Molecular Interaction Fields

Edited by
G. Cruciani

Related Titles

Methods and Principles in Medicinal Chemistry

Edited by R. Mannold, H. Kubinyi, G. Folkers
Editorial Board
H.-D. Höltje, H. Timmermann, J. Vacca, H. van de Waterbeemd, T. Wieland

Previous Volumes of this Series:

O. Zerbe (ed.)

BioNMR in Drug Research
Vol. 16

2002, ISBN 3-527-30465-7

P. Arloni, F. Alber (eds.)

Quantum Medicinal Chemistry
Vol. 17

2003, ISBN 3-527-30456-8

H. van de Waterbeemd, H. Lennernäs, P. Artursson (eds.)

Drug Bioavailability
Vol. 18

2003, ISBN 3-527-30438-X

H.-J. Böhm, S. S. Abdel-Meguid (eds.)

Protein Crystallography in Drug Discovery
Vol. 20

2004, ISBN 3-527-30678-1

Th. Dingermann, D. Steinhilber, G. Folkers (eds.)

Molecular Biology in Medicinal Chemistry
Vol. 21

2004, ISBN 3-527-30431-2

H. Kubinyi, G. Müller (eds.)

Chemogenomics in Drug Discovery
Vol. 22

2004, ISBN 3-527-30987-X

T. I. Oprea (ed.)

Chemoinformatics in Drug Discovery
Vol. 23

2004, ISBN 3-527-30753-2

R. Seifert, T. Wieland (eds.)

G Protein-Coupled Receptors as Drug Discovery
Vol. 24

2005, ISBN 3-527-30819-9

O. Kappe, A. Stadler

Microwaves in Organic and Medicinal Chemistry
Vol. 25

2005, ISBN 3-527-31210-2

Molecular Interaction Fields

Applications in Drug Discovery and
ADME Prediction

Edited by
Gabriele Cruciani

WILEY-VCH Verlag GmbH & Co. KGaA

Series Editors

Prof. Dr. Raimund Mannhold
Biomedical Research Center
Molecular Drug Research Group
Heinrich-Heine-Universität
Universtätsstrasse 1
40225 Düsseldorf
Germany
Raimund.mannhold@uni-duesseldorf.de

Prof. Dr. Hugo Kubinyi
Donnersbergstrasse 9
67256 Weisenheim am Sand
Germany
kubinyi@t-online.de

Prof. Dr. Gerd Folkers
Collegium Helveticum
STW/ETH Zürich
8092 Zürich
Switzerland
folkers@collegium.ethz.ch

Volume Editor

Prof. Dr. Gabriele Cruciani
University of Perugia
Department of Chemistry
06124 Perugia
Italy
gabri@chemiome.chm.unipg.it

■ All books published by Wiley-VCH are carefully produced. Nevertheless, authors, editors, and publisher do not warrant the information contained in these books, including this book, to be free of errors. Readers are advised to keep in mind that statements, data, illustrations, procedural details or other items may inadvertently be inaccurate.

Library of Congress Card No.: applied for

British Library Cataloguing-in-Publication Data
A catalogue record for this book is available from the British Library.

Bibliographic information published by Die Deutsche Bibliothek
Die Deutsche Bibliothek lists this publication in the Deutsche Nationalbibliografie; detailed bibliographic data is available in the Internet at <http://dnb.ddb.de>.

© 2006 WILEY-VCH Verlag GmbH & Co. KGaA, Weinheim

All rights reserved (including those of translation into other languages). No part of this book may be reproduced in any form – nor transmitted or translated into machine language without written permission from the publishers. Registered names, trademarks, etc. used in this book, even when not specifically marked as such, are not to be considered unprotected by law.

Printed in the Federal Republic of Germany.
Printed on acid-free paper.

Composition Kühn & Weyh, Satz und Medien, Freiburg
Printing betz-druck GmbH, Darmstadt
Bookbinding J. Schäffer GmbH i. G., Grünstadt

ISBN-13: 978-3-527-31087-6
ISBN-10: 3-527-31087-8

A Personal Foreword

I was a young organic chemist when I met Prof. Sergio Clementi, but from the very first moment I understood that his guide to the world of Chemometrics would have been a *brainwashing* for me. And indeed it was so. Sergio was a splendid teacher and what I know in the field of QSAR and Chemometrics is totally due to him.

From Sergio I learned the correct ways to produce mathematical models, and the tricks to interpret them. However, it was immediately clear that Chemometrics (and cheminformatics as well) can do very little when the numerical descriptors are poor or not related to the phenomena under study.

Few years later I had another *brainwashing* when I met prof. Peter Goodford at the European Symposium on QSAR in Sorrento (Italy). I was fascinated by his presentation and science, and I decided to learn more about. I spent in Oxford one year and Peter was a second scientific father for me. It was fantastic to complete my background working side by side with a scientist who did so much in the field of Structure-Based Drug Design. All I know on force-fields and numerical descriptions of complex phenomena such as (macro)molecular interactions is due to him. From that moment I never stopped to use his software GRID.

When ADME-attritions rate was large and *in silico* ADME procedures were still unknown, I went to Lausanne to learn pharmacokinetics working with Prof. Bernard Testa. Again another important man in my scientific life. Bernard pushed me deeply in the field of pharmacokinetics, and I was surprised to see how well Peter Goodford's GRID was working in such a different field.

My scientific career was guided and complemented by these scientists, and the reasons why my interests are so sparse depend on their enthusiasm and imprinting. However, one thing I have always used in all the problems I have encountered, or in all the procedure I developed. I have always used Molecular Interaction Fields to describe the structures of chemical and biological systems. After so may years of work, I'm still fascinated by the amount of information they contain. One never finishes to find new ways to extract information from them. Moreover, combining MIF with chemometric tools, is a powerful approach to all the fields of computer assisted drug development.

This led to the production of different algorithms, all reported in this volume and all based on Peter's GRID force field. It is noteworthy that GRID-MIF are currently applied in structure based, pharmacodynamic and pharmacokinetic fields, as well as in metabolism. Another proof, (although not necessary) of the versatility, flexibility and correct bio-parameterisation of Peter's GRID force field.

This volume reports the MIF theory, and several applications of MIFs in different arena of the drug discovery process. MIFs are decoding the common language of the (macro)molecules, the molecular interaction potential. Using MIF is simple, interpreting them straightforward.

It was a privilege to work on this volume with such a distinguished group of contributors, and I'm sure that this volume will open a window on the fascinating world of Molecular Interaction Fields.

Finally, I want to acknowledge my coworkers at Perugia University, and Prof. Raimund Mannhold and Prof. Hugo Kubinyi for their help, contribution and encouragement to produce this book.

Perugia, June 2005 *Gabriele Cruciani*

Contents

A Personal Foreword *V*

Preface *XIV*

List of Contributors *XVI*

I	**Introduction** *1*	
1	**The Basic Principles of GRID** *3*	
	Peter Goodford	
1.1	Introduction *3*	
1.2	Philosophy and Objectives *3*	
1.3	Priorities *4*	
1.4	The GRID Method *5*	
1.4.1	GRID Probes Are Anisometric *6*	
1.4.2	The Target "Responds" to the Probe *8*	
1.4.3	The Target is Immersed in Water *10*	
1.5	The GRID Force Field *10*	
1.5.1	The Lennard-Jones Term *11*	
1.5.2	The Electrostatic Term *11*	
1.5.3	The Hydrogen Bond Term *12*	
1.5.4	The Other Terms *12*	
1.6	Nomenclature *14*	
1.6.1	"ATOM" Records *14*	
1.6.2	"HETATM" Records *15*	
1.7	Calibrating the GRID Force Field *16*	
1.7.1	Checking the Calibration *17*	
1.7.2	Checking Datafile GRUB *17*	
1.8	The Output from GRID *18*	
1.8.1	GRID Maps from Macromolecules *19*	
1.8.2	GRID Maps from a Small Molecule *24*	
1.9	Conclusions *25*	

Molecular Interaction Fields. Edited by G. Cruciani
Copyright © 2006 WILEY-VCH Verlag GmbH & Co. KGaA, Weinheim
ISBN: 3-527-31087-8

2	**Calculation and Application of Molecular Interaction Fields** *27*
	Rebecca C. Wade
2.1	Introduction *27*
2.2	Calculation of MIFs *27*
2.2.1	The Target *27*
2.2.2	The Probe *28*
2.2.3	The Interaction Function *29*
2.2.3.1	Van der Waals Interactions *29*
2.2.3.2	Electrostatic Interactions *30*
2.2.3.3	Hydrogen Bonds *31*
2.2.3.4	Entropy *32*
2.3	Selected Applications of MIFs *33*
2.3.1	Mapping a Ligand Binding Site in a Protein *33*
2.3.2	Deriving 3D-QSARs *34*
2.3.3	Similarity Analysis of a Set of Related Molecules *36*
2.4	Concluding Remarks and Outlook *38*

II	**Pharmacodynamics** *43*

3	**Protein Selectivity Studies Using GRID-MIFs** *45*
	Thomas Fox
3.1	Introduction *45*
3.2	GRID Calculations and Chemometric Analysis *46*
3.2.1	Source and Selection of Target Structures *46*
3.2.2	Selection and Superimposition of Binding Sites *47*
3.2.3	Calculation of the Molecular Interaction Field *47*
3.2.4	Matrix Generation and Pretreatments *50*
3.2.4.1	Region Cut-outs *51*
3.2.5	GRID/PCA *51*
3.2.5.1	Score Plots *52*
3.2.5.2	Two-Dimensional Loading Plots *53*
3.2.5.3	Loading Contour Maps *54*
3.2.5.4	Problems of GRID/PCA *54*
3.2.6	GRID/CPCA *55*
3.2.6.1	Block Unscaled Weights *56*
3.2.6.2	CPCA *58*
3.2.6.3	Identification of Important Variable Blocks for Selectivity *59*
3.2.6.4	Contour Plots *59*
3.3	Applications *60*
3.3.1	DNA Minor Groove Binding – Compare AAA and GGG Double Helix *60*
3.3.2	Dihydrofolate Reductase *61*
3.3.3	Cyclooxygenase *61*
3.3.4	Penicillin Acylase *62*

3.3.5	Serine Proteases	63
3.3.5.1	S1 Pocket	64
3.3.5.2	P Pocket	64
3.3.5.3	D Pocket	66
3.3.6	CYP450	67
3.3.7	Target Family Landscapes of Protein Kinases	69
3.3.8	Matrix Metalloproteinases (MMPs)	70
3.3.9	Nitric Oxide Synthases	74
3.3.10	PPARs	75
3.3.11	Bile Acid Transportation System	75
3.3.12	Ephrin Ligands and Eph Kinases	76
3.4	Discussion and Conclusion	77
4	**FLAP: 4-Point Pharmacophore Fingerprints from GRID**	**83**
	Francesca Perruccio, Jonathan S. Mason, Simone Sciabola, and Massimo Baroni	
4.1	Introduction	84
4.1.1	Pharmacophores and Pharmacophore Fingerprints	84
4.1.2	FLAP	86
4.2	FLAP Theory	86
4.3	Docking	88
4.3.1	GLUE: A New Docking Program Based on Pharmacophores	89
4.3.2	Case Study	91
4.4	Structure Based Virtual Screening (SBVS)	92
4.5	Ligand Based Virtual Screening (LBVS)	94
4.6	Protein Similarity	95
4.7	TOPP (Triplets of Pharmacophoric Points)	97
4.8	Conclusions	101
5	**The Complexity of Molecular Interaction: Molecular Shape Fingerprints by the PathFinder Approach**	**103**
	Iain McLay, Mike Hann, Emanuele Carosati, Gabriele Cruciani, and Massimo Baroni	
5.1	Introduction	103
5.2	Background	104
5.3	The PathFinder Approach	105
5.3.1	Paths from Positive MIF	105
5.3.2	Paths from Negative MIF	107
5.4	Examples	109
5.4.1	3D-QSAR	109
5.4.2	CYP Comparison	112
5.4.3	Target–Ligand Complexes	112
5.5	Conclusions	115

6	**Alignment-independent Descriptors from Molecular Interaction Fields** *117*
	Manuel Pastor

6.1	Introduction *117*
6.1.1	The Need for MIF-derived Alignment-independent Descriptors *117*
6.1.2	GRIND Applications *119*
6.2	GRIND *120*
6.2.1	The Basic Idea *120*
6.2.1.1	Computation of MIF *121*
6.2.1.2	Extraction of Highly Relevant Regions *122*
6.2.1.3	MACC2 Encoding *124*
6.2.2	The Analysis of GRIND Variables *128*
6.3	How to Interpret a GRIND-based 3D QSAR Model *130*
6.3.1	Overview *130*
6.3.2	Interpreting Correlograms *131*
6.3.3	Interpreting Single Variables *133*
6.3.4	GRIND-based 3D QSAR Models are not Pharmacophores *134*
6.4	GRIND Limitations and Problems *135*
6.4.1	GRIND and the Ligand Conformations *135*
6.4.2	The Ambiguities *137*
6.4.3	Chirality *139*
6.5	Recent and Future Developments *139*
6.5.1	Latest Developments *139*
6.5.1.1	Shape Description *139*
6.5.1.2	Anchor GRIND *140*
6.5.2	The Future *140*
6.6	Conclusions *141*

7	**3D-QSAR Using the GRID/GOLPE Approach** *145*
	Wolfgang Sippl

7.1	Introduction *145*
7.2	3D-QSAR Using the GRID/GOLPE Approach *147*
7.3	GRID/GOLPE Application Examples *149*
7.3.1	Estrogen Receptor Ligands *149*
7.3.2	Acetylcholinesterase Inhibitors *158*
7.4	Conclusion *165*

III	**Pharmacokinetics** *171*
8	**Use of MIF-based VolSurf Descriptors in Physicochemical and Pharmacokinetic Studies** *173* *Raimund Mannhold, Giuliano Berellini, Emanuele Carosati, and Paolo Benedetti*
8.1	ADME Properties and Their Prediction *173*
8.2	VolSurf Descriptors *174*
8.3	Application Examples *179*
8.3.1	Aqueous Solubility *180*
8.3.2	Octanol/Water Partition Coefficients *184*
8.3.3	Volume of Distribution (VD) *190*
8.3.4	Metabolic Stability *192*
8.4	Conclusion *193*
9	**Molecular Interaction Fields in ADME and Safety** *197* *Giovanni Cianchetta, Yi Li, Robert Singleton, Meng Zhang, Marianne Wildgoose, David Rampe, Jiesheng Kang, and Roy J. Vaz*
9.1	Introduction *197*
9.2	GRID and MIFs *198*
9.3	Role of Pgp Efflux in the Absorption *199*
9.3.1	Materials and Methods *199*
9.3.1.1	Dataset *199*
9.3.1.2	Computational Methods *199*
9.3.1.3	ALMOND Descriptors *200*
9.3.2	Results *200*
9.3.3	Pharmacophoric Model Interpretation *202*
9.3.4	Discussion *203*
9.4	HERG Inhibition *204*
9.4.1	Materials and Methods *204*
9.4.1.1	Dataset *204*
9.4.1.2	Computational Methods *204*
9.4.2	Results *205*
9.4.2.1	Nonbasic Nitrogen Subset *205*
9.4.2.2	Ionizable Nitrogen Subset *206*
9.4.2.3	Interpretation of Pharmacophoric Models *208*
9.5	CYP 3A4 Inhibition *209*
9.5.1	Materials and Methods *209*
9.5.1.1	Dataset *209*
9.5.1.2	Computational Methods *210*
9.5.1.3	Ligand GRIND Descriptors *210*
9.5.1.4	Protein GRIND Descriptors *210*
9.5.1.5	Overlap of Structures *211*
9.5.2	Results *211*

9.5.2.1	Distances in the Protein Pocket 214
9.5.3	Discussion 215
9.6	Conclusions 216

10 Progress in ADME Prediction Using GRID-Molecular Interaction Fields *219*
Ismael Zamora, Marianne Ridderström, Anna-Lena Ungell, Tommy Andersson, and Lovisa Afzelius

10.1	Introduction: ADME Field in the Drug Discovery Process 219
10.2	Absorption 223
10.2.1	Passive Transport, Trans-cellular Pathway 225
10.2.2	Active Transport 227
10.3	Distribution 228
10.3.1	Solubility 228
10.3.2	Unspecific Protein Binding 230
10.3.3	Volume of Distribution 230
10.4	Metabolism 232
10.4.1	Cytochrome P450 Inhibition 233
10.4.2	Site of Metabolism Prediction 233
10.4.3	Metabolic Stability 234
10.4.4	Selectivity Analysis 235
10.5	Conclusions 242

11 Rapid ADME Filters for Lead Discovery *249*
Tudor I. Oprea, Paolo Benedetti, Giuliano Berellini, Marius Olah, Kim Fejgin, and Scott Boyer

11.1	Introduction 249
11.2	The Rule of Five (Ro5) as ADME Filter 250
11.3	Molecular Interaction Fields (MIFs): VolSurf 251
11.4	MIF-based ADME Models 253
11.5	Clinical Pharmacokinetics (PK) and Toxicological (Tox) Datasets 254
11.6	VolSurf in Clinical PK Data Modeling 256
11.7	ChemGPS-VolSurf (GPSVS) in Clinical PK Property Modeling 257
11.8	ADME Filters: GPSVS vs. Ro5 261
11.9	PENGUINS: Ultrafast ADME Filter 264
11.10	Integrated ADME and Binding Affinity Predictions 267
11.11	Conclusions 268

12	**GRID-Derived Molecular Interaction Fields for Predicting the Site of Metabolism in Human Cytochromes** *273*
	Gabriele Cruciani, Yasmin Aristei, Riccardo Vianello, and Massimo Baroni
12.1	Introduction *273*
11.2	The Human Cytochromes P450 *274*
12.3	CYPs Characterization using GRID Molecular Interaction Fields *275*
12.4	Description of the Method *279*
12.4.1	P450 Molecular Interaction Fields Transformation *280*
12.4.2	3D Structure of Substrates and Fingerprint Generation *281*
12.4.3	Substrate–CYP Enzyme Comparison: the Recognition Component *282*
12.4.4	The Reactivity Component *283*
12.4.5	Computation of the Probability of a Site being the Metabolic Site *284*
12.5	An Overview of the Most Significant Results *285*
12.5.1	Importing Different P450 Cytochromes *287*
12.6	Conclusions *289*
12.7	Software Package *289*

Index *291*

CD-ROM Information *305*

Preface

Volume 27 of our series "Methods and Principles in Medicinal Chemistry" is dedicated to "Molecular Interaction Fields" and their impact on current drug research.

In the early 1980s Peter Goodford developed the GRID force field for determining energetically favorable binding sites on molecules of known structure. The GRID force field has always been calibrated as far as possible by studying experimental measurements, and the calibration is then checked by studying how well GRID predicts observed crystal structures. Crystal packing is determined by free energy considerations rather than by enthalpy alone. The force field includes entropic terms; GRID can detect the hydrophobic binding regions which are so important when high-affinity ligands are being designed, and it can also detect sites for the polar groups which determine ligand selectivity. GRID may be used to study individual molecules such as drugs, molecular arrays such as membranes or crystals, and macromolecules such as proteins, nucleic acids, glycoproteins or polysaccharides.

Moreover GRID can be used to understand the structural differences related to enzyme selectivity, a fundamental field in the rational design of drugs. GRID maps can also be used as descriptor input in statistical procedures like CoMFA, GOLPE or SIMCA for QSAR or 3D-QSAR analyses.

The GRID force field represents the basis for several software packages specifically developed for application to pharmacodynamic aspects of drug research, including the programs ALMOND, Pathfinder, and FLAP or, in the ADME field, the programs VolSurf and MetaSite.

Correspondingly, the present volume is quite logically divided into three sections. An introductory section contains two chapters dealing with the theoretical background. The chapter of Peter Goodford, who originally developed the GRID software, focuses in detail on the basic principles of GRID, whereas the chapter by Rebecca Wade is dedicated to "Calculation and Application of Molecular Interaction Fields".

The second section refers to pharmacodynamic aspects and contains chapters on "Protein selectivity studies using GRID-MIF" by Thomas Fox, "The Complexity of Molecular Interaction: Molecular Shape Fingerprint by PathFinder Approach" by McLay, Hann, Carosati, Cruciani, and Baroni, "Alignment-Independent Descriptors from Molecular Interaction Fields" by Manuel Pastor, "FLAP: 4-point pharmacophore fingerprints from GRID" by Perruccio, Mason, Sciabola, and Baroni as well as a chapter on "3D QSAR using the GRID/GOLPE approach" by Wolfgang Sippl.

The third and last section is dedicated to pharmacokinetics including chapters on "Molecular Interaction Fields in ADME and Safety" by Cianchetta, Li, Singleton, Zhang, Wildgoose, Rampe, Kang, and Vaz, "MIF-based VolSurf descriptors in Physicochemical and Pharmacokinetic studies" by Mannhold, Berellini, Carosati, and Benedetti, "Progress in ADME prediction using GRID-Molecular Interaction Fields" by Zamora, Ridderström, Ungell, Andersson, and Afzelius, "Rapid ADME filters for Lead Discovery" by Oprea, Benedetti, Berellini, Olah, Fejgin, and Boyer and finally a chapter on "GRID-Derived Molecular Interaction Fields for Predicting the Site of Metabolism in Human Cytochromes" by Cruciani, Aristei, Vianello, and Baroni.

A remarkable peculiarity of this volume is the inclusion of a CD-ROM containing some software packages used in the three sections of the book.

The series editors believe that this book is unique in its topic and presentation and adds a fascinating facet to the series. We are indebted to all authors for their well-elaborated contributions and we would like to thank Gabriele Cruciani for his enthusiasm in organizing this volume. We also want to express our gratitude to Renate Doetzer and Frank Weinreich from Wiley-VCH for their valuable contributions to this project.

September 2005

Raimund Mannhold, Düsseldorf
Hugo Kubinyi, Weisenheim am Sand
Gerd Folkers, Zürich

List of Contributors

Lovisa Afzelius
AstraZeneca R&D Mölndal
DMPK and Bioanalytical Chemistry
431 Mölndal
Sweden

Tommy Andersson
AstraZeneca R&D Mölndal
DMPK and Bioanalytical Chemistry
431 Mölndal
Sweden

Yasmin Aristei
University of Perugia
Chemistry Department
Via Elce di Sotto 8
06123 Perugia
Italy

Massimo Baroni
Molecular Discovery
215 Marsh Road
HA5 5NE Middlesex – London
UK

Paolo Benedetti
University of Perugia
Chemistry Department
Via Elce di Sotto 8
06123 Perugia
Italy

Giuliano Berellini
University of Perugia
Chemistry Department
Via Elce di Sotto 8
06123 Perugia
Italy

Scott Boyer
AstraZeneca R&D Mölndal
Global Safety Assessment
43183 Mölndal
Sweden

Emanuele Carosati
University of Perugia
Chemistry Department
Via Elce di Sotto 10
06123 Perugia
Italy

Giovanni Cianchetta
Universitá di Perugia
Dipartimento di Chimica e Tecnologia
del Farmaco
Via del Liceo 1
06123 Perugia
Italy

Molecular Interaction Fields. Edited by G. Cruciani
Copyright © 2006 WILEY-VCH Verlag GmbH & Co. KGaA, Weinheim
ISBN: 3-527-31087-8

List of Contributors

Gabriele Cruciani
University of Perugia
Chemistry Department
Via Elce di Sotto 8
06123 Perugia
Italy

Kim Fejgin
Göteborg University
Department of Pharmacology
Institute of Physiology and
Pharmacology
Box 431
40530 Göteborg
Sweden

Thomas Fox
Boehringer Ingelheim Pharma KG
Department Chemical Research /
Structural Research
J51-00-01
88397 Biberach
Germany

Peter Goodford
Oxford University
Laboratory of Molecular Biophysics
South Parks Road
Oxford OX1 3Q
UK

Mike Hann
GlaxoSmithKline Medicines Research
Centre
Structural and Biophysical Sciences
Gunnels Wood Road
Stevenage
Hertfordshire SG1 2NY
UK

Jiesheng Kang
Sanofi-Aventis Pharmaceuticals
1041 Route 202/206 N
Bridgewater, NJ 08807
USA

Yi Li
Sanofi-Aventis Pharmaceuticals
1041 Route 202/206 N
Bridgewater, NJ 08807
USA

Raimund Mannhold
Heinrich-Heine-Universität
Department of Laser Medicine
Molecular Drug Research Group
Universitätsstr. 1
40225 Düsseldorf
Germany

Jonathan S. Mason
Pfizer Global R&D
Molecular Informatics, Structure &
Design
Sandwich, Kent CT13 9NJ
UK

Iain McLay
GlaxoSmithKline Research &
Development
Gunnels Wood Road
Stevenage, SG1 2NY
UK

Marius Olah
University of New Mexico
School of Medicine
Division of Biocomputing
MSC08 4560
Albuquerque, NM 87131-0001
USA

Tudor I. Oprea
Sunset Molecular Discovery LLC
1704 B Llano St., S-te 140
Santa Fe, NM 87505
USA

Manuel Pastor
IMIM/UPF
Computer-Assisted Drug Design
Laboratory
Passeig Marítim de la Barceloneta, 37-49
08003 Barcelona
Spain

Francesca Perruccio
Pfizer Global Research & Development
Ramsgate Road
Sandwich, Kent CT13 9NJ
UK

David Rampe
Sanofi-Aventis Pharmaceuticals
1041 Route 202/206 N.
Bridgewater, NJ 08807
USA

Marianne Ridderström
AstraZeneca R&D Mölndal
DMPK and Bioanalytical Chemistry
431 Mölndal
Sweden

Simone Sciabola
University of Perugia
Chemistry Department
Laboratory of Chemometrics
Via Elce di Sotto 10
06123 Perugia
Italy

Robert Singleton
Sanofi-Aventis Pharmaceuticals
1041 Route 202/206 N.
Bridgewater, NJ 08807
USA

Wolfgang Sippl
University of Düsseldorf
Universitätsstr. 1
40225 Düsseldorf
Germany

Anna-Lena Ungell
AstraZeneca R&D Mölndal
DMPK and Bioanalytical Chemistry
431 Mölndal
Sweden

Roy K. Vaz
Sanofi-Aventis Pharmaceuticals
1041 Route 202/206 N.
Bridgewater, NJ 08807
USA

Riccardo Vianello
Molecular Discovery Ltd.
215 Marsh Road
Pinner, Middlesex
HA5 5NE London
UK

Rebecca Wade
EML Research gGmbH
Molecular and Cellular Modeling Group
Schloss-Wolfsbrunnenweg 33
69118 Heidelberg
Germany

Marianne Wildgoose
Sanofi-Aventis Pharmaceuticals
1041 Route 202/206 N.
Bridgewater, NJ 08807
USA

Ismael Zamora
Lead Molecular Design
Francesc Cabanes i Alibau, 1-3 2-1
08190 Sant Cugat del Valle, Barcelona
Spain

Meng Zhang
Sanofi-Aventis Pharmaceuticals
1041 Route 202/206 N.
Bridgewater, NJ 08807
USA

I
Introduction

1
The Basic Principles of GRID
Peter Goodford

1.1
Introduction

One cannot go out and buy a computer program in the confident expectation that it will do its job exactly as expected. Of course there are some things, like a lawn mower, where a relatively quick and easy test can be made to discover if it is good enough. Can it cut long grass? Cut wet grass? Does it pick up all the clippings? Will it leave beautiful light and dark stripes on the lawn? However, a molecular interaction field (MIF) is a good deal more complicated than a lawn mower, and it is not at all easy to establish which MIF programs work in a satisfactory way. Each program must be assessed very carefully before deciding what software should be used for any particular task, and many different factors must be taken into account. Some are obvious, like the available computer hardware; its speed; the size of its memory; and the amount of disk space on the user's system. Some are less apparent, such as the objectives, priorities and overall philosophy of the people who wrote the software, and the way in which they devised and calibrated their MIF. The most important factor is to be certain in one's own mind about the precise jobs which one wants the program to do.

1.2
Philosophy and Objectives

Even the most superficial study of molecular interaction fields shows that each MIF has its own particular characteristics. This field may put great emphasis on the accurate computation of the individual atomic charges. A different MIF may give more attention to the way in which those charges are distributed between an atom and its bonds, and a third may place some of each atom's charge onto its lone pair electrons. Another MIF may attempt to make accurate predictions of the pK_a of every polar atom, in order to be certain that each one is appropriately protonated before the MIF computations begin. Some fields may require the system under investigation to have zero overall charge. Other fields will happily do com-

Molecular Interaction Fields. Edited by G. Cruciani
Copyright © 2006 WILEY-VCH Verlag GmbH & Co. KGaA, Weinheim
ISBN: 3-527-31087-8

putations on a couple of phosphate ions, for example, with none of the oxygens protonated and no counter cations so that the two anionic phosphates move remorselessly apart until they vanish at the edge of the universe! One field may always compute the local pH, and another may need pH information as part of the input data. This field may give detailed attention to estimating the local dielectric environment and how it changes from place to place, while that one may assume an arbitrary overall dielectric constant.

It is not only the electrostatic treatment which is different in each MIF, but also other molecular characteristics. Many fields require all the hydrogen coordinates to be defined, but some only need the location of hydrogen-bonding hydrogens, and others take no specific account of any hydrogen positions. Some fields use simple harmonic motion to describe bond vibrations, but others attempt to consider deviations from harmonicity. Some have dedicated computations which deal with hydrogen bonds, and others pay no particular attention to hydrogen bond geometry. Most fields do not allow for tautomeric changes, but some can take tautomerism into account and a few can cope with alterations in the hybridisation of an atom. Some deal exclusively with enthalpy, but others can take account of entropy which is a major component of the hydrophobic effect.

Whenever another research group begins to study MIFs, they introduce a new perspective and a new set of ideas, so the extension and improvement of force fields has been a matter of continuously improving approximations. There will never be an absolutely correct MIF, but even the very earliest work was surprisingly valuable. Huggins and Pauling [1] introduced their atomic radii seventy years ago, but they immediately extended the understanding of crystal packing and of many other properties. However no force field is perfect, and one can only hope that the approximations will continue to improve in the years ahead.

1.3
Priorities

The priorities of the people who create any MIF are a concrete manifestation of their scientific philosophy and overall objectives, and seven requirements seemed particularly important when the GRID force field [2] was being designed:

1. This force field was explicitly intended for use with the GRID method.
2. The overall objective was to predict where ligands bind to biological macromolecules, and so gain a better understanding of the factors involved in binding. However that improved understanding should also help in the design of improved ligands.
3. The GRID force field could have been calibrated either by using theoretical calculations, or by studying experimental observations, and after much discussion the experimental approach was adopted whenever possible.
4. The input data must always be thoroughly checked before every computation, and an associated program called GRIN was written to do this job.

5. The equations used for the computation must be reasonably straightforward, so that anybody working in the drug-discovery field (biologist, pharmacologist, medicinal chemist, crystallographer, clinician, statistician, patent expert, administrator etc.) could easily discover exactly how program GRID had calculated any particular result.
6. It must be relatively easy for anybody working in the field to interpret the output from GRID.
7. An annual reappraisal policy was established so that the worst features in the current version of GRID would always be identified, and could be dealt with appropriately in each successive year.

It is conventional to write impersonally about scientific research, but subjective decisions are made when one decides which features are worst, or which objectives are most important for a program. The personal pronoun "we" will therefore be used in this article, whenever it is appropriate to draw attention to subjectivity. "We" are still discussing priorities for the forthcoming release of GRID, but before describing the GRID force field in detail we must first describe how the GRID method actually works.

1.4
The GRID Method

There are many programs which can be used compute the electrostatic potential around a molecule. A computer model is first prepared from the x, y, z coordinates of the atoms, and this model is then surrounded by an imaginary orthogonal grid.

The next step is to compute the work needed to bring a unit electrostatic charge from infinity to the first point on the grid, and the total work required for this job is a measure of the electrical potential at that particular grid point. The same procedure is then repeated for each of the other grid points, including those which are actually inside the molecule, until the potential has been calculated for every position.

At this stage it would be possible to print out the individual potential values as a table of numbers for detailed study. The findings could then be used as input for further computations, but studying a printed data table would be a rather clumsy way of displaying the results, and a much better method is to create a three-dimensional computer plot showing a contour surface surrounding the molecule. This contour surface defines a single user-selected value of the electrostatic potential, and the final picture usually shows the molecule together with something looking rather like a child's balloon!

Program GRID works in very much the same way, but the objective is to obtain chemically specific information about the molecule. An electrostatic potential does not normally allow one to differentiate between favorable binding sites for a primary or a secondary or a tertiary amine cation, or tetramethyl ammonium or pyridinium or a sodium cation, and the GRID method is an attempt to compute analogous potentials which do have some chemical specificity. The generic name *"target"* is given to the molecule (or group of molecules) being studied by GRID, and the object used to measure the potential at each point is called a *"probe"*. The individual potential values are called *"GRID values"* and the final computer plot is called a *"GRID map"*. Many different probes can be used on the same target one after the other, and each probe represents a specific chemical group so that chemically specific information can be accumulated about the way in which the target might interact favorably with other molecules.

The GRID method differs in three critical ways from traditional programs which just display electrostatic potentials:

1. GRID probes are often anisometric.
2. The target "responds" when the probe is moved around it from place to place.
3. It is assumed that both the target and the probe are immersed in water.

These differences must now be considered in more detail.

1.4.1
GRID Probes Are Anisometric

Most GRID probes are anisometric because each probe represents an atom or a small group of atoms. For example a carbonyl oxygen probe is one oxygen atom with a couple of sp2 lone pairs. It has a size and a polarizability and an electrostatic charge, and each lone pair can accept one hydrogen bond. The center of the oxygen is placed at the first grid point, and a check is then made for unacceptably bad close contacts. If none is found the program then searches for nearby hydrogen-bond donor atoms on the target, and a list of those donors is made and sorted. Target atoms are rejected from the list if their donor hydrogens are pointing the wrong way, and the probe is then rotated (keeping its oxygen fixed at the grid point) so that its lone pairs will be oriented until they make the best possible hydrogen bonds to nearby target atoms. When this has been done the GRID force field is used to compute a GRID value for that particular probe at that particular point, and the whole process is repeated systematically until the potential for carbonyl oxygen is known for every grid point on the map.

An aromatic sp2 hydroxy probe differs in several ways from carbonyl oxygen. The oxygen atom of the hydroxy is placed at the grid point as before, but the probe has a larger polarizability and makes hydrogen bonds of a different strength. It can accept only one hydrogen bond, but the oxygen is bonded to a hydrogen atom which can donate. If both the donor and acceptor hydrogen bonds are made simultaneously they will be mutually constrained towards the sp2 angle of 120°. The bond length from the oxygen to its hydrogen is about 1 Å, and the probe's donor hydrogen moves round the grid point at this distance when the probe is rotated. Figure 1.1 shows the target with an sp2 hydroxy probe placed at the first point, ready for the computation to begin.

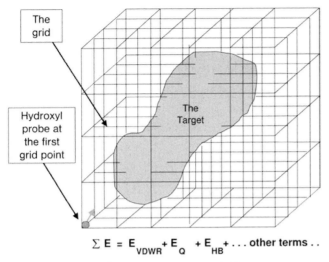

$$\Sigma E = E_{VDWR} + E_Q + E_{HB} + \ldots \text{other terms} \ldots$$

Figure 1.1. The set up for GRID. See text Section 1.4.1.

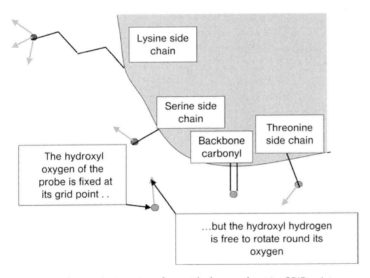

Figure 1.2. The initial orientation of an sp2 hydroxy probe at its GRID point.

The sp2 carboxyl oxygen probe differs from both sp2 carbonyl and sp2 hydroxy, having a much greater polarizability and much greater negative charge than either. The sp3 aliphatic hydroxy probe is distinguished by making its hydrogen bonds at the sp3 angle of 109° instead of 120°, and by accepting at two lone pairs instead of just one. "Multi-atom probes" can also be used, such as aromatic carboxylate which represents the anion of a complete benzoic acid molecule. This multi-atom probe has two sp2 carboxy oxygens both bonded to the carboxy carbon which is bonded to the aromatic ring. Each oxygen has a couple of lone pairs, all appropriately oriented and of appropriate strength. Both oxygens are deprotonated, and they both have a substantial negative charge which is partly counterbalanced by a modest positive charge on the carboxy carbon, so the whole probe has an overall charge of –1. Its oxygens are both identical, and one of them is fixed as usual at the grid point. The whole multi-atom probe is then rotated to find all the orientations in which it can make good hydrogen bonds to the target, and good electrostatic interactions, while avoiding steric clashes. The chosen oxygen always stays on its grid point, and the GRID potential for that point is computed when the best orientation of the whole multiatom probe has been established.

A multiatom probe usually finds pairs of minima which would correspond in this example to the two oxygens of the carboxylate group. Of course the computation for a multi-atom probe takes somewhat longer than the map for a simpler probe, but the force field was written explicitly for the program and so GRID computations are never particularly time consuming.

1.4.2
The Target "Responds" to the Probe

Figure 1.2 shows in more detail how an aromatic sp2 hydroxy probe might be placed on its grid point at the start of a cycle of computation. In this figure the hydrogen of the probe happens to be pointing by chance towards a nearby serine residue of the target. The orientation of the probe is completely random at this early stage of the job, but with a slight readjustment GRID can make the probe's hydrogen point directly at the serine's side chain sp3 hydroxy oxygen. A hydrogen bond could then be formed and that would be quite a good arrangement, but a better one is shown in Fig. 1.3. Program GRID has to search and find the better alternative, and must do three things to make this happen:

1. GRID has to rotate the probe, while keeping its oxygen firmly anchored at the grid point, until its hydrogen is redirected towards the nearby backbone sp2 carbonyl oxygen as shown in Fig. 1.3.
2. The probe then has to spin about an imaginary sp2–sp2 axis (A in Fig. 1.3) which links it to the backbone carbonyl oxygen, until the probe's own lone pair points as directly as possible towards the sp3 hydroxy oxygen of the serine.
3. The sp3 hydroxy group of the serine must finally spin about bond B which links it to the to the rest of the protein, until its hydrogen points as well as possible towards the probe's lone pair.

This rotation of the serine oxygen is called the "response of the target to the probe", and finding the best response is often a much more complicated job than it appears in Fig. 1.3. There are usually many different hydrogen bonding groups on or near the surface of the target, reasonably close to the probe, as shown in Fig. 1.4, and they must all be taken into account.

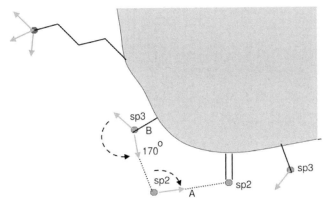

Figure 1.3. Rotational adjustments of the probe. See text.

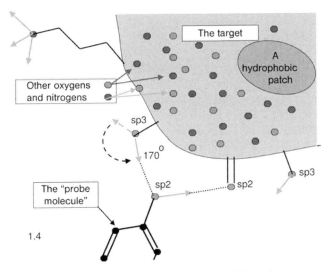

Figure 1.4. The final position of the probe showing additional features of the binding site. See text.

Methods are provided in GRID so that the user can adjust the size of the "response." For instance he could prevent the serine hydroxy from rotating on its axis, if he knew that it was already making another strong hydrogen bond which would be broken if the probe interacted as described above. There is also a lysine side chain shown near the top of Figs. 1.2, 1.3 and 1.4, but the NH_3^+ group of that lysine cannot reach the probe at its grid point as things are shown in the figures. However resetting one of the directives would allow the lysine side chain to swing down, and perhaps make a useful hydrogen-bond interaction with the probe. The directives are always set by default so that things like this do not happen, and long side chains like the lysine do not normally search around during a regular GRID run, unless the user has made a positive decision to release them. That kind of decision can only be taken after a thorough study of the binding site of the protein. The user must understand some or all of its properties, and this enhancement of the user's understanding was a major objective when program GRID was being written.

1.4.3
The Target is Immersed in Water

The concept of electrical potentials was developed by physicists in the 19th century, and they quite naturally took a vacuum as their reference state. The dielectric constant of a vacuum is 1.0 by definition, and many of the early experiments on electrostatics were made in air which has a dielectric constant very close to unity. However biological systems are full of water, and biologists must invoke a dielectric constant of up to 80 in order to make traditional electrostatic calculations. It is therefore hardly surprising that MIF computations in biological systems tend to give unstable results, when such a large dielectric correction factor must be used.

The GRID force field was designed on a more appropriate basis for biology. It is assumed a priori that the environment surrounding the target has a bulk dielectric of 80, and that the dielectric diminishes towards 4 in the deep center of a large globular macromolecule. These are the default values which were used in calibrating the MIF, but of course each user can alter them to any reasonable alternative during his own GRID runs. It has been reported [3] that a value between 10 and 20 gives results which agree better with experiments on small molecules.

Some years ago a large oil company wanted to use program GRID for calculations on zeolites. These are minerals, and it was first necessary to calibrate several elements such as silicon which had not previously been used in GRID runs. Preliminary computations were then started, but the results from zeolites were misleading. A bulk dielectric of 80 would clearly be inappropriate in this case, because zeolites are used at approximately 300 °C for oil refining and are therefore completely dry. However it was impossible to find any dielectric values which yielded satisfactory results for zeolites, and this seems to demonstrate that one should not expect a single MIF to work for every system. Each force field should be calibrated for the job in hand, and much more sophisticated methods are needed if one wishes to study all 100 elements in all experimental conditions. GRID and its force field must be restricted to the wet biological environment for which they were calibrated.

1.5
The GRID Force Field

The target is always prepared and checked by an associated program called GRIN which is used before the actual GRID run begins, and perhaps the most important job of GRIN is the amalgamation of every nonpolar hydrogen atom of the target with its neighboring heavy atom to give an "extended atom". Consider, for example, a very small target H_3C–CH_3 consisting of one ethane molecule. GRIN will represent this by two extended methyl atoms instead of two carbons plus six hydrogens, and this condensation of eight real atoms into a pair of extended atoms allows the GRID programme to run much faster. Of course there is some loss of accuracy, but real targets for GRID are usually much more complicated than ethane. Real targets usually have conformationally flexible side chains, and it is very easy to place too much emphasis on the exact hydrogen coordinates of a biological macromolecule when those hydrogens have not even been observed by the X-ray crystallographer.

Programme GRIN also checks the target for errors, and the GRID run then begins. The GRID energies are usually computed pairwise between the probe at its grid point and each extended atom of the target, one by one. Recent releases of the program include more terms, but early versions used only three energy components for each pairwise energy E_{PAIR}:

$$E_{PAIR} = E_{LJ} + E_Q + E_{HB} \tag{1}$$

1.5.1
The Lennard-Jones Term

The E_{LJ} term in Eq. (1) is the well-known "Lennard-Jones energy", and is computed as the sum of two terms:

$$E_{LJ} = (Ad^i - Bd^j) F \tag{2}$$

in which $i = -12$, $j = -6$ and $F=1$. A and B are positive constants which are chosen so that E_{LJ} will be calculated in kcal mol^{-1}, and d is the distance between the probe at its grid point and the extended atom of the target. The first term Ad^i is always positive, and represents the repulsion of the atoms for each other if they are unacceptably close together. The second term $-Bd^j$ is negative and measures their induction and dispersion attractions for each other.

1.5.2
The Electrostatic Term

E_Q is an electrostatic term computed as $E_Q = q_1q_2/dD$ where q_1 is the charge of the probe, q_2 the charge of the extended target atom, and D is the dielectric constant value to be used when their pairwise electrostatic interaction is calculated. Computing D is a slow business [2] because a square-root calculation is always required, and many atom pairs must be studied in a GRID run. D must be estimated individually for each pair, and extensive tests have shown that acceptable results cannot be obtained reliably unless all pairwise values of D and E_Q are worked out. Of course it is very tempting to ignore E_Q if the interacting atoms are more than 20 (or 30 or 40) Å apart, but this attractive short cut is unacceptable because it often gives rise to significant errors. The method finally adopted [2] to compute D is based on classical electrostatics [4] with the important additional assumption that one is dealing with a system of two homogeneous phases separated by a flat planar surface. It is easy to construct models in which the assumption of a flat surface can give rather misleading results, but in practice this does not seem to happen very often and the general approach for calculating D seems a reasonable approximation.

1.5.3
The Hydrogen Bond Term

E_{HB} is a hydrogen bond term [5–7] which is used only when one of the interacting atoms can donate a hydrogen bond and the other can accept. Equation (2) is again used but the constants A and B now have values which depend on the chemical nature of the interacting atoms, and the function F depends on their hybridisation and the relative positions of the interacting atoms and their bonded neighbors. E_{HB} and E_{LJ} both define relatively short-range effects, and are set to zero if the interacting atoms are more than a few Angstroms apart.

1.5.4
The Other Terms

The E_{LJ}, E_Q and E_{HB} functions are very simple, but they are also very well known which gives them one particularly important advantage: everybody understands them and can judge and criticise them for themselves. Moreover GRID displays by default the individual E_{LJ}, E_Q and E_{HB} and dielectric D values for every pairwise interaction, and so the source of any suspected error can usually be discovered very easily. After careful analysis it may then turn out that GRID did not make the suspected error, and that the user's worries were misplaced. Irrespective of the final outcome, the user may gain an enhanced understanding of the system by checking things like this for himself, and this enhanced comprehension was a major objective when the program GRID was first being written.

As mentioned above, there is an ongoing policy to search continually for the worst features in the current release of GRID, and then to take account of them. Later versions of the programme therefore include many extra terms which were not present in the original E_{PAIR} function. For instance:

1. When there is a rather close contact between a target and a probe atom, the computed E_{LJ} value may be strongly positive, suggesting mutual repulsion. However, if E_{HB} is simultaneously negative the atoms may actually be close together because they are making a hydrogen bond to each other, and GRID must detect when this happens and then allow E_{HB} to override E_{LJ}.
2. An adjustment must be made for the effect of an atom's charge upon the strength of the hydrogen bonds which it makes. For example hydrogen atoms bonded to an aliphatic carbon do not normally participate in hydro-

gen bonding. However GRID must take special account of the alpha carbon atom at the N-terminal of a protein chain, because this carbon sometimes donates a hydrogen bond as it can pick up positive charge from the nearby cationic N-terminal nitrogen.
3. Metals now receive special attention in GRID according to their hardness or softness.
4. Some water molecules in a biological system appear to make four tetrahedral sp3 hydrogen bonds. Others donate two hydrogen bonds but accept only one, making these three interactions in roughly the same plane. GRID must therefore be able to deal with both the flat and the tetrahedral arrangements.
5. The input programme GRIN always checks the overall electrostatic charge of the target, and expects nucleic acids to be surrounded by a cloud of counterions which maintain overall electroneutrality. However the ions were not mobile in early releases of GRID, and GRID maps of DNA were therefore full of holes which surrounded each counterion. They looked rather like a Swiss cheese, but GRID Probes can now nudge the counterions out of the way and thus generate a GRID map without misleading holes.
6. Some water molecules near a target may be so strongly bound that they almost behave like a part of the target itself. GRID must therefore treat each of these waters in a way which depends on its particular environment. For example, a water already bound to two carboxy groups would normally be donating a hydrogen bond to each carboxy oxygen, and would be much more likely to accept a hydrogen bond from the probe than to donate. On the other hand, a water already accepting from a couple of arginine guanidinium NH_2 groups would be most likely to donate to the probe. GRID must therefore be able to examine each water of the target and take its local environment into account.
7. The force field now incorporates entropy terms. For instance a lysine side chain of the target can adopt only one or two conformations when fully extended towards a distant probe, but can assume more conformations to reach a probe which is nearer, and GRID must be able to allow for this.
8. There is now a "hydrophobic probe" which detects hydrophobic regions on the surface of the target, and this probe must also take account of entropy.

The list of interesting special cases could be extended indefinitely. GRID will deal automatically with some of them, and directives are provided so the user can decide how to deal with others.

1.6
Nomenclature

Several different types of oxygen atom have already been mentioned including sp2 carbonyl oxygen, sp2 carboxyl oxygen, sp2 hydroxy, sp3 hydroxy, phosphate oxygens, aliphatic ether oxygen and furan oxygens. All of these have the same chemical symbol O, but each type of oxygen has its own specific properties. One needs to have a straightforward list of this detailed information for each kind of atom, showing all its characteristics on one line across the page or visual display screen. In order to reduce the amount of data on each line we therefore decided to give each atom a "*Type*" number which would specify its hybridisation and other electronic properties. Carbonyl oxygen for example, is a Type 8 oxygen.

Many different ways of tabulating the necessary information have been proposed by scientists working in various international agencies, national laboratories, universities and companies. The system adopted for GRID is based on the methods of the Protein Data Bank (PDB). Their nomenclature was agreed after extensive international discussions between workers in many fields, and it divides atoms into two distinct categories: "ATOM" and "HETATM".

1.6.1
"ATOM" Records

ATOM records are used to specify molecules which occur frequently in biological systems. These are called the "known molecules" and include amino acids, heme, cofactors, some of the "unnatural" amino acids used by medicinal chemists and a variety of molecules of general interest to GRID users. Here is a Protein Data Bank ATOM record:

ATOM   234 NZ LYS   28   21.361 29.854 65.530 1.00 81.36

and some aspects of this record require a brief explanation:
1. The PDB nomenclature uses the first six characters on a line in order to define different kinds of record, and this is an "ATOM" record.
2. The PDB nomenclature defines the sequence in which the atoms of a protein must be specified, and this happens to be the 234th atom of its protein.
3. The abbreviated name of each amino acid is also defined, and LYS is the abbreviated name for this amino acid which is lysine.
4. The name of each atom of an amino acid is defined, and this is "nitrogen zeta" of the lysine. The abbreviated name for this atom is defined as NZ.
5. This lysine is the 28th amino acid residue along the protein chain.

The next three numbers are the x, y, and z coordinates of this particular atom, measured in Angstrom using orthogonal reference axes. They are essential input data for any MIF, but the last two numbers 1.00 and 81.36 which represent the relative occupancy and isotropic temperature factor of the atom in the protein crystal, are not used by GRID.

A valuable characteristic of PDB format is that so much relevant information can be condensed into about 60 characters at the start of a line. The rest of the line can then be used by GRID for all the other data which are needed to specify the properties of an atom. The input program GRIN prepares all this data automatically, and writes it at the end of the line after the first 60 characters.

1.6.2
"HETATM" Records

Of course GRID users also need to study all sorts of molecules as well as proteins, and PDB format provides "HETATM" records as an easy way for doing this. If the user wanted for some special reason to define his protein structure using HETATM records, the same atom in the same molecule would appear like this:

HETATM   25 N3+ MOL   1   21.361 29.854 65.530 1.00 81.36

Notice the similarities and differences between the ATOM and HETATM records:
1. All the numbers and symbols are lined up in the same columns as before, but the record now begins with the word HETATM instead of ATOM. This name HETATM is an abbreviation for "heteroatom".
2. The protein is no longer being treated as a string of amino acids, but as a single molecule which can be given any molecule number (in this case 1).
3. The molecule is now called MOL instead of using the specific name LYS which was required by PDB format for lysine in a protein. GRID can accept any three-letter name for a molecule when HETATMs are being used, except names such as LYS which are reserved for "known molecules".
4. The sequence of ATOM records in a molecule is specified by PDB format. However HETATM records can be listed in any sequence, and the nitrogen has been moved to the 25th row of the new HETATM file. There is nothing special about its new position, and the user could just as easily have moved it to the first or last row of the file, or left it where it was.
5. There is a convention for HETATM names in GRID. They indicate the structure of the atom, so this N3+ nitrogen is a HETATM with three bonded hydrogens and it is positively charged as indicated by the + sign. Note in particular that 3 is the count of bonded hydrogens, and not the hybridisation which by coincidence is also sp3 in this case. Hybridisation and other electronic properties are defined by the type of an atom which is determined by the input programme GRIN when it prepares the data.
6. The x, y, z coordinates are unchanged, and the last two numbers 1.00 and 81.36 have not been altered. They will be overwritten by GRIN when it prepares the input data.

1.7
Calibrating the GRID Force Field

It is often convenient to think of drugs and proteins in terms of their chemical formulae and three-dimensional structures. However, an alternative interpretation is to regard the structure as nothing more than a set of frictionless rods and levers which transmit forces from one part of the system to another. This is the philosophy which underlies GRID, and it puts the main emphasis onto thermodynamics rather than structure. However it does raise a number of problems:

1. When the thermodynamic viewpoint has been adopted, it is the free energy of the system rather than the chemical structure of the molecules which needs the most careful study. Free energies can be most conveniently computed for reversible equilibria, and so the results from GRID should apply, strictly speaking, only to equilibrium systems. GRID has been found in practice to give useful predictions [8–11], but it is not easy to estimate the size of any errors caused by deviations from equilibrium.
2. Some free energy changes in biology are almost vanishingly small, while others may be greater by several orders of magnitude. The biggest changes often correspond to covalent reactions which break the "rods and levers", and these can completely swamp the weaker effects. We therefore decided to study only the ground state at body temperature, and so the GRID force field is not applicable to ligands which bind covalently to their receptor. In many cases this may just be another way of saying that GRID predictions are restricted to reversible equilibrium systems.
3. It is the differences and not the similarities between one drug molecule and another which are important, and the calibration of the GRID force field must be sensitive enough to differentiate between similar yet different atoms. For instance it would have been easy to assign the same parameters to the oxygen of an aliphatic ether and the oxygen of furan, but GRID would not have been able to differentiate between those two kinds of oxygen atom if this had been done. We therefore decided *not* to restrict the number of atom types in the force field, and we always welcome suggestions from GRID users, although the calibration of a new atom type is a nontrivial job which may take some considerable time. However, more than 10 different types of oxygen atom and 20 types of nitrogen have now been calibrated for GRID as a result of this policy.
4. One of the earliest decisions was to calibrate the GRID force field whenever possible by using experimental measurements rather than theoretical computations, and calorimetric measurements were therefore needed for the initial calibration in order to differentiate the enthalpic and entropic contributions to the overall free energy. However, only a very little calorimetric data was readily available at that time, about well characterised biological systems in which the structures of the interacting ligand and macromolecule were both known, and so a different approach was initially needed.

Fortunately several other kinds of experimental data were available for calibrating the GRID force field. Crystallographic measurements provided values for the Van der Waals radii of many atoms in all sorts of molecules, and corresponding but shorter radii were estimated for atoms making hydrogen bonds to each other. Many experimental determinations of atomic polarizabilities have been reported, and these were used together with the number of outer-orbital electrons in an atom to predict its Lennard-Jones interaction energy E_{LJ}. The observed structure of a molecule allows one to determine the bond order and hybridisation of its atoms, and hence to predict the maximum number of hydrogen bonds which each atom can donate and accept. Atomic charges can, in principle, be deduced from accurate X-ray data, but relatively few X-ray observations are precise enough for this job and theoretical methods were therefore used to estimate atomic charges. This only left the hydrogen-bond strength as an undetermined variable to be fitted to the observations.

All the necessary data was collected together in a file called GRUB which is revised whenever a new version of the program GRID is released. The first part of the GRUB file contains data values for ATOMS in known molecules (The natural amino acids, heme, cofactors, etc.). For example, there is an entry for the NZ ATOM of lysine. The second part of GRUB has individual HETATM values, and so it has an entry for N3+.

1.7.1
Checking the Calibration

Very many crystallographic observations have been reported on ligands bound to macromolecules. The structure of these complexes is usually measured to within a fraction of an Angstrom, and the formation of the crystals is determined by free energy. We therefore decided to use these readily available, crystallographically observed, ligand-macromolecule structures in order to check and refine the GRID force field after the initial calibration. In the absence of appropriate calorimetric measurements one cannot know whether enthalpy and entropy each make their appropriate contributions to the overall energy values computed by GRID. However as crystal structures were used to check and refine the force field, it seemed reasonable to hope that GRID would be able to predict the location of favorable binding sites, and this is indeed the case [8,11].

1.7.2
Checking Datafile GRUB

Users may often want to edit their copy of GRUB, or copy it from one directory to another, and of course mistakes may be made. It is therefore easy for errors to creep into the stored parameters, and so various kinds of check are made:
1. The input programme GRIN always analyses the input data before each new GRID computation. It checks both the structure of the target and the integrity of datafile GRUB, and warns the user about any doubtful features

in either file. In particular, it always reports the overall electrostatic charge of the target, because MIF computations can give very misleading results if the total charge of the whole system is significantly different from zero.

2. We frequently check GRID maps prepared from high resolution X-ray structures, because significant calibration errors in datafile GRUB would cause a systematic bias in the output. Many mistakes in the datafile were corrected in this way when GRUB was first being prepared, but such changes are not required so often now.

3. Regular users would quickly detect errors in the output from GRID by visual inspection of their maps, and we have had valuable feedback from users for many years. They would let us know very quickly if probes were being systematically predicted in the wrong place. In the 1986 release of GRUB, for example, something was wrong with the amino acid histidine and it was a great help to learn about this from a user.

4. Another kind of check comes from people who use the results generated by GRID as input data for further computations. For instance, we learned in 1997 that GRID was giving statistically biased results for compounds which contained acetylenic carbon atoms, and we had not been aware of this until a GRID user informed us. A reappraisal showed that relatively little information about acetylenes had been available when the GRID force field was first being parametrised, and a slight adjustment brought acetylene into line with the rest of the calibration data once we knew about the bias. This shows that the statistical analysis of GRID results can make an important contribution to the improvement of the force field.

Particular emphasis must be placed on the importance of checking all the input data before beginning any MIF computation. Of course error checking is not a satisfying job, but more problems seem to occur because of input errors than for any other reason. Program GRIN always checks the input thoroughly by default, and this helps to diminish the workload, but some users ignore error messages or try to save a little time by altering directive LEVL to a low value which turns off checking altogether. This is not recommended, and for our own research we never set LEVL in programme GRIN below the default value of 3.

1.8
The Output from GRID

The GRID method was explicitly designed in order to get selective information about binding sites, and the output can be used in two quite distinct ways:

1. To prepare GRID maps which are intuitively easy to understand, and can therefore provide a focal point for discussions between people with backgrounds in different fields of science, and indeed for people with little formal scientific training.

2. To generate matrices of numerical data which can be analysed statistically.

This article is not the place in which to consider statistical methods in detail, but the use of GRID maps to interpret interesting features of molecular structures will now be described.

1.8.1
GRID Maps from Macromolecules

Figure 1.5 shows an amphipathic alpha helix whose structure as part of a large globular protein was observed by X-ray crystallography. No hydrogens are displayed, and the helix has been separated from the rest of the protein in order to have an uncluttered figure. The side of the helix which faces towards the bottom of the page contains alanine, leucine and similar hydrophobic amino acids. The opposite side of the helix (the top side in the figure) has a prominent lysine side chain and other polar groups. This type of amphipathic helix has often been observed floating on the outside surface of globular proteins, with its hydrophobic amino acids facing towards the globular centre and its polar side chains in the surrounding water phase.

The flexible side chain of the lysine (CH_2–CH_2–CH_2–CH_2–NH_3^+) is displayed in an all-trans conformation in Fig. 1.5 because it was assigned all-trans coordinates by the X-ray crystallographer. An all-trans structure like this is often reported when the atoms of a side chain are in such vigorous dynamic thermal motion that they cannot be detected by X-ray methods. Arbitrary all-trans coordinates are then assigned by default, because the crystallographer knows that the amino acid is lysine from the DNA sequence although he cannot observe the side chain atoms himself.

The GRID map in Fig. 1.5 was deliberately prepared in order to demonstrate how easy it is to obtain misleading results when inappropriate directives are thoughtlessly used for an MIF computation. GRID would never normally generate such a deceptive map, and a special set up was needed in order to force it to prepare Fig. 1.5 at all. The blue sphere marks the terminal N3+ group of the lysine side chain, and GRID was deliberately used on the implausible and unrealistic assumption that the helix and its side chains were all completely rigid. There are three energy minima (colored red in the figure) corresponding to the three hydrogen atoms of the cationic nitrogen, and these minima misleadingly suggest that an incoming ligand would be able to make particularly favorable interactions in these three highly localised positions. This must be an incorrect conclusion if the side chain is actually sweeping backwards and forwards in vigorous motion across a relatively wide region.

Figure 1.6 shows what happens when slightly more appropriate settings are used for the GRID run. Torsional rotation of the sp3 amino group is now allowed round the terminal CH_2–NH_3^+ bond of the side chain, and so a halo is generated. The halo in Fig. 1.6 is not a uniform ring because there would be eclipsing between the CH_2 and NH_3^+ hydrogens at some torsion angles, and eclipsing would be energetically unfavorable. This description of the NH_3^+ interactions might be reasonable if the methylene groups of the side chain were buried within the bulk

1 The Basic Principles of GRID

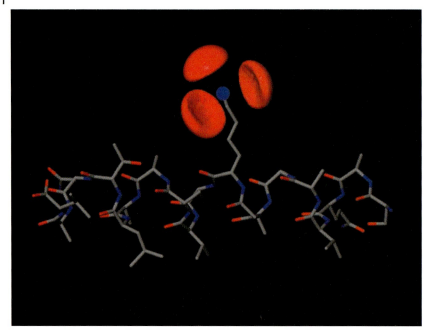

Figure 1.5. An alpha helix with a lysine side chain. The terminal NH_3^+ group of the lysine is marked with a blue sphere. GRID was deliberately misused to prepare this figure on the unrealistic assumption that the helix and its side chains were all completely rigid. This GRID map is therefore misleading, and this figure demonstrates how important it is to use MIF programs with great care because it is very easy to obtain deceptive results by misusing any MIF program. See text.

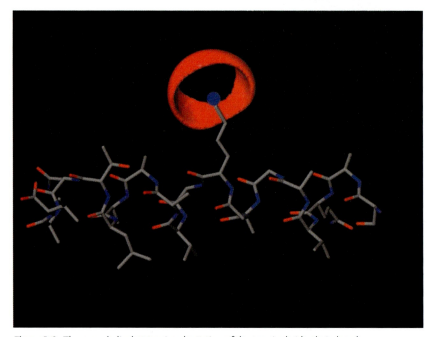

Figure 1.6. The same helix, but torsional rotation of the terminal side chain bond is now permitted, and this map is slightly more realistic than Fig. 1.5. See text.

1.8 The Output from GRID

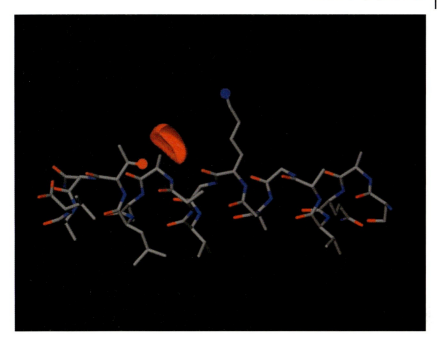

Figure 1.7. The same helix when the whole side chain is allowed to move freely. GRID now detects a favorable binding site where the hydroxy group (red sphere) of a threonine side chain and the terminal NH_3^+ group (blue sphere) of the lysine can both interact simultaneously with the probe. See text.

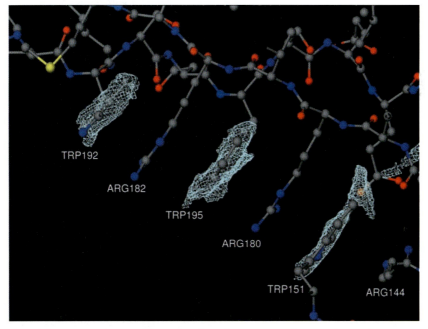

Figure 1.8. Some residues near the surface of another protein. GRID is used to elucidate why the polar arginine and nonpolar tryptophan side chains pack so closely together. See text.

of the protein, as they often are in some proteins because methylene is a hydrophobic moiety. Movement of the buried methylenes would then be restricted, and the most significant torsional rotation in the lysine side chain might be around the CH_2–NH_3^+ bond. However, the methylene groups of the lysine in Fig. 1.6 are not deeply buried, and the side chain with its nitrogen is actually free to move over a wide region, so Fig. 1.6 is nearly as misleading as Fig. 1.5.

Programs GRIN and GRID always search the target for any small parts which can move freely. However the user must set a dedicated directive (called MOVE) if he wants this feature to be used in his computations. This directive never allows total flexibility because the whole structure of the target might then unravel. Domains are always treated as rigid units, and when MOVE has been set the helix backbone is still treated as a rigid domain. However side chains can now move [12], and the resulting GRID map is shown in Fig. 1.7.

A small red sphere now marks the hydroxy group of a nearby threonine which is on the same side of the helix as the NH_3^+ of the lysine, and directive MOVE alerts GRID to the proximity of these two polar groups. GRID then tests whether the flexible lysine side chain could reach far enough in the direction of the threonine, so that a probe might be able to interact with both the NH_3^+ and the threonine hydroxy group at the same time. In this example the geometry is acceptable, but the torsional flexibility of the lysine would be restrained by its interaction with the probe, and an entropic allowance must be made for this. However the enthalpic benefit of two good hydrogen bonds outweighs any entropic penalty for torsional restraint, and so this is a particularly appropriate place for an incoming ligand.

The red region in Fig. 1.7 shows where a probe would be located when interacting with both the threonine hydroxy and the lysine NH_3^+ groups. However, it was necessary to contour this GRID map at a slightly more negative energy level than Fig. 1.6 in order to show the result clearly, and Fig. 1.7 was therefore contoured at an energy roughly corresponding to a pair of hydrogen bonds. At this energy level there is no blurring due to the weaker interactions which the NH_3^+ group would make as it searched through wide regions alone at the end of its side chain, and the absolute minimum near the threonine is unmistakable in the GRID map.

A completely different application of GRID is illustrated in Fig. 1.8 which shows several amino acids in the cytokine binding region of the human protein gp130. The surface of the protein faces towards the bottom left corner of the figure, and there is a sandwich structure in this part of the macromolecule where alternate tryptophan (TRP) and arginine (ARG) side chains lie one above each other like slices in a loaf of bread. The close-packed TRP–ARG relationship is unexpected because arginine is one of the most polar amino acids, while the side chain of tryptophan consists almost entirely of nonpolar hydrocarbon groups.

GRID maps suggest an interpretation of this structure. Tryptophan has two aromatic rings with nine CH and CH_2 groups but only one nitrogen which can make one hydrogen bond and no more. Arginine is very polar because it has a permanent cationic charge and three nitrogen atoms in a guanidinium group which can donate up to five hydrogen bonds. The cationic charge of guanidinium tends to increase the strength of its donated bonds, but arginine would never accept a hydrogen bond from a tryptophan side chain. Moreover a glance at Fig. 1.8 shows that the observed geometry does not permit the hydrogen bonding of either ARG180 or ARG182 to the nearby tryptophan rings, and it is most surprising that such very polar residues should be squashed between the hydrophobic tryptophans in such an apparently unfavorable position. Arginines also have long flexible side chains which are frequently found in vigorous motion, like the lysine in Fig. 1.7, but the X-ray findings from gp130 show that the side chain methylene groups of ARG180 and ARG182 are not moving much more than the adjacent main-chain alpha-carbon atoms. This is another surprising feature of the observed structure, because there must be an entropic penalty when these flexible side chains are so firmly pinned down.

The PDB structure for this protein (PDB Reference: 1BQU) was therefore edited to remove the side chains of TRP192, TRP195 and TRP151, and thus make way for GRID probes to explore the volume normally occupied by these bulky moieties. The edited file was then used to prepare a target for GRID, and the contours in Fig 1.8 were generated using the hydrophobic probe on this target. The tryptophan side chains are shown in their observed positions (although the map itself was generated *when they were absent*), and the contours show that GRID predicts a hydrophobic region roughly surrounding each tryptophan ring. Further examination shows that the extended methylene chains of arginines 180 and 182 are almost ideally arranged for making hydrophobic interactions with the tryptophans, although 144 has a less favorable crumpled conformation.

However, it is not only the arginine methylene groups which generated the hydrophobic contours. GRID also predicts that the top and bottom faces of the arginine guanidinium groups can make favorable hydrophobic interactions, because the hydrogen bonds of guanidinium are so very firmly constrained to the plane of the guanidinium system. This is particularly well shown by the contours surrounding TRP195, which extend well beyond the reach of the methylene groups in the side chains of ARG180 and ARG182.

One must conclude that hydrophobic interactions may stabilise the multilayer TRP–ARG sandwich of gp130, in spite of the different character of these two amino acids, and in spite of the entropic penalty mentioned above. However the gp130 crystals themselves came from a solution which contained glycerol molecules and sulfate ions, and both of these components were trapped in the crystals where they may have helped to stabilise the observed protein structure. It would not be altogether surprising if some alternative conformation or conformations of gp130 may also occur *in vivo*, if those somewhat unphysiological substances are not present in the human body in sufficient concentrations to stabilise the structure as observed.

1.8.2
GRID Maps from a Small Molecule

Leucine is an amino acid, and is one of the building blocks of proteins. It has a nitrogen atom which is shown as a blue sphere in Fig. 1.9, a pair of oxygens both shown in red, and a small cluster of hydrophobic groups shown as yellow spheres towards the left of the figure. Hydrogen atoms are not displayed in order to keep the picture as clear and simple as possible, but the N3+ nitrogen (blue) has three bonded hydrogens and is therefore cationic. The carboxy oxygens have no bonded hydrogens and are negatively charged, so the molecule taken as a whole is electrically neutral.

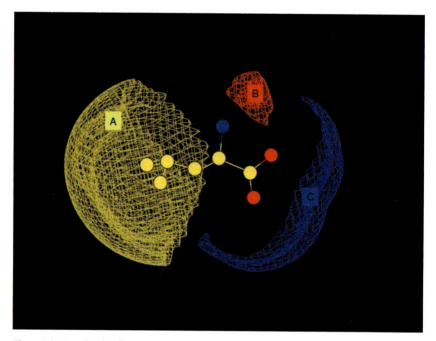

Figure 1.9. A molecule of leucine with GRID maps for a hydrophobic probe (A, yellow); a multiatom cis-amide probe (B, red); and an sp3 NH_3^+ probe (C, blue). See text.

This very simple target was chosen in order to demonstrate the selectivity of the GRID method. The yellow contours (A) were generated using the hydrophobic probe, and they show that one part of the amino acid is nonpolar and very hydrophobic. Binding clefts on biological macromolecules often expose a hydrophobic surface, and it is very important to detect the hydrophobic surfaces of ligands if one wishes to design high affinity molecules.

The blue contours (C) in Fig. 1.9 were generated by the N3+ cationic amine probe which makes good hydrogen bonds to the (red) carboxy oxygen atoms of the target. When generating these blue contours GRID takes account of the fact that both the interacting atoms are charged, and can therefore make a particularly strong hydrogen bond to each other. GRID also takes account of the local dielectric, and the electrostatic attraction between the target's oxygens and the cationic probe is therefore attenuated towards the right of the figure, because this is where the probe would be most exposed to the higher dielectric of the surrounding bulk water.

It is important to give careful consideration to apparently small details in a GRID map. The carboxy group in Fig. 1.9 may appear to be symmetrical, but the blue contours are stronger round the oxygen at the bottom of the figure. This difference between the oxygens may be caused by two quite distinct influences:

1. The upper oxygen is closer to the cationic nitrogen N3+ of the leucine, and so the cationic N3+ probe may experience an unfavorable electrostatic repulsion when it is close to the leucine nitrogen's cationic charge.
2. The bottom oxygen in the figure is partly shielded from bulk water by the hydrophobic moiety (yellow spheres) of the target, and so its dielectric environment may tend to favor the electrostatic attraction of a cationic probe.

The red contours (B) at the top of Fig. 1.9 were generated by a multiatom amide probe CO.NH which was arranged cis so that its hydrogen and oxygen are both on the same side of the CN axis. This multiatom probe therefore detects regions where it can donate a hydrogen bond from its nitrogen and can accept at its carbonyl oxygen. It sits between the N3+ group and a carboxy oxygen of the target, and is at a slightly awkward angle because the hydrogen bonding atoms of the target do not line up perfectly with those of the multiatom probe.

There are many other selective probes which can be used to elucidate the properties of a target. For instance the "amphipathic probe" finds boundary surfaces where a part of the target with polar characteristics touches neighboring hydrophobic regions. It would draw attention to a boundary of this type which runs up the middle of Fig. 1.9 where it separates the polar nitrogen and oxygen atoms on the right from the hydrophobic carbons towards the left of the figure. Each additional probe provides qualitatively different information, and competing research groups may reach interestingly different conclusions when studying the same set of molecules simply because they select different probes for their investigations.

1.9
Conclusions

The design of molecular interaction fields has been a matter of continuously improving approximations, and no force field is perfect. It is therefore critically important to choose the right MIF for each particular job, and the GRID force field was explicitly designed for use with the GRID method. This is an approach which generates selective information about binding sites on proteins, therapeutic agents and other important biological molecules of known structure. The output is intuitively easy to understand, and can provide a focal point for discussions between people with backgrounds in different fields of science. Results from GRID can also be analysed statistically.

In this chapter some emphasis has been placed on the subjective influences which can often modify the results of force field computations, and examples of subjective decision making have been provided. However the need to make decisions can enhance the GRID user's intuitive understanding of noncovalent interactions between molecules, and this enhancement was a prime objective when the GRID method was first being devised. Figures 1.5, 1.6 and 1.7 show how important it is to have a proper understanding of the system under investigation, if one wants to obtain meaningful results.

The GRID force field has always been calibrated as far as possible by studying experimental measurements, and the calibration is then checked by studying how well GRID predicts observed crystal structures. Crystal packing is determined by free energy considerations rather than by enthalpy alone, and recent versions of the force field include entropic terms. GRID can detect the hydrophobic binding regions which are so important when high-affinity ligands are being designed, and it can also detect sites for the polar groups which determine ligand selectivity. The GRID method is being systematically extended, and new versions are issued from time to time.

Acknowledgment

To print a list of the teachers and colleagues who have helped me would occupy too much space here. My debt to all of them is immense, and is only matched by my gratitude to each one for their generosity, and to my wife for all her support and encouragement.

References

1 M.L. Huggins, L.Pauling *Z. Kryst.* **1934**, *87*, 205–238.
2 P. J. Goodford *J. Med. Chem.* **1985**, *28*, 849–857.
3 G. Cruciani, K. A. Watson *J. Med. Chem.* **1994**, *37*, 2589–2601.
4 L. D. Landau; E. M. Lifshitz *Course of Theoretical Physics*, Englis (ed.), Pergamon Press, Oxford, **1960**, Vol. 8.
5 D. N. A. Boobbyer, P. J. Goodford, P. M. McWhinnie, R. C. Wade *J. Med. Chem.* **1989**, *36*, 1083.
6 R. C. Wade, J. Clark, P. J. Goodford *J. Med. Chem.* **1993**, *36*, 140–147.
7 R. C. Wade, P. J. Goodford *J. Med. Chem.* **1993**, *36*, 148–156.
8 M. von Itzstein et al. *Nature (London)* **1993**, *363*, 418–423.
9 P. J. Goodford *J. Chemometrics* **1996**, *10*, 107–117.
10 T. Langer *Quant. Struct.-Act. Relat.* **1996**, *15*, 469–474.
11 A. Berglund; M. C. De Rosa, S. Wold *J. Comput-Aided. Mol. Design* **1997**, *11*, 601–612.
12 P. J. Goodford **1998** Rational Molecular Design in Drug Research, in *Alfred Benzon Symposium 42*, Munksgaard, Copenhagen, Liljefors (ed.), **1998**, 215–226.

2
Calculation and Application of Molecular Interaction Fields
Rebecca C. Wade

2.1
Introduction

Molecular interaction fields (MIFs) can be calculated for any molecule of known three-dimensional (3D) structure. A MIF describes the spatial variation of the interaction energy between a molecular target and a chosen probe. The target may be a macromolecule or a low molecular weight compound or a molecular complex. The probe may be a molecule or a fragment of a molecule. MIFs can be applied in many ways [1]. They can guide the process of structure-based ligand design, their original intended application in the GRID program [2]. They may be used to dock ligands to macromolecules [3–5]. They are frequently used to derive quantitative structure–activity relationships (QSARs) for low molecular weight compounds [6] but can also be used to study the structure–activity relationships of macromolecules [7, 8]. MIFs can also be applied to the prediction of pharmacokinetic properties, such as in the VolSurf methodology [9].

In this chapter, I will first describe how MIFs are computed and then give selected examples of how MIFs can be applied. MIFs will be described primarily with reference to their calculation with the GRID program [10]. Other programs may be used to compute MIFs; these have different energy functions and parametrizations.

2.2
Calculation of MIFs

2.2.1
The Target

The starting point for a MIF calculation is provided by the atomic coordinates of the target molecule. These may have been determined experimentally or theoretically. In many calculations of MIFs, the target is treated as a rigid structure. How-

Molecular Interaction Fields. Edited by G. Cruciani
Copyright © 2006 WILEY-VCH Verlag GmbH & Co. KGaA, Weinheim
ISBN: 3-527-31087-8

ever, in applications, it is often important to treat target flexibility, at least partially. There are several strategies for doing this:
- Compute MIFs for multiple conformations of the target; the conformations may come from an NMR ensemble or from conformational searches or molecular dynamics simulations.
- Permit adaptation of the position of some atoms in the target to optimize the interaction energy of the probe during calculation of the MIFs. In the GRID program, this is routinely done for rotatable hydrogen atoms. It is also possible for the user to do calculations with specified side chains containing several nonhydrogen atoms treated as movable in response to the probe position.

The target is also usually considered to have a single titration state and to be unaffected by the position of the probe. However, the GRID program does allow for probe-induced switching between histidine tautomers.

The target may consist of a single molecule or a complex of molecules or molecules and ions, such as e.g. for metalloproteins. Well ordered water molecules may also be considered part of the target. GRID also permits the possibility for a water-bridged target–probe interaction to be considered without a priori defining the position of the water molecule. The remaining solvent molecules are treated as a continuum that modulates the interaction energy between probe and target and may also have an entropic effect on the probe–target interaction.

2.2.2
The Probe

MIFs are computed for positions of the probe at points on a rectilinear grid superimposed on the target. It is this grid that gives the GRID program its name.

Grids of target–probe interaction energy values can be read into many molecular graphics programs which can display the MIFs as isoenergy contours or project the energies onto molecular surfaces.

The probe is placed at each grid point in turn to compute a MIF. At its simplest, the probe is a unit positive charge representing a proton; in this case the MIF is the molecular electrostatic potential (MEP). Most probes are spheres parametrized to represent a specific atom or ion type. Hydrogen atoms are usually treated implicitly. For example, a carbonyl carbon atom and a methyl CH_3 group are both treated as spherical probes but the methyl probe has a larger radius because of the hydrogen atoms bonded to the carbon atom. Polar hydrogen atoms that can make hydrogen bonds can also be treated implicitly but the directional character of their hydrogen bonds must also be modelled (see below).

Nonspherical probes containing more than one nonhydrogen atom may also be used. In this case, the relation between the probe position and orientation and the grid point is not completely symmetric and has to be defined. In the GRID programme, a carboxylate group, for example, is treated as a 3-point probe with one of the oxygen atoms centered on the grid point. At each grid point, the probe is rotated around the grid point oxygen to energetically optimize its orientation.

2.2.3
The Interaction Function

Above, MIFs have been described as fields of probe–target interaction energies. The MIFs thus provide information on where the favorable and the unfavorable locations for the probe around the target are. The probe–target interaction function is typically an empirical molecular mechanics energy function. Its functional form is chosen to represent the underlying physical interactions and the function is parametrized against experimental observations. The energy function may contain explicit entropic terms or be parametrized such that it represents an interaction free energy rather than just an interaction energy. More recently, functions, sometimes referred to as potentials of mean force or knowledge-based functions, derived from statistical analysis of molecular structures have been derived to map out favorable and unfavorable locations for a probe with respect to a target molecule [11–15]. These functions, unlike an energy function, are not required to take a physically meaningful form. Given enough experimental data, they may reproduce experimental observations better than empirical energy functions but this can be at the disadvantage of a lack of analytical differentiability and of a transparent physical interpretation. They are most often used as scoring functions for ligand docking but some can be used to compute MIFs.

The empirical energy functions used to compute MIFs can consist of the sum of one or more terms. In the GRID program, the energy function is given by the following terms:

$$E = E_{VDW} + E_{EL} + E_{HB} + S \tag{1}$$

where E_{VDW} is the van der Waals energy, E_{EL} is the electrostatic energy, E_{HB} is the hydrogen-bond energy and S is an entropy term.

2.2.3.1 Van der Waals Interactions

Atomic repulsion and induced dipole–induced dipole dispersive attraction are typically described by a Lennard-Jones function [16, 17]:

$$E_{VDW} = \sum_{pt} \frac{A}{r_{pt}^{12}} - \frac{C}{r_{pt}^{6}} \tag{2}$$

This is summed over all probe (p)–target (t) atom pairs and is a function of the distance, r_{pt}, between the atoms in the pair, which have van der Waals radii, R_p and R_t. $A = 0.5C(R_p + R_t)^6$ and C is given by the Slater–Kirkwood formula [18] and is dependent on atomic polarizability and the number of effective electrons per atom.

The functional form of the repulsive part is purely empirical and was chosen by Lennard-Jones to fit experimental data for rare gases and because of its computational convenience. For some uses of MIFs, it is useful to make the repulsion less abrupt by using a lower power of distance, e.g. r_{pt}^8. This has the advantage that the MIF calculation is less sensitive to the atomic positions and thus more robust to errors in atomic coordinates or to the effects of atomic motions.

2.2.3.2 Electrostatic Interactions

Electrostatic interactions have long-range character. This makes them very important for MIF calculations. The simplest way to compute the electrostatic interactions is to compute a Coulombic energy:

$$E_{el} = \sum_{pt} \frac{q_p q_t}{4\pi\varepsilon_0 \, \varepsilon_r r_{pt}} \quad (3)$$

where ε_0 is the permittivity of free space, ε_r is the relative dielectric constant of the surrounding medium, and q_p and q_t are partial atomic point charges in the probe and target respectively.

The difficulty with computing electrostatic calculations is that molecular systems are heterogeneous media composed of molecules with different dielectric properties. The solvent is usually treated implicitly as a dielectric continuum. While water has a value of ε_r of about 80, proteins have been assigned values of ε_r ranging from 2 to 80, varying according to location and type of calculation. To account for the environment, ε_r may be assigned a constant value or treated as a function of interatomic distance, r_{pt}. To account for the dielectrically discontinuous boundary between molecules and the implicit solvent, the method of images may be used (as in GRID), or the Poisson(–Boltzmann) equation solved.

The method of images treats the dielectric boundary as an infinite plane whose effect can be modelled by the addition of image charges. Computation of the electrostatic energy using the method of images requires estimation of the depths of the atoms in the target. In GRID, this is done by counting the number of neighboring target atoms within a specified distance and translating this into a depth using a precalibrated scale [2].

Numerical solution of the Poisson–Boltzmann equation permits the irregular shape of the dielectric boundary to be accounted for. Finite difference and multigrid methods on grid(s) superimposed on the target molecule are the most commonly used methods to solve the Poisson–Boltzmann equation. Details of solution of the Poisson–Boltzmann equation for biomacromolecules are given in references [19–22]. To solve the Poisson–Boltzmann equation, a suitable value of the dielectric constant for the molecular interior and the definition of the dielectric boundary must be chosen. These are adjustable parameters to be fitted in the context of the complete energy function, the treatment of molecular flexibility, and the properties to be computed [23, 24]. The dielectric boundary can be chosen as the van der Waals surface, the solvent accessible molecular surface (mapped by a solvent probe surface) or the solvent accessible surface (mapped by a solvent probe center). The choice of dielectric boundary definition can significantly impact the magnitude of electrostatic binding free energies [25, 26].

Suitable partial atomic charges must be assigned in order to compute the electrostatic energy. As for all other molecular mechanics calculations, it is important that all the charges are consistent with each other and with all the other parameters and the complete energy function. The charges used in different empirical energy functions or force fields may vary considerably but this difference may be compensated in other parts of the energy functions. Charges may be derived from *ab initio* or semiempirical quantum mechanical calculations, or simpler approaches such as electronegativity equalization. Overall, the energy function may be parametrized to reproduce structural properties and/or energetic properties such as solvation energies.

Specific polarization effects, beyond those modelled by a continuum dielectric model and the movement of certain atoms, are neglected in MIF calculations. Many-body effects are also neglected by use of a pair-wise additive energy function. Polarizable force fields are, however, becoming more common in the molecular mechanics force fields used for molecular dynamics simulations, and MIFs could be developed to account for polarizability via changes in charge magnitude or the induction of dipoles upon movement of the probe.

2.2.3.3 Hydrogen Bonds

Hydrogen bonds are important for MIF calculations because of their specificity. Hydrogen bonds are short-range, directional interactions that have distance and angular dependences on the arrangements of the atoms involved. The strength of hydrogen bonds is determined by a combination of attractive electrostatic, charge-transfer, polarization and dispersion components and a repulsive electron-exchange component, all of whose relative magnitudes vary. In many molecular mechanics force fields, hydrogen bonds are modelled implicitly by means of Coulombic and van der Waals terms. However, in some, including GRID, there is a separate hydrogen-bond term [27–29]. A special hydrogen-bond term is particularly important for a polar probe like a water molecule that is modelled as a sphere with two implicit hydrogen atoms but for which, nevertheless, the geometry of the hydrogen bonds that it makes must be treated explicitly [29].

In GRID, the hydrogen-bond energy is the product of three terms:

$$E_{hb} = E_{r_{pt}} E_t^\theta E_p^\phi \tag{4}$$

The functional form and parametrization of these three terms is empirical and constructed to reproduce experimentally observed structures in the context of the complete energy function. Initially, these were mostly small molecule structures from the Cambridge Structural Database (CSD). As the number of protein structures determined experimentally has grown, the Protein Databank (PDB) has become the primary source of structural data for parametrizing hydrogen-bond functions. For example, in recent work it was possible to develop a hydrogen-bond angular function for fluorine atoms occurring in ligands in GRID based on observations in the Protein Databank [5].

$E_{r_{pt}}$ is dependent on the separation between target and probe nonhydrogen atoms participating in the hydrogen bond. It has the form:

$$E_{r_{pt}} = \frac{M}{r_{pt}^m} - \frac{N}{r_{pt}^n} \tag{5}$$

where M and N depend on the chemical nature of the hydrogen-bonding atoms. Possible values of the m and n parameters are $m = 6$, $n = 4$; $m = 8$, $n = 6$ (as in GRID); $m = 12$, $n = 10$.

The angular terms take different functional forms depending on the chemical types of the hydrogen bonding atoms and whether they are in the probe or the target. The angular dependence differs for the same atom type in the probe and in the target. This is because the target includes the interactions of the hydrogen-bonding atom's neighbors whereas these are absent for the probe. The probe also has more freedom to rotate to an optimal orientation than a hydrogen-bonding atom in the target. At each grid point, the probe is rotated to the orientation that results in an optimal hydrogen-bond energy and, when there are multiple hydrogen bonds possible, the best combination is found by systematic search or analytically. In GRID, the target hydrogen atoms and lone pairs of electrons in the target may also be allowed to move to make the most favorable hydrogen bonds with the probe. For an sp3-hybridized hydroxy group, for example, the hydrogen assumes the position on a circular locus perpendicular to the C–O bond and subtending a tetrahedral angle to the C–O bond that results in the most energetically favorable interaction with the probe.

2.2.3.4 Entropy

When a probe binds a target molecule, it may displace ordered water molecules and it may result in ordering of a flexible part of the protein. Both these events have entropic effects. To account for these, an entropy term has been included in more recent versions of GRID and takes different forms according to the type of calculation. It is computed when parts of the target are treated as flexible, when the probe interactions are compared to that of water, and for the hydrophobic probe, known as the 'DRY' probe.

The entropic cost of reducing the flexibility of the target by binding a probe can be estimated by a term proportional to the reduction in the number of accessible target rotamers when the probe–target interaction is energetically favorable. It should be born in mind that this is a very approximate way to estimate the entropic change due to a change in mobility upon binding. Binding can be entropically unfavorable or favorable due to changes in the motions of a target protein upon binding a ligand [30]. Experimental [31, 32] and theoretical [33] studies suggest that protein backbone entropy can increase upon ligand binding, particularly to stabilize binding of a small hydrophobic ligand, and that the increased protein mobility can be remote from the ligand binding site. These effects are difficult to account for in MIF calculations in which the probe is only a fragment of a ligand; the rotamer-dependent entropic term provides a measure of entropic cost local to the probe position.

In GRID [10], the favorable entropic contribution due to the displacement of one water molecule from a hydrophobic surface is assumed constant. Its value is obtained by comparing the possible hydrogen bond combinations that the water molecule can form when at the hydrophobic surface and when in bulk water. In bulk water, the water molecule is assumed to make three hydrogen bonds from the possible four and there are four possible combinations of three hydrogen bonds (1,2,3; 1,2,4; 1,3,4; 2,3,4). The entropy change is simply given by $RT\ln(4) = -0.848\,\text{kcal mol}^{-1}$. This value is larger in magnitude than a typical attractive Lennard-Jones interaction between two atoms (about $-0.2\,\text{kcal mol}^{-1}$) but smaller in magnitude than a typical hydrogen-bonding interaction (about -2 to $-4\,\text{kcal mol}^{-1}$).

The hydrophobic effect can be ascribed to entropic and enthalpic effects. The 'DRY' probe detects hydrophobic regions. It can be considered to be like an 'inverse' water probe. It makes a Lennard-Jones interaction in the same way as a water probe. It is also neutral like a water probe and has no electrostatic interaction term. The hydrogen-bond energy is however inverted to reflect the fact that polar parts of the target that are able to make hydrogen bonds will not be energetically favored next to a hydrophobic probe. In addition, the constant entropic term of $-0.848\,\text{kcal mol}^{-1}$ is added to the total interaction energy. The 'DRY' hydrophobic probe is very useful for detecting hydrophobic patches on proteins and can also be used in ligand docking.

2.3
Selected Applications of MIFs

2.3.1
Mapping a Ligand Binding Site in a Protein

The classic way to use MIFs is to identify energetically favorable binding sites on a macromolecular target for probes and use these to design better binding ligands. This was done, for example, by Mark von Itzstein and colleagues, when they identified an energetically favorable site for an amino or guanidino group in the active site of the influenza virus neuraminidase using GRID [34]. This enabled a transition-state analogue lead compound to be modified by adding a guanidine substituent and this resulted in the drug, Relenza, which is used clinically against influenza. The application of GRID MIFs to the design of anti-influenza agents is reviewed in [35].

The computation of MIFs is generally very helpful for identifying regions where substituents could be added to known ligands and then using the MIFs in conjunction with docking or *de novo* design tools to design further compounds [36, 37]. MIFs may also be applied to detecting selective regions by visual comparison of MIFs with different probes or with different targets, or in a systematic way by the GRID/PCA procedure [38]. GRID is also often applied to the location of binding sites for ordered water molecules in biomacromolecules, particularly water molecules bridging protein–ligand interactions, see e.g. [39].

2.3.2
Deriving 3D-QSARs

Probably the most widespread application of MIFs is the derivation of 3D-QSARs by the CoMFA [6] or GRID/GOLPE [40, 41] approach for low molecular weight compounds. A series of compounds for which experimental measurements of binding or activity have been made is modelled and the structures are aligned. MIFs are computed for each molecule and then the values of the MIFs at the grid points are correlated with activities using PLS (Partial least squares projection to latent structures). Important regions around the molecules for explaining differences in activity in the series are detected and these guide the further design of new molecules.

That this approach is also applicable to macromolecules was demonstrated recently in a study of WW domains [7]. WW domains are small domains (ca. 40 amino acids) occurring in a wide variety of proteins, with a characteristic sequence that includes two tryptophan (W) residues. They form a three-stranded beta sheet and are stable without the formation of disulfide bridges or the presence of cofactors. WW domains bind to peptides that are often proline rich. There are several classes of WW domain, each having different peptide binding specificity. To compare and classify the binding properties of a set of WW domains, MIFs were computed for the WW domains. MEPs were compared by PIPSA (see below). MIFs were correlated with measurements of binding to one peptide, a proline-rich peptide containing tyrosine, which one class of WW domains recognises. A set of 23 WW domains was used to build the 3D-QSAR models, including WW domains that recognise tyrosine-containing peptides, as well as WW domains that can be considered representative of 'nonbinders' to tyrosine-containing peptides.

3D-QSAR models were made with both the GRID/GOLPE combination (GRID used for computing MIFs and GOLPE [42] used for chemometric analysis) and with CoMFA (with computation of MIFs and chemometric analysis as implemented in SYBYL (Tripos)). For the GRID/GOLPE model, several probes (CH_3, DRY, H, PO_4 and combinations of these) were tested. The methyl (CH_3) probe was found to describe the surface properties (mainly the shape properties) relevant for explaining the tyrosine recognition activity best and was therefore used to build the final GRID/GOLPE PLS model. A four latent variable (component) model was obtained with excellent fitting (R^2 = 0.99) and predictive performance (Q^2 = 0.88, standard deviation of errors of prediction SDEP = 0.03) determined by leave-one-out analysis. The most important regions for binding specificity to tyrosine-containing peptides detected in the GRID/GOLPE model are shown by the labeled contoured regions in Fig. 2.1. With CoMFA, molecular electrostatic and steric fields were computed for the region of the peptide binding site. In the CoMFA QSAR model, which gave values of Q^2 = 0.6 and SDEP =0.07, the same three regions also contributed most. The region of positive coefficient was detected not only by the steric field but also by the electrostatic one, consistent with the large residues being positively charged at residue 35 of the WW domains.

The MIF-based QSAR models fit well with the known peptide specificities of the WW domains, and aid identification of the important determinants of binding specificity.

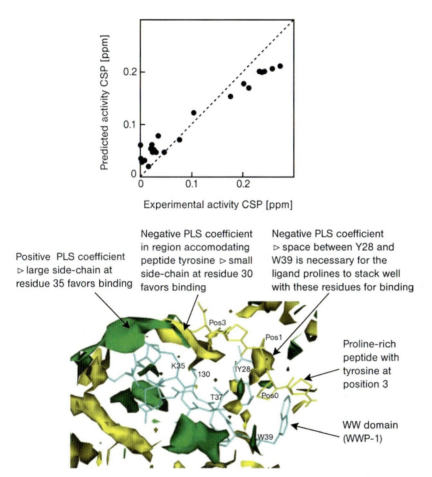

Figure 2.1. Results of chemometric analysis by the GRID/GOLPE method for 23 different WW domains. (a) Plot of the predicted vs. experimental NMR chemical shift perturbation (CSP). (b) Important contributions to recognition of tyrosine-containing peptides are shown. The WW domain (WWP3-1) is shown in blue, the peptide in yellow, and the PLS coefficients for the GRID/GOLPE model for a methyl probe are shown by the contours. Positive weighted PLS coefficients (contour level 0.00005) are colored green, negative PLS coefficients (contour level −0.00005) are colored yellow, respectively. Adapted from [7].

2.3.3
Similarity Analysis of a Set of Related Molecules

For one or several targets, MIFs can be analysed visually. However, it is not unusual to need to analyse the interaction properties of tens to thousands of molecules. In this case, computed MIFs must be analysed automatically. This can be done by computing similarity indices (SI) that permit the similarity of the MIFs of two molecules to be described by one number. If $SI = 1$, the MIFs are identical. If $SI < 1$, the MIFs differ. There are a number of expressions for computing MIFs. Commonly used definitions are the Carbo and Hodgkin indices. These were first developed for comparing the quantum mechanically computed electron densities and potentials of small organic compounds [43]. They can equally well be used to compare the MEPs and other MIFs of macromolecules, as well as small molecules.

Similarity indices of MIFs can be used to aid classification of compounds according to activity. This is exemplified by a similarity analysis of auxin plant hormones and related compounds [44, 45]. In a training set, about 50 compounds related to auxins were divided, on the basis of a biological assay and similarity analysis, into four classes: active auxin, weak auxin activity with weak antiauxin behavior, inactive and inhibitory. Further auxin-related compounds from a test set were then classified by a procedure to compare similarity indices for MIFs for the compounds with representative training set compounds for each of the classes. MIFs were computed with the GRID program for four probes: H_2O, $-NH_2^+$, $-CH_3$ and $=O$ (carbonyl oxygen). The molecules were aligned either by optimizing the MIF similarity at the molecular surface or by the similarity of atomic properties, with biasing towards electrostatic or steric properties. All the three-dimensional structures of the compounds, as well as the similarity data, can be explored interactively in [45], which is freely accessible on the web. This MIF-based similarity analysis was recently combined with estimation of log P and log D values, using VolSURF [9], which is also based on GRID MIFs [46]. This led to improvement in predictions of biological activity for certain types of auxin-like compound, showing that their biological activity depends both on receptor binding and lipophilicity. The recent solution of the crystal structure of an auxin binding protein (ABP1) permitted docking studies for the auxin compounds. These yielded computed binding modes in which the carboxyl group of the auxin-like compounds coordinates a zinc ion and the rest of the compound is accommodated in a rather hydrophobic pocket of the protein binding site. The presence of the zinc ion was unknown at the time of the original MIF-based similarity analysis and no metal ion probe was used for the MIF calculations. Nevertheless, the binding modes were consistent with the 'ligand-only' alignment and similarity analysis. The docking showed that both auxins and inhibitors could bind in the active site but that the antiauxins did so in a way that could hinder subsequent signal transfer and auxin response.

For proteins, a similarity analysis is typically most revealing for the MEP but can be computed for other MIFs. The MIF similarity analysis is implemented in PIPSA (Protein Interaction Property Similarity Analysis) [8, 47] (The PIPSA software can be obtained via http://projects.villa-bosch.de/mcm/). In the PIPSA procedure, the proteins are aligned using a structure-based alignment, MIFs are computed on a grid superimposed on the molecules, similarity indices are computed by summation over all points within a defined skin around the proteins and within a user-defined arc. The arc may be chosen to cover a specific binding region or the active site or may include the whole protein skin. The matrix of pairwise similarity indices obtained may be subjected to clustering, principal component analysis or tree analysis. PIPSA has been applied to diverse types of proteins including PH domains [47], WW domains [7], cupredoxins [48] and E2 ubiquitin conjugating enzymes [49].

In application to 33 cupredoxin electron transfer proteins, both the MEP and the GRID 'DRY' hydrophobic probe MIF were used [48]. The DRY probe results correlate better with sequence comparison of the proteins than the MEP PIPSA results. This is because the MEP at a given grid node is governed by residues discontiguous in space and extending some distance from the node. The hydrophobic probe energy is, on the other hand, governed by short-range interactions for which contiguous sequence plays a greater role (see Fig. 2.2). The analysis showed

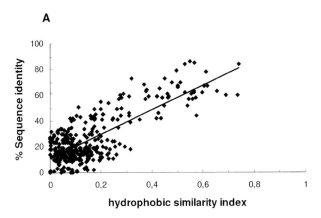

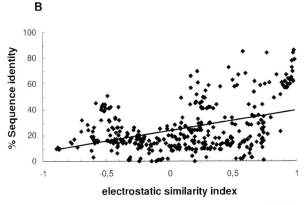

Figure 2.2. Pairwise percentage sequence identity for 33 cupredoxins plotted against (A) hydrophobic MIF similarity index and (B) electrostatic potential similarity index. The linear regression correlation coefficients are respectively, $r^2 = 0.61$ and $r^2 = 0.17$. The electrostatic potential similarity index is not well correlated with sequence identity because it is determined by relatively long range interactions. Adapted from [48].

how proteins in the same sequence subfamily could have different binding specificities and how proteins in different sequence subfamilies could have similar recognition properties resulting in isofunctionality. It also showed how proteins with low sequence identity could have sufficient similarity to bind to similar electron donors and acceptors while having different binding specificity profiles. Both the presence of a hydrophobic patch near the site of electron transfer and the electrostatic potential distributions were found to be important in determining the electron transfer specificity of this family of proteins.

Application of similarity analysis to WW domains was most revealing for the MEP [7]. This showed how a set of 42 WW domains could be classified according to MEP, in accordance with known peptide binding specificities, see Fig. 2.3.

In the case of E2 ubiquitin conjugating enzymes, the interaction properties of about 200 protein structures were compared [49]. The pairwise similarity matrix was visualized as a dendrogram and a kinemage projection to three-dimensional space (see www.ubiquitin-resource.org). The analysis revealed relations between functional groupings and electrostatic properties at specific parts of the protein structure.

2.4
Concluding Remarks and Outlook

MIFs provide a very useful way to analyse the interaction properties of a known molecular structure, be it of a macromolecule or a small molecule. A number of different software packages can be used to compute MIFs, each using different definitions of MIFs and parametrizations. The user needs to make a careful choice of probe type(s), and accordingly the interaction function for computing the MIF. MIFs can be used in many ways ranging from visual analysis of one MIF to automated comparison and correlation of the data in many MIFs. MIF computations can guide and complement other types of calculation such as molecular docking and design.

While MIF functions have been developed over about two decades, there is still scope for improvement. Improvements are motivated by the increasing amount of experimental data on the structures of macromolecule–ligand complexes, and the need to account more completely for the physical properties governing probe–target interactions. The flexibility of the target, while partially accounted for, is still incompletely treated, both in terms of conformational sampling and in terms of the energetics, and particularly the entropic contributions. Solvent effects are similarly incompletely treated. Changes in protonation and polarization upon binding also need to be better described. Nevertheless, the current functions for computing MIFs generally give good agreement with experiment and are very valuable components of the molecular designer's toolkit.

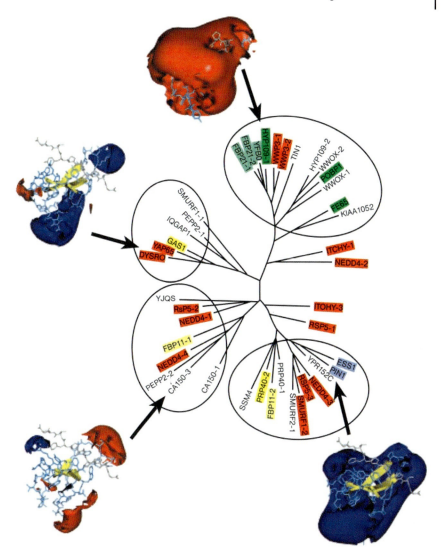

Figure 2.3. Electrostatic potential dendrogram constructed from an electrostatic potential distance matrix for a set of 42 WW domains showing four clusters. WW domains are colored according to their peptide ligand binding preferences. Four WW domains representative of the clusters are shown with positive (blue) and negative (red) electrostatic isopotential contours. In the cluster of predominantly positive potential WW domains are WW domains binding phosphorylated peptides; in the cluster of predominantly negative potential WW domains are WW domains binding peptides containing a positively charged residue. The WW domains in the other two clusters show specificity towards neutral peptides. Adapted from [7].

Acknowledgments

I am indebted to Dr Peter Goodford for his excellent guidance through my doctoral studies on GRID and its applications, and for continuing support. I thank all my former and present group members whose work is referred to in the references and applications. Financial support from the Klaus Tschira Foundation is gratefully acknowledged.

References

1 R. C. Wade, Molecular Interaction Fields, in *3D QSAR in Drug Design. Theory, Methods and Applications*, H. Kubinyi (ed.) ESCOM, Leiden, **1993**, pp. 486–505.
2 P. J. Goodford, A computational procedure for determining energetically favorable binding sites on biologically important macromolecules. *J. Med. Chem.* **1985**, *28*, 849–857.
3 G. Wu, D. H. Robertson, C. L. Brooks, Detailed analysis of grid-based molecular docking: A case study of CDOCKER- A CHARMm-based MD docking algorithm, *J. Comput. Chem.* **2003**, *24*, 1549–1562.
4 G. M. Morris, D. S. Goodsell, R. Huey, A. J. Olsen, Distributed automated docking of flexible ligands to proteins: parallel applications of AutoDock 2.4. *J. Comput. Aided Mol. Des.* **1996**, *10*, 293–304.
5 E. Carosati, S. Sciabola, G. Cruciani, Hydrogen Bonding Interactions of Covalently Bonded Fluorine Atoms: From Crystallographic Data to a New Angular Function in the GRID Force Field, *J. Med. Chem.* **2004**, *47*, 5114–5125.
6 R. D. Cramer, D. E. Patterson, J. D. Bunce, Comparative molecular field analysis (CoMFA). 1. Effect of shape on binding of steroids to carrier proteins, *J. Am. Chem. Soc.* **1988**, *110*, 5959–5967.
7 K. Schleinkofer, U. Wiedemann, L. Otte, T. Wang, G. Krause, H. Oschkinat, R. C. Wade, Comparative Structural and Energetic Analysis of WW Domain-Peptide Interactions, *J. Mol. Biol.* **2004**, 865–881.
8 R. C. Wade, R. R. Gabdoulline and F. De Rienzo, Protein Interaction Property Similarity Analysis, *Int. J. Quantum Chem.* **2001**, *83*, 122–127.
9 P. Crivori, G. Cruciani, P. A. Carrupt, B. Testa, Predicting blood-brain barrier permeation from three-dimensional molecular structure, *J. Med. Chem.* **2000**, *43*, 2204–2216.
10 P. J. Goodford, GRID, Molecular Discovery, www.moldiscovery.com.
11 J. Gunther, A. Bergner, M. Hendlich, G. Klebe, Utilising structural knowledge in drug design strategies: applications using Relibase, *J. Mol. Biol.* **2003**, *326*, 621–636.
12 M. I. Zavodsky, P. C. Sanschagrin, R. S. Korde, L. A. Kuhn, Distilling the essential features of a protein surface for improving protein-ligand docking, scoring, and virtual screening, *J. Comput. Aided Mol. Des.* **2002**, *16*, 883–902.
13 D. R. Boer, J. Kroon, J. C. Cole, B. Smith, M. L. Verdonk, SuperStar: comparison of CSD and PDB-based interaction fields as a basis for the prediction of protein-ligand interactions, *J. Mol. Biol.* **2001**, *312*, 275–287.
14 H. Gohlke, M. Hendlich, G. Klebe, Knowledge-based scoring function to predict protein-ligand interactions, *J. Mol. Biol.* **2000**, *295*, 337–356.
15 M. Böhm, G. Klebe, Development of New Hydrogen-Bond Descriptors and Their Application to Comparative Molecular Field Analyses, *J. Med. Chem.* **2002**, *45*, 1585–1597.
16 J. E. Lennard-Jones, *Cohesion Proc. Phys. Soc.* **1931**, *43*, 461–482.

17 J. E. Lennard-Jones, On the determination of Molecular Fields. II. The equation of state of a gas. *Proc. R. Soc. London Ser. A.* **1924**, *106*, 463–477.

18 J. C. Slater, J. G. A. Kirkwood, *Phys. Rev.* **1931**, *37*, 682–686.

19 F. Fogolari, A. Brigo, H. Molinari, The Poisson–Boltzmann equation for biomolecular electrostatics: a tool for structural biology, *J. Mol. Recognit.* **2002**, *15*, 377–392.

20 N. A. Baker, Poisson–Boltzmann methods for biomolecular electrostatics, *Methods Enzymol.* **2004**, *383*, 94–118.

21 D. Bashford, Macroscopic electrostatic models for protonation states in proteins, *Front. Biosci.* **2004**, *9*, 1082–1099.

22 M. Neves-Petersen, S. Petersen, Protein electrostatics: a review of the equations and methods used to model electrostatic equations in biomolecules – applications in biotechnology, *Biotechnol. Annu. Rev.* **2003**, *9*, 315–395.

23 E. Demchuk, R. C. Wade, Improving the Continuum Dielectric Approach to Calculating pKas of Ionizable Groups in Proteins, *J. Phys. Chem.* **1996**, *100*, 17373–17387.

24 C. Schutz, A. Warshel, What are the dielectric "constants" of proteins and how to validate electrostatic models? *Proteins* **2001**, *44*, 400–417.

25 T. Wang, S. Tomic, R.R. Gabdoulline, R.C. Wade, How Optimal are the Binding Energetics of Barnase and Barstar? *Biophys. J.* **2004**, *12*, 1563–1574.

26 F. Dong, M. Vijayakumar, H.-Y. Zhou, Comparison of calculation and experiment implicates significant electrostatic contributions to the binding stability of barnase and barstar, *Biophys. J.* **2003**, *85*, 49–60.

27 D. N. A. Boobbyer, P. J. Goodford, P. M. McWhinnie, R. C. Wade, New Hydrogen-bond Potentials for Use in Determining Energetically Favourable Binding Sites on Molecules of Known Structure, *J. Med. Chem.* **1989**, *32*, 1083–1094.

28 R. C. Wade, K. Clark, P. J. Goodford, Further Development of Hydrogen Bond Functions for Use in Determining Energetically Favorable Binding Sites on Molecules of Known Structure. 1. Ligand Probe Groups with the Ability To Form Two Hydrogen Bonds, *J. Med. Chem.* **1993**, *36*, 140–147.

29 R. C. Wade, P. J. Goodford, Further Development of Hydrogen Bond Functions for Use in Determining Energetically Favorable Binding Sites on Molecules of Known Structure. 2. Ligand Probe Groups with the Ability To Form More Than Two Hydrogen Bonds, *J. Med. Chem.* **1993**, *36*, 148–156.

30 M. J. Stone, NMR relaxation studies of the role of conformational entropy in protein stability and ligand binding, *Acc. Chem. Res.* **2001**, *34*, 379–388.

31 L. Zidek, M. V. Novotny, M. J. Stone, Increased protein backbone conformational entropy upon hydrophobic ligand binding, *Nat. Stuct. Biol.* **1999**, *6*, 1118–1121.

32 S. Arumugam, G. Gao, B. L. Patton, V. Semenchenko, K. Brew, S. R. Van Doren, Increased backbone mobility in beta-barrel enhances entropy gain driving binding of N-TIMP-1 to MMP-3, *J. Mol. Biol.* **2003**, *327*, 719–734.

33 B. Tidor, M. Karplus, The contribution of vibrational entropy to molecular association. The dimerization of insulin, *J. Mol. Biol.* **1994**, *238*, 405–414.

34 M. von Itzstein et al., Rational design of potent sialidase-based inhibitors of influenza virus replication, *Nature* **1993**, *363*, 418–423.

35 R. C. Wade, 'Flu' and Structure Based Drug Design, *Structure* **1997**, *5*, 1139–1146.

36 M. T. Pisabarro, A.R. Ortiz, A. Palomer, F. Cabre, L. Garcia, R.C. Wade, F. Gago, D. Mauleon, G. Carganico, Rational Modification of Human Synovial Fluid Phospholipase A2 Inhibitors, *J. Med. Chem.* **1994**, *37*, 337–341.

37 W. Bitomsky, R. C. Wade, Docking of Glycosaminoglycans to Heparin-Binding Proteins: Validation for aFGF, bFGF, and Antithrombin and Application to IL-8, *J. Am. Chem. Soc.* **1999**, *121*, 3004–3013.

38 M. A. Kastenholz, M. Pastor, G. Cruciani, E.E.J. Haaksma, T. Fox, GRID/CPCA: A New Computational Tool To Design Selective Ligands, *J. Med. Chem.* **2000**, *43*, 3033–3044.

39 M. Fornabaio, F. Spyrakis, A. Mozzarelli, P. Cozzini, D. J. Abraham, G. E. Kellogg, Simple, intuitive calculations of free energy of binding for protein-ligand complexes. 3. The free energy contribution of structural water molecules in HIV-1 protease complexes, *J. Med. Chem.* **2004**, *47*, 4507–4516.

40 M. Pastor, G. Cruciani, K. A. Watson, A Strategy for the Incorporation of Water Molecules Present in a Ligand Binding Site into a Three-Dimensional Quantitative Structure-Activity Relationship Analysis, *J. Med. Chem.* **1997**, *40*, 4089–4102.

41 A. R. Ortiz, M. Pastor, A. Palomer, G. Cruciani, F. Gago, R. C. Wade, Reliability of comparative molecular field analysis models: effects of data scaling and variable selection using a set of human synovial fluid phospholipase A2 inhibitors, *J. Med. Chem.* **1997**, *40*, 1136–1148.

42 M. Baroni, G. Costantino, G. Cruciani, D. Riganelli, R. Valigi, S. Clementi, Generating optimal linear PLS estimations (GOLPE): an advanced chemometric tool for handling 3D-QSAR problems, *Quant. Struct.-Act. Relat.* **1993**, *12*, 9–20.

43 C. Burt, W.G. Richards, P. Huxley, *J. Comput. Chem.* **1990**, *11*, 1139–1146.

44 S. Tomic, R. R. Gabdoulline, B. Kojic-Prodic, R. C. Wade, Classification of auxin plant hormones by interaction property similarity indices, *J. Comput. Aided Mol. Des.* **1998**, *12*, 63–79.

45 S. Tomic, R. R. Gabdoulline, B. Kojic-Prodic, R. C. Wade, Classification of auxin related compounds based on similarity of their interaction fields: Extension to a new set of compounds, *Internet J. Chem.* **1998**, *1*, 26. http://www.ijc.com/articles/1998v1/26/.

46 B. Bertosa, B. Kojic-Prodic, R. C. Wade, M. Ramek, S. Piperaki, A. Tsantili-Kakoulidou, S. Tomic, A New Approach to Predict the Biological Activity of Molecules Based on Similarity of Their Interaction Fields and the log$P$ and log$D$ Values: Application to Auxins, *J. Chem. Inf. Comput. Sci.* **2003**, *43*, 1532–1541.

47 N. Blomberg, R. R. Gabdoulline, M. Nilges, R. C. Wade, Classification of protein sequences by homology modeling and quantitative analysis of electrostatic similarity, *Proteins* **1999**, *37*, 379–387.

48 F. De Rienzo, R. R. Gabdoulline, M. C. Menziani, R. C. Wade, Blue copper proteins: a comparative analysis of their molecular interaction properties, *Protein Sci.* **2000**, *9*, 1439–1454.

49 P. J. Winn, T. L. Religa, J. D. Battey, A. Banerjee, R. C. Wade, Determinants of Functionality in the Ubiquitin Conjugating Enzyme Family, *Structure* **2004**, *12*, 1563–1574.

II
Pharmacodynamics

3
Protein Selectivity Studies Using GRID-MIFs
Thomas Fox

3.1
Introduction

During the course of a chemical compound from initial hit in a screening campaign to a marketable drug, a number of obstacles usually have to be overcome and various characteristics have to be optimized. These are, amongst others, the affinity to the primary target, physicochemical properties like lipophilicity and solubility, as well as its ADME (absorption, distribution, metabolism, elimination) and toxicology profile. Another important aspect is ligand selectivity: a given ligand should interact selectively with only one biomolecule, as often undesired side effects are due to interaction of the ligand with biomolecules other than the primary target.

In the search for new, selectively interacting compounds, knowledge about the three-dimensional structure of the desired target provides extremely useful information enabling one to focus on ligand–protein interaction and selectivity regions. However, the intermolecular interaction between protein and ligand is a complex phenomenon. Therefore, it is difficult to extract enough information from these (static) structures that can be used in the design of selective ligands, and sometimes these (static) structures do not give sufficient useful information at all.

In silico techniques have gained wide acceptance as a tool to support the drug discovery and optimization process. Binding mode predictions via docking, affinity predictions via QSAR and CoMFA, or the prediction of ADME(T) properties are routinely applied [1–3].

Far less experience is available for selectivity predictions, which may be at least partly due to the fact that the underlying experimental data are less accurate (being the combination of binding affinities to two or more targets). In most cases, ligand-based QSAR or 3D-CoMFA methods and their variants have been used with the binding affinity ratio as the objective function to train the mathematical model [4]. However, such methods are restricted to the chemical space of the training set, and become rather awkward if more than two targets have to be analyzed simultaneously.

Molecular Interaction Fields. Edited by G. Cruciani
Copyright © 2006 WILEY-VCH Verlag GmbH & Co. KGaA, Weinheim
ISBN: 3-527-31087-8

To complement the ligands' view on affinity and selectivity, it would be desirable to use the information about the nature of the binding sites contained in the three-dimensional (3D) structures of the targets. Traditionally, this has been done by visual inspection of the protein structures. However, this is error prone, as it is too easy to overlook an important difference, or to identify a potential source of selectivity which cannot be confirmed experimentally.

In the last few years, a number of publications have demonstrated that the GRID/PCA or GRID/CPCA methods can be successfully applied to characterize the structural differences between protein binding sites, and to identify differences in the protein–ligand interactions as well as the regions on the target enzymes which mediate highly selective interactions [4–17].

On the basis of the 3D structures of the proteins, the GRID/CPCA method analyzes the selectivity differences from the viewpoint of the target and is therefore independent of the availability of appropriate ligand binding data for a ligand-based QSAR analysis.

Briefly, the methodology involves five major steps: (i) retrieval of adequate 3D structures for each of the target proteins; (ii) superimposition of the regions in which the ligands interact with the target (binding sites); (iii) multivariate characterization of the binding site by the energies of interaction between the target macromolecules and different small chemical groups; (iv) summary of the results by means of principal component analysis (PCA) or consensus PCA (CPCA); (v) graphical analysis and chemical interpretation of the results.

At the same time, this approach also allows one to make a straightforward classification of protein binding sites which depends only on the interaction patterns in the regions of interest, and not on the sequence alignment and differences in the amino acid composition.

In this chapter, we will first review the basic principles of GRID/PCA and GRID/CPCA, highlighting some important technical aspects necessary for their successful application. Some space will be given to the discussion of the differences between PCA and the (hierarchical) CPCA methods, before their application to various selectivity problems will be summarized.

3.2
GRID Calculations and Chemometric Analysis

3.2.1
Source and Selection of Target Structures

In most cases, three-dimensional target structures have been taken from crystallographic analysis, however, NMR structures [14], homology models [9, 11, 12, 15] and snapshots from MD trajectories [16] have also been employed. In the GRID/PCA method, only one structure per target protein is used, thus a careful selection is necessary. In general, high resolution structures with suitable ligands occupying the same binding sites should be chosen. Thus, conformational changes due to ligand binding are somewhat accounted for, and the analysis concentrates on the differences in the protein–ligand complexes and not on possible induced-fit phenomena. In the GRID/CPCA approach, it is possible to use several structures for each target, thus including some information about the conformational degrees of freedom of the protein. Therefore, as many structures as possible should be analyzed; keeping in mind that obvious outliers, which would dominate the PCA analysis, should be excluded.

3.2.2
Selection and Superimposition of Binding Sites

The proper alignment of the ligand binding sites is crucial for the success of the analysis. In particular, the goal of the superimposition is not to compare the whole protein but only the binding site, i.e. those positions where known ligands interact and also those that a potential new ligand could reach. In some cases previous investigations point to a set of important residues that interact with substrates or inhibitors and represent the parts of the binding site involved in the ligand recognition. Alternatively, starting from the protein–ligand complex, one can select those residues which either have a direct contact with the ligand by visual inspection, or – in a more automated fashion – select residues which are within a certain distance from the ligand. Here, a distance threshold of 3–4 Å provides a useful definition of the binding site [14, 18]. These selected residues, or rather their Cα atoms, are used for the optimal superimposition of the proteins via a least-squares method.

A somewhat more elaborate scheme was proposed by Matter and Schwab [4]. Starting from a global 3D alignment, in an iterative procedure they focus on those amino acids which, after superimposition, show a low RMS deviation. Only these residues are considered in the next round of structure alignment. In this way they arrive at an unbiased superimposition of a large number of proteins focusing on the most conserved parts of the structure.

However, in many cases the proteins considered in the analysis are very similar and show only little structural variation. In these cases a superimposition of structurally conserved regions with standard protein homology modeling tools has yielded satisfactory protein superimpositions suited for the GRID/PCA or GRID/CPCA analysis [8,9,13,17].

3.2.3
Calculation of the Molecular Interaction Field

In principle, any method which is able to produce a molecular interaction field (MIF) could be applied to the characterization of the targets. However, almost exclusively the GRID program [19] has been used. This is due to the size of the problem which excludes the use of standard *ab initio* or semiempirical methods to produce interaction maps. It also reflects the many successful applications of the GRID force field in the characterization of protein active sites and the interpretation of protein–ligand interactions [20–24]. Additionally, GRID is well linked to GOLPE [25, 26], a software that can be used for the analysis and visualization of the interaction energy data.

In the GRID procedure, the interaction energy between the target and a probe is calculated at each node of a three-dimensional grid which contains the chosen binding site. Several probes, which are parameterized to represent various small chemical groups, are available. The results of these calculations are a collection of three-dimensional matrices, one for each probe–target interaction. A detailed description of the GRID program and the underlying force field parameters can be found elsewhere [20–22].

The grid size and location should be chosen in such a way that the grid box encloses all the positions around the binding site in which atoms of a potential ligand could be found. This can be done by visual inspection, or one can identify all residues which are able to interact with the ligand or are within a certain distance from the ligand, and select the box size such that all these residues are included.

Using a 1 Å spacing between the grid points seems to be a good compromise between a sufficient mapping of the binding site and creating many variables without real information content. With larger grid spacings, one risks missing important differences in the binding sites, whereas a finer grid not only significantly increases the computation time and file sizes, but also bears the risk of introducing noise which hampers the chemometric analysis.

Complications arise when the proteins under investigation are differently charged, as the charge differences and the resulting differences in the electrostatic interactions may well mask more subtle differences in the binding sites. Therefore, the systems should be neutralized. This is usually accomplished by preparing MIFs for counter ions and then placing the appropriate number of counter ions around the proteins, using the utility programs MINIM and FILMAP [19]. It is obvious, that the counter ions should be placed outside the binding site box.

In the original GRID/PCA approach, all available GRID probes were usually used, unless there was some indication that certain interaction types (e.g. metal ions) were not present or would show very unusual behavior [4,7]. In this way, a detailed and comprehensive description of possible ligand–target interactions is achieved. In the GRID/CPCA application, only a subset of five to ten probe types are used, usually including the DRY, C3, N1, O, and OH probes, to get a manageable yet diverse set of interactions in the analysis (see Table 3.1).

Since GRID version 19, protein side chains may be treated as flexible by setting the MOVE directive to values greater 0. In most analyses to date, the MOVE directive has been set to 0, i.e. the protein was considered rigid. In some applications, the results of rigid and flexible protein treatment were compared; these comparisons yielded mixed results: in some cases only little differences were observed [11], sometimes additional residues important for selectivity could be identified with flexible side chains [9], or largely different results were reported [7]. Generally, it seems that the calculated binding site cavities are larger, and MIF differences caused by slight differences in side chain conformations disappear when side chain flexibility is incorporated.

Table 3.1 Overview of PCA/CPCA models described in the literature and discussed in this review.

Author	Year	System	Method	Probes	MOVE	Ref.
Cruciani & Goodford	1994	DNA	PCA	31 probes	0	5
Pastor & Cruciani	1995	DHFR	PCA	41 probes	0	6
Matter & Schwab	1999	MMP	PCA	38 probes, no metals	0	4
Filipponi et al	2000	COX	PCA	42 probes	0,1	7
Braiuca et al.	2003	Penicillin Acylase	PCA	DRY,C3,C1,N2:,N2,N3+,OH,ON,O:,O	1	12
Braiuca et al.	2004	Penicillin Acylase	PCA	DRY,C3,C1=,OH,O1,S1,N2:,O,O::,ON,BR,CL,F	0,1	17
Kastenholz et al	2000	Serine Protease	CPCA	DRY,C3,N1,N:,NH=,N1+,NM3, O,OH,OS,O::,COO$^-$	0	8
Ridderström et al.	2001	CYP2C	CPCA	DRY,C3,N1,N:,NH=,N1+,NM3,O,OH,COO$^-$	0,1	9
Afzelius et al.	2004	CYP2C	CPCA	DRY,C3,N1,N1+,O$^-$,O,OH$_2$	0	16
Naumann & Matter	2002	Protein Kinases	CPCA	DRY,O,N1,(C3),(OH$_2$)	0	10
Terp et al.	2002	MMP	CPCA	DRY,N1,N:,NH=,N1+,NM3,O,OH,OS,O::	0,1	11
Ji et al.	2003	NOS	CPCA	DRY,C3,N1,N:,NH=,N1+,NM3,O,OH,COO$^-$	0,1	13
Matter & Kotsonis	2004	NOS	CPCA	?	?	59
Pirard	2003	PPAR	CPCA	DRY,C3,N1,O,OH	0	64
Kurz et al.	2003	ILBP/FABP	CPCA	DRY,C3,N1,O,OH	0	14
Myshkin & Wang	2003	EphKinases	CPCA	DRY,C3,N1,O	0	15

Recently, FLOG [27] has been used for the characterization of protein binding sites [28]. Conceptually, the maps produced by FLOG are similar to those obtained by GRID, with the advantage that FLOG does not require atomic charges, and allows for additional freedom in the placement of polar hydrogens. By default, only five probe types are available: donor/cation, acceptor/anion, polar, hydrophobic, and van der Waals. Moreover, the information from these maps seems to be somewhat redundant, so a protein characterization could be obtained with only a subset of three probe types (donor/cation, acceptor/anion, and hydrophobic).

3.2.4
Matrix Generation and Pretreatments

For the chemometric analysis, the three-dimensional MIFs obtained from GRID are rearranged as one-dimensional vectors. In the GRID/PCA approach [6], one such vector is obtained for each MIF, and the vectors are used to build a two-dimensional *X* matrix, in which the rows are the probe–target interactions (the objects) and each column contains the variables that describe energetically these interactions at a given grid point. The process used to obtain the *X* matrix is illustrated in Fig. 3.1.

A careful pretreatment is necessary to focus on the relevant variables. Often variables with low absolute values (<0.01 kcal mol^{-1}) and those with low standard deviations (<0.02–0.03 kcal mol^{-1}) are removed in order to eliminate noise. Autoscaling is not recommended, since all the data comes from the same source (GRID probe–target interaction energies) and all the data are expressed in the same units (kcal mol^{-1}). Thus, autoscaling might introduce noise in the model,

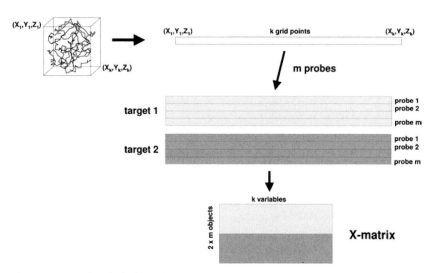

Figure 3.1. Procedure for building the *X* matrix in GRID/PCA. The analysis of the interaction energies of the *m* probes with the 2 target proteins produces 2 × *m* three-dimensional matrices. These are unfolded to obtain 2 × *m* one-dimensional vectors from which the two-dimensional *X* matrix is built.

increasing the importance of insignificant regions (i.e. rows in the X matrix with little variance).

Usually, any positive interaction energy present in the X matrix is then set to 0 kcal mol^{-1}. Thus one focuses on the negative, favorable interaction energies and removes the information about small protein shape differences. Additionally, only using negative interaction energies allows a straightforward interpretation of the results.

3.2.4.1 Region Cut-outs

One of the disadvantages of the original GRID/PCA approach was due to the fact that the whole rectangular box was used in the analysis. Thus data were also collected for regions that do not lie within binding pockets. Since the homology between targets tends to be higher in the binding site than further away, these regions are often only badly superimposed, and the PCA finds significant structural differences that hide more relevant but more subtle differences in the region of primary interest. During the development of GRID/CPCA [8], a 'cutout tool' was implemented in GOLPE, which allows one to select user-defined, irregularly shaped regions within the original GRID box. This not only leads to a significant reduction of the x variables but also facilitates the overall analysis. Indeed, most authors use the 'cut-out tool' to focus their selectivity analysis on individual subsites.

3.2.5
GRID/PCA

The X data matrix which contains all information describing the probe–target interactions can be analyzed by PCA [29, 30]. PCA is a multivariate projection method which allows one to extract the systematic information which is contained in the data matrix and to present it in a simplified form. The original number of variables is reduced to a few factors called principal components (PCs). The result of such an analysis can then be visualized by means of two informative plots which allow a straightforward interpretation of the problem. In this way, PCA provides an understanding of similarities and dissimilarities between the different protein binding sites with respect to their interaction with potential ligands.

Briefly, PCA summarizes the variation of a data matrix as a product of two lower-dimensional matrices, the score matrix T and the loadings matrix P. With n objects and k grid points, the $(n \times k)$ data matrix is decomposed into the $(n \times a)$ score matrix and the $(a \times k)$ loadings matrix plus a $(n \times k)$ 'error' matrix of residuals E:

$$X = T \cdot P' + E \qquad (1)$$

As long as the number of principal components a is small, a considerable simplification of the problem is achieved. In their derivation, the PCs or score vectors (the columns in T) are sorted in descending importance.

The score matrix gives a simplified picture of the objects (probe–target interactions), represented by only a few, uncorrelated new variables (the PCs). Score plots, i.e. plots of the score vectors against each other, are a summary of the relationships between the objects and reveal the essential data patterns of the objects. Thus, objects which behave similarly have similar scores and are close in the score plot. In our context, score plots can be used to identify clusters of objects according to the different kind of targets (macromolecules) and probes (ligand chemical groups) involved.

The loadings matrix gives a similar summary of the variables. The loadings indicate how the original variables are linearly combined to form the scores. In the loading plot, the distance of a point from the origin is important: variables most relevant for the model are found at the periphery of the loading plot. Conversely, uninfluential variables show up near the origin of the plot.

Loading plots are a means to interpret the patterns seen in the score plot. The two plots are complementary and superimposable; a direction in one plot corresponds to the same direction in the other. Thus, a pattern seen in the score plot can be interpreted by looking along this interesting direction in the loading plot.

If most of the variation of the original data can be described by the first few PCs, a much simpler data structure exists. In GRID/PCA, typically about 60–75% of the variance can be explained with the first two PCs; this allows an interpretation based on the 2D plots of the leading components.

3.2.5.1 Score Plots

A typical score plot from a GRID/PCA analysis is shown in Fig. 3.2, taken from an analysis for thrombin (thr) and trypsin (try) [31]. In this plot every point represents a single probe–target interaction ('object'). The figure clearly shows two clusters of objects, one referring to interactions with thrombin, the other to interactions with trypsin. PC 1 discriminates the two target proteins, thus, it can be associated with selectivity. On the other hand, PC 2 describes general, nonselective ligand–target interactions. Therefore, this PC ranks the probes according to their general ability to interact chemically with residues of the binding site. Due to the focusing on negative energies, there is an inverse relationship between the strength of the binding and the PC 2 scores – stronger interactions are related to smaller (i.e. more negative) interaction energies. Therefore, the probes with lower interaction energies show up in the top part of the score plot in Fig. 3.2, and the probes that interact more strongly (including all the multi-atom probes) are in the bottom part. Moreover, the points spread from top to bottom, indicating that the probes which interact strongly with both targets should also be the most interesting from the point of view of selectivity.

Thus, groups represented by probes at the bottom left and right areas of Fig. 3.2, when properly placed in the binding site, may be able to increase the selectivity of the interaction towards one of the targets.

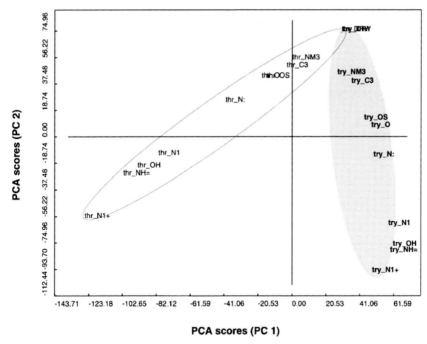

Figure 3.2. Typical PC 1 vs. PC 2 score plot obtained with GRID/PCA (t1 vs. t2) (T. Fox, unpublished work). The points in the plot represent the objects of the **X** matrix: the interactions of a particular probe with one of the two targets, in this example thrombin (thr) and trypsin (try).

3.2.5.2 Two-Dimensional Loading Plots

Figure 3.3 shows the corresponding loading plot of the PCA model. In this plot every point represents a position in the grid box where probe–target interactions were computed (variables). Based on the correspondence of the score and loading plots, the interpretation of the loading plot is straightforward.

The horizontal axis represents PC 1; therefore, variables with high absolute PC 1 loadings represent regions in the binding site where the probes show extremely different behavior in their interaction with the two targets. As only negative (favorable) interactions are considered, these regions will reveal positions where a chemical group can bind loosely with one of the targets and tightly with the other. The further out a variable (and therefore the corresponding 3D lattice point) is along the PC 1 axis, the more important is this binding site region for discriminating between the two targets.

The vertical axis relates to PC 2 loadings; thus points in the top part indicate positions where the probes interact strongly with both targets, whereas points with a PC 2 contribution close to zero represent regions with weak interactions.

Therefore, one may distinguish three types of variables and thus positions in the binding site: (i) Points near the origin: positions where the probes only estab-

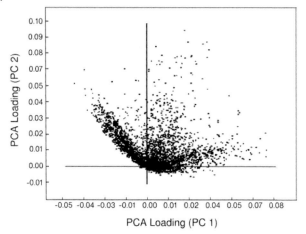

Figure 3.3. Typical PC 1 vs. PC 2 loading plot from a GRID/ PCA analysis. The points in the plot represent the variables of the X matrix, i.e. the positions in the grid space.

lish weak, nonselective interactions. Most of the binding site positions fall into this category. (ii) Low absolute values for PC 1 loadings and high PC 2 loadings: positions where the probes interact strongly with both targets. They are not interesting from the point of view of the selectivity but might be exploited to increase the affinity for both targets. (iii) High absolute values for PC 1 loadings and (at least) intermediate PC 2 loadings: positions of selective interactions. Adequate groups located in these positions would induce or increase the selectivity of a ligand.

3.2.5.3 Loading Contour Maps

Due to the nature of the PCA variables in the X matrix, it is possible to relate the points in the loading plot back to the 3D region in the binding site. In GOLPE, this can be accomplished by loading contour maps. They highlight those active site regions identified by the statistical model which interact strongly or selectively with ligands. Hence, by selecting the appropriate PC and contour level it is possible to display the regions of the binding site most relevant for selective binding and common affinity.

3.2.5.4 Problems of GRID/PCA

Whereas the GRID/PCA approach has been successfully used in a number of selectivity problems (see below), this procedure has several shortcomings:

First, since PCA is quite sensitive to the scale of interaction energies, the information given by probes representing weak nonbonded interactions (van der Waals and hydrophobic) is masked by the effect of probes representing stronger interactions (coulombic and hydrogen bonds). Thus, although hydrophobic or van der Waals interactions are important for ligand binding and selectivity, they are almost never identified during a standard GRID/PCA calculation. Generally, the relevance assigned by the method to the different probes can be unreliable and misleading.

Another important problem is that PCA generates a 'general' model – the binding site regions selected in the loading plot represent generally important regions in the active site, but are not linked to a particular probe. For the design of selective ligands, it is much more useful to identify also the exact type of interaction which is important for selectivity at a certain region in the binding site.

Moreover, the GRID/PCA method relies on the visual inspection of the score plot to identify the PC which predominately discriminates the clustered objects. This can be difficult when there are more than two clusters (e.g. arising from more than two targets), since two or more PCs might be relevant for selectivity. Therefore, the analysis is practically limited to two target proteins at a time; the analysis of three or more targets is very awkward.

3.2.6
GRID/CPCA

To overcome some of the limitations of GRID/PCA, GRID/CPCA was developed [8]. Again, the 3D structures representing the targets are analyzed using GRID, but the MIFs obtained for the different probes are organized differently: After rearranging the 3D MIFs, the resulting one-dimensional vectors for the different probes are combined side-by-side, adding new variables to the same object (Fig. 3.4). Thus, the matrix describing the systems has a row for each 3D structure

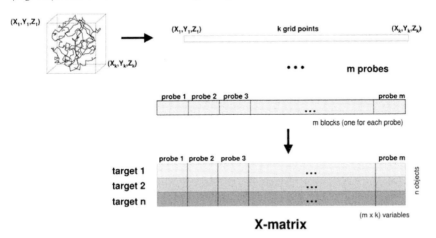

Figure 3.4. Data collection for GRID/CPCA: Starting from the GRID calculations for one probe, a one-dimensional vector containing all interaction energies at the k grid points is constructed. Then the vectors for the m probes are compiled into one long vector which contains $(k \times m)$ data points. The final **X** matrix is built by stacking these vectors for every target protein.

studied and $m \times k$ columns, corresponding to m probes × k variables present in a single MIF.

The PCA of this matrix produces a score plot where each target 3D structure is represented by a single point (see Fig. 3.5). When several different structures are used to represent each target protein, the score plots should show them clustered, indicating that the differences between the 3D structures of the same target are less important than the differences between the proteins.

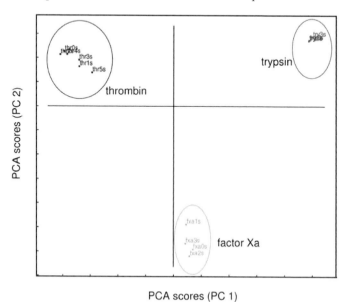

Figure 3.5. Typical example of a CPCA score plot, here for the serine proteases thrombin, trypsin, and factor Xa. Each point in the plot represents one input protein structure. The different structures for each protein are clustered. PC 1 discriminates thrombin and trypsin, PC 2 distinguishes factor Xa from thrombin and trypsin.

This approach has a number of additional advantages. First, several protein targets can be included in the analysis and selectivity profiles between groups of targets can be generated. In addition, many 3D structures representing the same target protein can be used. This minimizes spurious results originating from minute differences in the crystal structures of the targets. Actually, the method highlights the differences between the common features of the targets. In some sense, using more than one target structure can be considered as a means to incorporate information about conformational flexibility in the analysis.

3.2.6.1 Block Unscaled Weights

One of the main drawbacks of the original method results from the different overall range of the interaction energies obtained for different probes. This cannot be corrected by autoscaling of the *X* matrix as each column vector contains all probe information at one single grid point. In GRID/CPCA, the different probes are organized in different blocks of variables. Therefore, it is now possible to apply a weighting procedure called 'block unscaled weights' (BUW). BUW considers all

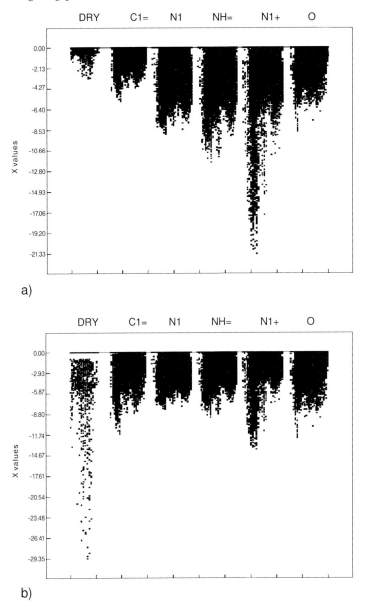

Figure 3.6. Effect of BUW on the distribution of the *x* variables (i.e. the interaction energies obtained with GRID) for each probe: (a) distribution before and (b) distribution after BUW to unit variance.

variables within a single block and scales them to unit variance, and each variable block is scaled separately. Thus, the relative scales of single variables within each block remain unchanged, whereas each probe gets the same importance within the model. Figure 3.6 illustrates a typical BUW procedure, showing the initial energy distribution of the x variables for each probe and the normalized distribution after the variable block weighting.

3.2.6.2 CPCA

As the X matrix produced by this problem formulation is structured in meaningful blocks, hierarchical PCA methods provide interesting additional insight regarding the relative importance of the different blocks (i.e. probes) in the analysis.

Multiblock methods have been introduced in cases where the number of variables is large and additional information is available for blocking the variables into conceptually meaningful blocks [32, 33]. Several algorithms have been described including hierarchical PCA (HPCA) or consensus PCA (CPCA), which differ slightly in the underlying algorithms [34].

Within GOLPE, the CPCA approach is implemented. In Fig. 3.7 the usual 'arrow scheme' for CPCA is shown. The data matrix is divided into m blocks X_1, ...X_m given by the m probes used in the GRID calculation. A starting consensus or super score is selected as a column of one of the blocks. On each block, this vector is multiplied by X_i to get an approximation of the block loadings p_b. Then the p_b is normalized and multiplied back by X_i to get a new score t_b. This is repeated until convergence of t_b and is the usual NIPALS algorithm for PCA. Once the block scores t_b for all blocks have been calculated, they are combined into a super block T. The super score t_T is then regressed on the super block to give the super weight w_T of each block score to the super score. The super weight is normalized and a

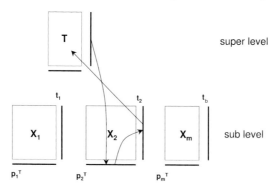

Figure 3.7. Principle of CPCA: A start super score t_T is regressed on all blocks X_i to give the block loadings p_b. Then the block loadings are normalized to length one and then multiplied back through the blocks to give the block scores t_b. The block scores t_b are combined into a super block T. Then a PCA round on T is performed to give the super weight w_T and a new super score. This is repeated until convergence of t_T [taken from Westerhuis et al. [32]].

new t_T is calculated. A new iteration starts until the super score converges. The super score is derived using all variables, whereas the block scores are derived using only the variables within the corresponding block. The super weight gives the relative importance of the different blocks X_i for each dimension. After convergence, all the blocks are deflated using the super score, and the next super score, orthogonal to the first, can be determined by repeating the described iteration on the residual matrix.

Therefore, CPCA uses exactly the same objective function as PCA: it tries to best explain the overall variance of the X matrix, but the analysis is made on two levels: the block level, which considers each of the probes, and the super level, which expresses the "consensus" of all blocks. CPCA provides a solution on the super level that is identical to a solution found in regular PCA, i.e., the same T and P matrices are obtained. Additionally, the method produces block scores t_b and block loadings p_b for each of the probes and a weight matrix which gives the contribution of each block to the overall scores. The block scores represent a particular point of view of the model given by a certain probe and provide unique information not present in regular PCA. Object distances in the block scores can be used to assess the relative importance of the different probes in their discrimination.

We note that on the block level, an objective function is used to obtain the scores which is different from the standard PCA: the principal components should also reproduce the values obtained in the overall PCA level. As a consequence, the percentage of the variance explained by PC 1 and PC 2 varies between 20 and 30% compared to the much higher values in the GRID/PCA approach. Also, within the blocks the scores are not necessarily ordered in decreasing importance. However, the separation of the objects and the interpretability of a model should be considered as more important criteria for the quality of a model than the percentage of the variance explained by PC 1 and PC 2.

3.2.6.3 Identification of Important Variable Blocks for Selectivity

Plots like those in Figs. 3.6 and 3.9 (below), which show, for each probe separately, the distribution of the x values, convey additional information. They can be used to identify probes that are able to distinguish between the different targets: if especially high interaction energies are found in one of the targets, this is a clear indication for the importance of this interaction type for selectivity.

3.2.6.4 Contour Plots

As for PCA, CPCA loadings can be translated into contour plots describing the interaction fields between a GRID probe and the target protein structure. For a selectivity study one is interested in the loadings discriminating different target proteins. Using 'active plots' in GOLPE, one draws a vector linking pairs of objects in a 2D score plot which is then translated into isocontour plots that identify those variables which contribute most to distinguishing the selected objects. To obtain such isocontours, GOLPE calculates the difference between the two points for the first and second principal component and projects these differences back into the original space (a pseudo-field) using the PCA loadings. The result is a grid plot of the differences in the pseudo-fields that highlight the object differences for the corresponding probe. This allows one to identify both the regions and the interaction type (i.e. probe) that can produce selective interaction with respect to the start and end points of the drawn vector (which e.g. could connect a pair of protein targets).

3.3
Applications

In the following, we will discuss applications, which use the GRID/PCA and the GRID/CPCA methods for selectivity analyses between several macromolecular targets. The serine proteases and the matrix metalloproteases were the first applications of the CPCA method, and therefore will be used to describe the approach in some detail. For the other studies, we can only give a short overview, highlighting aspects that we find especially interesting or that constitute a development beyond the original formulation. For more details we refer to the original literature.

3.3.1
DNA Minor Groove Binding – Compare AAA and GGG Double Helix

The first study to investigate selectivity profiles using MIFs in connection with a chemometric analysis was published about 10 years ago [5]. There, the behavior of all 64 possible DNA triplets with respect to 31 GRID probes was studied with GRID/PCA.

In the comparison of the double-stranded TTGGGTT and TTAAATT base pair sequences, the first PC distinguishes between different probes – despite neutralization of the system with counter ions, the high charge density in DNA dominates the electrostatic interactions and thus the analysis. This is especially evident for negatively and positively charged probes which end up at different extremes of PC 1. This is intuitively clear – cations should interact far more favorably with DNA than anions. The second PC discriminates between the AAA and GGG DNA; the largest separation is achieved for positive probes.

This publication is the first example to demonstrate that MIFs can be used to differentiate between affinity and selectivity, and that, using the appropriate probe, selectivity for one of two targets may be achieved.

3.3.2
Dihydrofolate Reductase

In 1995, Pastor and Cruciani investigated the differences between human and bacterial dihydrofolate reductase (DHFR) [6]. This was the first study specifically aimed at the investigation of structural and energetic differences between macromolecules using GRID MIFs. The authors showed that the analysis of GRID MIFs with PCA can successfully highlight regions important for selectivity and activity, and allows one to focus attention on the important parts of the active site.

Starting from the X-ray structures of one bacterial and one human DHFR, the GRID/PCA analysis shows that PC 1 distinguishes between the two target proteins, clustering the objects into two groups, while PC 2 ranks the probes.

Transferring back the interesting regions of the loading plot into the 3D space of the active site, three regions in the binding site could be identified as important for selectivity. In the lower part of the binding site, differences were highlighted which stem mostly from conformational changes of a rather flexible loop, and to some extent to an Asp in the human enzyme which would allow hydrogen bonding to a ligand. In the middle of the hydrophobic pocket, differences in size and orientation between two side chains (Leu vs. Phe) are highlighted in great detail. However, as conformational flexibility was not taken into account in the analysis, small changes in the side chain conformations will produce a very different picture.

A third region at the top of the active site showed markedly different interactions with the probes. While most differences can be attributed to interactions with the side chains of an unconserved amino acid, another area in this region is produced by different backbone carbonyl orientations. Due to the presumably lower conformational flexibility of the protein backbone compared to side chain atoms, the authors speculate that this area would be especially promising for the design of selective ligands.

Complementary to regions of selective interactions, the authors identified areas with high PC 2 loadings where ligands should have strong interactions with both enzymes. Indeed, these regions coincide with ligand atoms that interact with conserved acidic residues known to be crucial for ligand binding.

3.3.3
Cyclooxygenase

Filipponi et al. [7] studied the differences in the active sites of the two isoforms of cyclooxygenase (COX), an enzyme involved in the biosynthesis of pro-inflammatory prostaglandins. Two isoforms of COX exist: a constitutive cyclooxygenase-1 and an inducible cyclooxygenase-2, and it is believed that COX-2 selective inhibitors provide anti-inflammatory agents with a superior safety profile [35].

The two enzymes are very similar. The only amino acid difference at the core of the binding site is the mutation Ile523 in COX-1 to the smaller Val523 in COX-2, which increases the size of the binding site in COX-2 and additionally opens a second pocket, which is inaccessible in COX-1.

Using X-ray structures of COX-1 and COX-2, the GRID calculations were performed both with the protein held rigid (MOVE = 0) and with flexible side chains (MOVE = 1). While the calculation with flexible protein produced a significantly larger binding site cavity of COX-2 than with the rigid protein, the identified selectivity regions were very similar.

Selective probe-COX-2 interactions occur in a region within the second pocket of COX-2 which is determined by Val523. The most selective probes, which can donate more than one hydrogen, have a high degree of affinity for COX-2 in this region.

The comparison with experimental data is somewhat complicated by the fact that the entry to the second pocket may be closed by a salt bridge which has to be broken to gain access to the pocket. Nevertheless, good agreement of the selectivity analysis with the selectivity profile of known inhibitors is found. In this case, the authors come to the conclusion, that taking protein flexibility into account with MOVE = 1 better reproduces the experimental data.

3.3.4
Penicillin Acylase

In two recent studies, Braiuca et al. applied GRID/PCA to the investigation of substrate selectivity of different forms of penicillin acylase (PA), an important enzyme in the β-lactam antibiotics industry [12, 17]. Several microbiological sources of PA exist, the enzymes differing in selectivity, activity, or stability. The authors used GRID-MIFs to explain the differences in PA from different sources, *E. coli* (PA-EC), *P. rettgeri* (PA-PR), and *A. faecalis* (PA-AF). GRID/PCA was employed to focus on the important parts in the active site and to reduce the noise in the untreated MIFs.

An important aspect of this analysis was that the authors decided to build up to four different sub-models for different probe types (e.g. donor, acceptor, hydrophobic, and halogen probes) to circumvent the known problem of underestimating hydrophobic interactions in the GRID/PCA approach.

The first study compared PA-EC with PA-AF [12]. The mutation of S67A in one part of the active site leads to weaker interactions with H-bonding probes in PA-AF. Several other amino acid differences between the enzymes translate into different interaction strengths or even structural differences of the protein backbone, which are reflected in the shape of the MIFs and the interaction energy maxima. Together with docking calculations of model substrates, the authors were able to explain the experimental selectivity profile and the enantioselectivity of the enzymes.

The second publication analyzes the differences between PA-EC and PA-PR [17]. In the score plots of all sub-models the second component discriminates the enzymes, while the first component differentiates the probes, indicative of very similar enzymes.

Indeed, the major difference is due to a mutation of Met142 (PA-EC) into a Leu (PA-PR), leading to a smaller binding site in PA-PR, consistent with the observed substrate selectivity.

Taking side chain flexibility into account with MOVE = 1, this steric hindrance is less evident for all probe groups due to a higher conformational freedom of the Met side chain. Interestingly, PA-PR also presents a larger MIF for halogens in the flexible model, and this suggests that it is able to bind larger halogenated species, such as bromine derivatives.

In addition, the flexible model identified a selectivity region which is not seen in the rigid model. The difference is due to mutations further away from the active site, which limit the flexibility of two amino acids in PA-PR directly involved in substrate recognition and thus cause a more restricted interaction region. The difference is noticeable for all probes but appears to be energetically particularly important for the polar ones. The region of this variation is crucial for the PA–β-lactam interactions, so the authors speculate that this might explain the lower rate of hydrolysis of some substrates in PA-PR compared to PA-EC.

3.3.5
Serine Proteases

Kastenholz et al. published the first application of the CPCA method. It dealt with the three serine proteases thrombin, factor Xa, and trypsin [8]. The score plot of the analysis which employed a total of 13 X-ray structures is shown in Fig. 3.5. PC 1 distinguishes between the thrombin and trypsin structures, while PC 2 separates factor Xa from the remaining two enzymes. The usual ligand binding region, which is very similar in all three enzymes, can be subdivided into three sub-pockets: S1, P and D [36] (see Fig. 3.8). With the cut-out tool, each of the three sub-pockets was investigated separately.

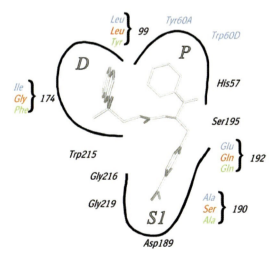

Figure 3.8. Schematic drawing of the active sites of thrombin, trypsin, and factor Xa. The crystallographic binding mode of the inhibitor NAPAP is shown to indicate the important sub-pockets S1, P, and D. Amino acids important for inhibitor binding and those differing between the enzymes are indicated (thrombin: blue; trypsin: red; factor Xa: green: common residues: black).

3.3.5.1 S1 Pocket

The deep hydrophobic S1 pockets are very similar in all three proteins. All residues are conserved except for an A190S mutation in trypsin which makes its pocket smaller and slightly more hydrophilic. While this is reflected by the pseudo-contour fields, the authors doubt that this difference can be exploited to design selective ligands.

3.3.5.2 P Pocket

The most striking difference between thrombin and the other two proteases is its insertion loop Tyr60A-Trp60D that rests as a lid on the active site and forms the P pocket. Consequently, as can be seen from the distribution of the x-variables (GRID interaction energies) after BUW, both the hydrophobic DRY and C3 probes, and to a lesser extent the cationic probes N1+ and NM3, are most important in the thrombin pocket (Fig. 3.9). Therefore, a ligand with either hydrophobic or positively charged functional groups in the region of the P pocket should improve selectivity toward thrombin. This can be visualized by the CPCA differential plot between thrombin and trypsin for the DRY probe (Fig. 3.10). Right below the thrombin insertion loop residues, the large yellow contour indicates that hydrophobic groups in a ligand at that position would increase its selectivity for thrombin. This is in excellent agreement with experimental data: most thrombin inhibitors point lipophilic groups in this direction [37, 38].

Figure 3.11 illustrates the advantage of using a chemometric description of the protein differences over a simple comparison of the MIFs: if one considers the simple difference of the DRY MIFs for thrombin and trypsin, both positive (yellow) and negative (cyan) regions show up and it is not clear whether a hydrophobic group in the ligand could be used to increase selectivity towards one of the targets.

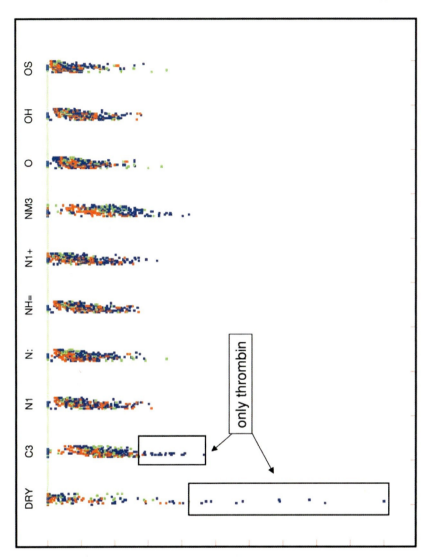

Figure 3.9. *x*-Variable distribution for the 10 grid probes in the P pocket after BUW. Blue dots indicate energies in thrombin, red dots in trypsin, and green dots in factor Xa.

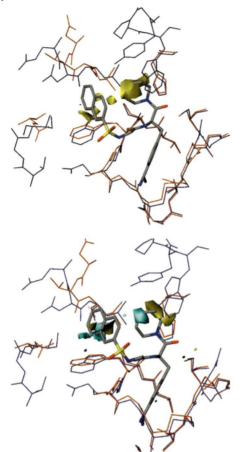

Figure 3.10. CPCA contour plot showing field differences between thrombin and trypsin for the DRY probe within the P and D pockets. Thrombin is drawn in blue, trypsin in red, the thrombin inhibitor NAPAP in gray. The yellow contours indicate regions where a hydrophobic group in a ligand would increase selectivity towards thrombin.

Figure 3.11. Contour plot of the difference of the DRY MIFs of thrombin and trypsin in the P and D pockets. Thrombin is drawn in blue, trypsin in red, the thrombin inhibitor NAPAP in gray. The contours indicate regions where the difference of the DRY interaction exceeds a threshold – cyan regions favor thrombin, yellow regions trypsin No clear picture emerges whether a hydrophobic group in the ligand would be favorable for one of the proteins.

3.3.5.3 D Pocket

The D pocket forms a second hydrophobic area, in factor Xa it is lined by the residues Phe174, Tyr99, and Trp215 which form an 'aromatic box' able to accommodate hydrophobic and positively charged functional groups. Indeed, the DRY, C3, and the NM3 probes exhibit the highest interaction energies for factor Xa. Figure 3.12 shows the CPCA pseudo-difference field plot between factor Xa and thrombin for the cationic NM3 probe. Large cyan contours for the NM3 probe in the hydrophobic box indicate that the introduction of positively charged or polarized groups should increase selectivity for factor Xa over thrombin or trypsin. Indeed, a number of highly active and specific factor Xa inhibitors have positively charged groups directed toward the contour blobs in the D pocket [39]. A similar picture is obtained for the hydrophobic DRY probe. The introduction of hydrophobic groups in a potential ligand that reach into the D pocket has been a design principle for selective factor Xa ligands [39].

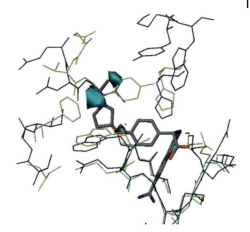

Figure 3.12. CPCA contour plot of the field differences of thrombin and factor Xa in the D pocket for the NM3 probe. The cyan contours indicate regions where interactions of the NM3 type would enhance selectivity towards factor Xa. Thrombin is drawn in blue, factor Xa in green, and the selective factor Xa inhibitor DX9065a [65] in gray.

3.3.6
CYP450

The family of cytochrome P450 enzymes (CYPs) plays a central role in the metabolism of a wide variety of xenobiotics including clinically important drugs. A number of approaches such as 3D-QSAR and pharmacophore modeling have been used to predict inhibitory potential and metabolism of drug candidates [40–45].

In a study by Ridderström et al., the GRID/CPCA strategy was applied to four human cytochrome P450 2C homology models (CYP2C8, 2C9, 2C18, and 2C19), all based on the X-ray structure of CYP2C5 [9].

In the score plots obtained, the first component discriminates between CYP2C8 and the other three enzymes, meaning that CYP2C8 is the most different among the CYP2C enzymes. The second component discriminates CYP2C18, and the third component CYP2C9 from the rest of the proteins, respectively.

The analysis of the rigid and flexible molecular interaction fields revealed that the hydrophobic regions along with shape differences of the active sites were the most important determinants for the selectivity among the CYP2C subfamily. Additionally, amino acids were identified which infer selectivity to one of the family members. The comparison with experimental mutagenesis data as well as the observed selectivities towards substrates of the CYP2C family partly confirmed the highlighted amino acids. However, in some cases experiments identified additional residues as important which do not show up in the chemometric analysis. Vice versa, for some important areas identified by the GRID/CPCA analysis no experimental evidence could be found. The authors ascribed these discrepancies to the following reasons: First, they could be partly a result of using homology models, and also using calculated binding modes for the analysis of ligand–protein interactions. It might also be possible that side chain movements taken into account by GRID are not sufficient for these flexible proteins. Finally, using the whole GRID box in the analysis might highlight regions as important which are not accessible for CYP2C inhibitors.

Nevertheless, the regions conferring selectivity towards CYP2C9 could be used to construct a receptor-pharmacophore model. This model agreed nicely with the calculated binding mode of diclofenac pointing its aromatic 4' position towards the heme. Hydroxylation of this position is specific for CYP2C9.

More recently, Afzelius and co-workers used GRID/CPCA for a comparative analysis of protein structures of CYP2C9 and CYP2C5 from different sources: crystal structures, homology models, and snapshots from molecular dynamics simulations [16]. The evaluation of molecular dynamics simulations by means of GRID/CPCA is an especially interesting new aspect in their publication.

In a first step, five available crystal structures of CYP2C5 and CYP2C9 were compared. The resulting score plot shows that PC 1 discriminates between the two enzymes. The second component separates substrate-free and substrate-bound CYP2C5. Using several probes, the loading contour plots highlighted a number of important differences in the binding sites of CYP2C5 and CYP2C9. These areas are often close to regions where visual inspection and distance measurement showed changes between the crystal structures of the two proteins. Moreover, many of the involved amino acids had been previously identified as important, e.g. by mutagenesis experiments.

In a second step, Afzelius et al. studied the results of cross-homology modeling, i.e. CYP2C5 models built from CYP2C9 crystal structures and CYP2C9 models based on CYP2C5 structures. Not surprisingly, the statistical analysis of the active site shows that the models are close mimics of their templates. That is, the CYP2C9 homology models closely resemble the CYP2C5 crystal structure from which they were built and not the target, the CYP2C9 crystal structure, and vice versa. Some improvement was possible, if the model was built from multiple targets, including bacterial CYP structures.

The last step incorporated snapshots from molecular dynamics simulations of CYP2C9 and CYP2C5 crystals in explicit water. They were analyzed to determine changes upon substrate binding and to investigate which parts of the cavity were more flexible and could participate in substrate recognition and access.

The main conclusion drawn from the MD simulations is that the proteins are highly flexible. The parts of the proteins that have high B-factors in the crystal structure also show great flexibility in the dynamics. The same regions are flexible in both runs, but the internal correlations of movements differ. This is reflected in the CPCA score plot: the snapshots of each of the two CYP2C9 runs and the X-ray structures showed up in a different quadrant and did not overlap at any time point of the simulation. Thus, the molecular dynamics simulations cover a different CPCA space from the crystal structures with and without substrate bound, independent of the different starting structures.

Moreover, when the homology models are added to the analysis, they do not occupy the same regions in the CPCA score plot as the molecular dynamics simulations or the crystal structures. Actually, inclusion of the homology models shows that they are even more diverse than the crystal structures compared with the molecular dynamics simulations. Finally, MD simulations snapshots, homology models, and crystal structures of both CYP2C5 and CYP2C9 were analyzed together, again highlighting substantial differences among structures obtained from experiment and from various calculation methods.

3.3.7
Target Family Landscapes of Protein Kinases

The potential of CPCA superscore plots to classify protein families according to the interaction patterns between protein active sites and GRID probe atoms was first explored by Naumann and Matter [10], leading to so-called 'target family landscapes'. This protein classification is solely based on a 3D interaction pattern in the binding site region, computed using GRID MIFs, and not on protein 2D similarity considerations via sequence alignment.

Naumann and Matter used a set of 26 X-ray structures of eukaryotic protein kinases, which were classified into subfamilies with similar protein–ligand interactions in the ATP binding site. As can be seen in Fig. 3.13, which shows the CPCA score plot, PC 1 separates CDK and MAP/receptor kinases on the left from the family of PKA kinases. The CDK family is represented by two distinct clusters in the target family landscape, formed by two different ATP binding site conformations. They correspond to the activated and inactivated kinase conformations

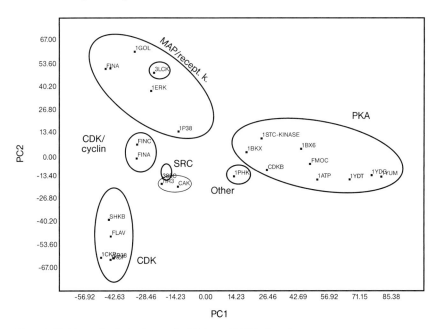

Figure 3.13. Target family landscape for 26 kinase ATP binding sites, illustrating the differences of several kinase families in chemometric space. A clear distinction between PKA, CDK and MAP kinases is seen in PC 1 and PC 2. With kind permission from Naumann and Matter [10].

depending on the binding of cyclin. The second PC separates MAP and other receptor kinases with positive scores in PC 2 from the CDK family showing negative PC 1 and PC 2 scores.

Interpretation of the structural features responsible for this landscape reveals that the main differences along PC 1 (i.e. between PKA and MAP/CDK kinases) are found in the purine and hinge binding region, whereas the discrimination in PC 2 is mainly driven by structural differences in the phosphate binding area.

To illustrate the use of the target family landscape for understanding kinase selectivity profiles, Naumann and Matter used a series of 86 2,6,9-substituted purines. These selective CDK inhibitors bind to the kinase ATP binding site [46]. A detailed comparison with experimental selectivity profiles showed good agreement with the chemometric analysis.

An important observation by Naumann and Matter is the identification of additional opportunities to achieve selective interactions in the kinases phosphate binding region, whereas most work so far has focussed on the purine binding regions.

3.3.8
Matrix Metalloproteinases (MMPs)

The matrix metalloproteinases (MMPs) are a large family of endopeptidases, responsible for degradation of a variety of extracellular matrix components in both normal tissue remodeling and pathological states [47]. The active site is a cavity spanning the entire enzyme, with three subsites on each side of the scissile bond (S3–S3', see Fig. 3.14). Most of the known MMP inhibitors so far exert their function by coordinating to residues in the primed side [48], only little effort has been put into exploring the unprimed sites [49, 50].

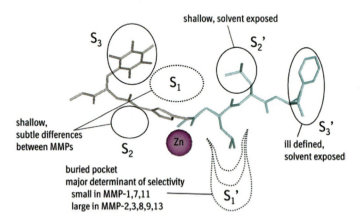

Figure 3.14. Schematic drawing of the active site of the MMPs. The crystallographic binding modes of the primed-side inhibitor PD-140798 (cyan) [66] and the unprimed-side inhibitor PNU-142372 (gray) [49] and the catalytic zinc ion (magenta) are shown to indicate the sub-pockets S3' to S3.

The MMP family has been the goal of a number of studies which tried to rationalize the differences between the family members, explain experimental selectivity profiles of known ligands, or to indicate new opportunities for selectivity design [4, 11, 51, 52].

In an unpublished study, Fox analyzed 46 X-ray structures of seven different MMPs from the RCSB Protein Data Bank [53] with GRID/CPCA [31]. The resulting score plot is shown in Fig. 3.15, with the structures colored according to family membership. The 'target family landscape' clearly shows a clustering of the individual MMP family members. Three large clusters encompass MMP1, MMP3, and MMP8, smaller clusters can be seen for MMP2, MMP7, MMP13, and MMP14. It is not clear how much the size of the clusters in the score plot reflects the inherent flexibility of the individual proteins, or is just a consequence of how many experimental structures for each MMP were available for the analysis. Nevertheless, especially the wide distribution of the MMP3 structures along PC 1 is some indication for conformational plasticity.

Inspecting the score plot, one can deduce that it should be possible to develop selective inhibitors for either MMP1, MMP3, MMP7, or MMP8, as their interaction pattern yields separate clusters in the plot. On the other hand, the scores for MMP2, MMP13, and MMP14 are very similar – here the GRID/CPCA analysis predicts that inhibitors cannot distinguish between these three MMPs if only the subsites S3–S3′ are considered in the calculations. Indeed, many experimental SAR show parallel trends between MMP2 and MMP13 [54, 55]. Mapping back the

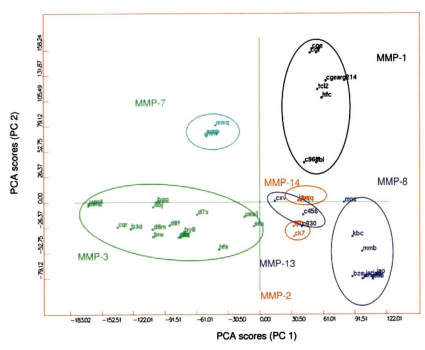

Figure 3.15. Target family landscape of 46 MMPs, colored according to family membership. Color scheme: MMP14 = orange, MMP13 = magenta, MMP7 = cyan.

loadings into the active site reveals that changes along PC 1 are distributed over the whole binding site. Changes along PC 2 map the well-known size difference in the S1' pocket – MMPs with positive PC 2 scores fall into the class with small S1' pockets, whereas a negative PC 2 indicates a large pocket. This is highlighted in Fig. 3.16, which shows the pseudo-difference plot between MMP1 and MMP8 for the C3 probe. Here, the yellow area indicates interactions which can only be reached in MMP8 with its large S1' pocket. Conversely, a bulky group in the cyan region is favorable for MMP1 with its relative short pocket. An interesting opportunity for selectivity design is shown by the DRY pseudo-difference field between MMP3 and MMP8: in the S3 pocket, a large yellow contour favoring MMP3 can be seen (Fig. 3.17). Indeed, some of the few inhibitors targeting the unprimed sites direct a phenyl ring into this region [49].

At about the same time, Terp et al. [11] used GRID/CPCA to analyze 10 MMPs with the intention of highlighting regions that could be potential sites for obtaining selectivity. Some of the structures were retrieved from the RCSB protein data bank [53], others were obtained through homology modeling [56]. To facilitate the analysis, the authors used the cut-out tool to focus on each of the six subsites in turn.

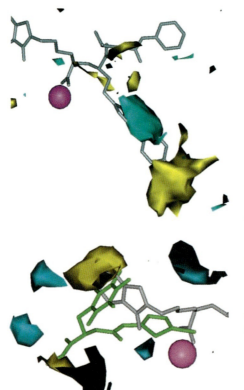

Figure 3.16. CPCA contour plot of the field differences of MMP1 and MMP8 in the S1' pocket for the C3 probe. The cyan contours indicate regions where interactions would enhance selectivity towards MMP1 with its small S1' pocket, interactions in the yellow region would favour MMP8 with its rather extended S1' subsite. The MMP inhibitor L764,004 [67] is shown in gray, the catalytic zinc ion in magenta.

Figure 3.17. CPCA contour plot of the DRY pseudo-field differences between MMP3 and MMP8 on the unprimed side of the active site. The big yellow contour indicates regions where hydrophobic interactions would enhance selectivity towards MMP3 – presumably via interaction with a Tyr. For comparison, the MMP-3 selective inhibitor PNU-142372 [49] is shown in green, the catalytic zinc ion in magenta, and parts of L764,004 [67] in gray.

Only small differences were found for the S3', S2', and S1 sub-pockets. Moreover, as these pockets are rather shallow and solvent exposed, it is doubtful that observed differences could be exploited for selectivity design.

The selectivity analysis for the S1' pocket is complex as it is surrounded by a loop. Its length and amino acid composition differs between the individual MMPs, leading to different shapes and interaction patterns for this subsite. Here, computational techniques like GRID/CPCA are especially advantageous, as they allow an automated, unbiased view on the interactions and an abstraction from a discussion of differences in single amino acids. They address the sum of all interactions at once, and the distances in the score plot allow one to somewhat quantify the differences among the proteins.

The S1' pocket is broad and elongated in all structures except for MMP1 and MMP7. Simulation of protein flexibility to some extent by using the MOVE option in GRID shows that the pocket in MMP1 is able to change its shape, making it accessible to more bulky substituents, in accordance with X-ray structures which show induced-fit behavior [57].

Terp et al. could identify a number of additional characteristics that further distinguish the different MMPs both at the top and bottom of the S1' pocket. MMP2, MMP3, MMP9, and MMP13 seem to have totally open S1' pockets and do not display any significant interactions. Nevertheless, it has been possible to derive inhibitors that show some selectivity between MMP9 and MMP13, utilizing differences in the S1' pocket. The authors attribute this to different backbone conformational changes induced by the inhibitors which were not taken into account in their analysis.

At the S2 pocket, the interactions of the DRY probe are most favorable with MMP7 and MMP20. In MMP2 and MMP14, residue Glu210 is a major determinant for selectivity: favorable interactions of the OH, N1, N1+, and NM3 probes, but unfavorable interactions with the O probe clearly distinguish these two MMPs from the remaining. A further distinction at this subsite is provided by the pocket size.

In the S3 pocket, H-acceptors are more favorable for MMP1, MMP8 and MMP14 than for the remaining MMP. In addition, MMP1 and MMP8 might also be distinguished from MMP14 by use of more sterically demanding, polar substituents at this site.

The differences found in the CPCA analysis mirror the experimental attempts to obtain selectivity, which concentrated mostly on the S1' site. Especially differences in the size and shape of the S1' subsite have been utilized, however, the GRID/CPCA calculations point to additional interactions in the S1' pocket that could be used to distinguish between several of the MMPs. In the S2' and S3' subsites, a wide range of substituents are tolerated and modifications there have often been used to optimize oral bioavailability and solubility.

These results are in partial agreement with a recent study by Lukacova et al. [51]. They mapped the binding sites of 24 human MMPs (either from experimental structure determination or from homology modeling) by calculating the interaction energies between probes and the protein. Then they used linear regression analysis to directly compare the interaction energies of the various MMPs and within the six subsites. The results reiterate the high similarity of the S2 and S3' pockets in all MMPs. Moreover, the well-known differences in the S1' pocket were observed, and the S3 pocket was also identified as being rather dissimilar among the MMP members and thus an interesting region for selectivity considerations.

In an earlier GRID/PCA study, Matter and Schwab [4] presented a detailed comparison of MMP3 and MMP8. There, the main selectivity difference was attributed to differences in the S1' pocket: the identified contour regions were in the vicinity of amino acid differences between the two enzymes. This analysis was supported by parallel CoMFA and CoMSIA analyses, which produced a consistent picture explaining the experimental affinity and selectivity of a series of MMP3 and MMP8 inhibitors.

3.3.9
Nitric Oxide Synthases

Nitric oxide synthases (NOS) catalyze the biosynthesis of nitric oxide (NO) using L-arginine as the substrate. Three isoforms of NOS have been identified: neuronal NOS (nNOS), endothelial NOS (eNOS), and inducible NOS (iNOS) [58]. Most of the inhibitors described interact with the substrate-binding site, some of them showing significant isoform selectivity. In addition, some pterin-based analogues were developed which target the tetrahydrobiopterin (H4Bip) binding site.

In a study by Ji et al., the X-ray structures of nine NOS were analyzed with GRID/CPCA [13]. In the score plot, the three NOS isoforms are well separated and cluster in different regions. Translating back the loadings into the binding site, the authors discussed several regions which might be explored to yield selective ligands. Dividing the active site into four sub-pockets, the analysis showed that in three of them, especially in the so-called M, C1, and C2 pockets, there are considerable differences among the structures, and these sites are predicted to be very important in the design of selective ligands.

In addition, Ji et al. docked 25 NOS inhibitors into the active site to understand the interaction mode of the inhibitors with the active site and also to understand the importance of the MIF differences in the active site for the isoform selectivity. In general, the experimental selectivities of the ligands agreed well with the derived selectivity regions. However, as the authors point out, the selectivity of some inhibitors with unusual kinetic behavior or even covalent binding to the target cannot be explained with the CPCA method.

This investigation is complemented by a study by Matter and Kotsonis [59] who studied the H4Bip binding site differences between the different NOS isoforms. In addition to the available crystal structures, they also employed a homology model of human nNOS in the analysis. They also found that the grouping of the different isoforms and species in the score plot agrees with the NOS isoform classification, and they could identify a number of interactions which might be explored to enhance selectivity for nNOS.

3.3.10
PPARs

A GRID/CPCA analysis of the three subtypes of Peroxisome Proliferator-Activated Receptors (PPARs) was reported by Pirard [18] using three PPARα, eight PPARγ, and three PPARδ X-ray structures of the ligand binding domain (LBD).

The CPCA superweight plot showed that hydrophobic and steric interactions contribute mostly to the discrimination between the three PPAR subtypes. From the superweight plot, the authors selected three representative probes (DRY, C3 and OH) for the subsequent analysis.

Due to the position of the subtypes in the score plot, the pseudofield contours in PC 1 distinguished PPARγ from PPARδ, whereas PC 2 highlighted the differences between PPARα and PPARγ/δ. Interestingly, all PPARγ LBDs crystallized with an agonist occupied the same region in the score plot, while the *apo* form and the one with a partial agonist were very close to the PPARα LBDs.

Most of the known PPAR agonists, which occupy only a fraction of the large T-shaped binding cavity, share a hydrophilic head group involved in key hydrogen bonds with several protein side chains, a central hydrophobic part and a flexible linker to the tail. The analysis of the main regions of the binding site (i.e. the head region, the left and right distal cavities, and the linker between the head and the distal cavities) shows that the differences in the distal pockets are most important, in agreement with experimental SAR. Especially within the left distal cavity, the different distribution of hydrophobic and bulky side chains can be exploited to modulate the selectivity for any of the three PPAR subtypes.

Although the three remaining subsites of the PPAR LBD exhibited less variation than the distal pockets, the GRID/CPCA calculations revealed some scope for selective interactions, in particular with the OH and C3 probes within the linker and head regions. This analysis agreed well with site-directed mutagenesis experiments, as well as the selectivity of known inhibitors.

Using the CPCA differential plots, the authors localized the structural differences that were responsible for the separation between the PPARγ LBDs with a bound agonist and the outliers. They found that this separation resulted from conformational changes in all five regions of the LBD, both from side chain and backbone atom movements. It is also noteworthy that conformational changes of helix 12, which represent the main difference between the agonist-bound and antagonist-bound forms of PPARα, cause a smaller separation between PPARα structures in the score plot than do these less localized conformational changes in the various PPARγ structures.

3.3.11
Bile Acid Transportation System

Ileal lipid-binding protein (ILBPs) and other fatty acid binding proteins (FABPs) were the targets of a GRID/CPCA study by Kurz et al. [14]. ILBP is a cytosolic lipid binding protein that binds both bile acids and fatty acids and presumably aids their reabsorption in the intestine.

For a first model, Kurz et al. used 30 conformers of human and pig ILBP obtained from NMR investigations. In the CPCA score plot, this resulted in a clear grouping; PC 1 revealed a separation of complexed from uncomplexed ILBPs, whereas PC 2 separated human and porcine ILBP conformers.

In the corresponding CPCA differential plots, the binding site interactions responsible for this grouping could be identified. The comparison of complexed and uncomplexed binding cavities for both ILBPs showed a conformational rearrangement upon ligand binding, in agreement with experimental distance constraints in the inner core of ILBP. In addition, significant differences between human and pig ILBP/bile acid complexes were identified.

A second model was built from 91 X-ray and NMR-derived conformers of 9 different FABPs, including the 30 ILBP conformers from the previous model. The resulting score plot showed a complex target family landscape of the lipid binding proteins. PC 1 separated all complexed ILBPs from uncomplexed ILBPs and from the remaining FABPs, PC 2 differentiated the various FABPs.

This analysis not only revealed significant differences in binding site interactions between FABPs and ILBPs. It also provided a classification of lipid binding proteins by their 3D interaction pattern. Thus, it is a first step towards the identification of interaction motifs aiding the understanding of recognition preferences of particular lipid binding proteins.

3.3.12
Ephrin Ligands and Eph Kinases

Eph receptor tyrosine kinases play a crucial role in intercellular processes such as angiogenesis, neurogenesis, and carcinogenesis. Based on their affinity for ephrin ligands, the Eph kinases are divided into two subfamilies, EphA and EphB, which bind to ephrinA and ephrinB, respectively [60].

Myshkin et al. [15] used the GRID/CPCA approach to simultaneously characterize the binding sites of ephrin ligands and Eph receptors. Starting from the crystal structure of the ephrinB2-EphB2 kinase complex, they built 3D models of the other 8 ephrins and 13 Eph kinase ligand binding domains and subjected them to a CPCA analysis.

In the resulting score plot, the first PC discriminated between the ephrins and the Eph kinases, which is in agreement with the complementary nature of interactions between these proteins. A separate classification of the kinases and ligands clearly showed two clusters corresponding to the A and B subfamilies. Kinases with a unique biological interaction profile showed up as outliers in the score plot.

Subsequently, the authors analyzed the structural basis of the eph–Eph kinase interaction. The differential plots for the cluster of Eph kinases and the cluster of ephrins, highlighted the complementary binding regions, i.e. areas of favorable interaction between kinase and ligand. Additionally, the differences between the

EphA and EphB classes were identified, as were structural features which characterize the outliers.

However, a number of biological findings could not be explained by the CPCA analysis. The authors attribute this to the fact that the ephrin–Eph kinase interactions are best described by an induced fit mechanism. The necessary computational assumption of keeping the protein structure rigid did not reflect the flexibility of the protein interfaces and their structural adaptability. The homology models, on the other hand, provided a rigid protein structure that is biased by the template protein EphB2.

3.4
Discussion and Conclusion

Despite much progress over the years, the complete understanding of ligand–protein interaction remains an elusive goal. Characterization of the binding site of proteins with GRID-MIFs and subsequent analysis with chemometric methods like PCA or CPCA may give some clues to answer the questions 'what substituent would improve the selectivity of a given compound and where to place this substituent?'.

As the examples published and briefly summarized above show, the GRID/PCA and GRID/CPCA methods can be used as tools that provide ideas for favorable structural features in newly designed ligands. In contrast to ligand-based methods, the analysis of the interactions in the binding sites of the macromolecules gives a receptor-based view of the problem, evaluated quantitatively and with an assignment of the relative importance of the possible interactions.

The close integration of MIF calculation, chemometrical analysis and visualization in GOLPE is an additional advantage which allows a straightforward application of the methodology. With GRID MIFs from different probes, one achieves a comprehensive description of possible interactions in the binding site. Additionally, PCA is a very effective tool for the comparison of the structures on a rational basis.

A comparison of two 3D structures carried out only on an empirical basis (simply by observation of differences) would be both more time consuming and less reliable since the chemometrical approach is able to dispose of the 'noise' of insignificant differences and to focus on the significant variations in chemical and structural terms.

Compared to the original GRID/PCA approach, the GRID/CPCA method offers a number of important advantages. First, not only the regions of selective interactions, but also the nature of these interactions can be identified, which leads to a much more precise design of selective ligands. Second, more than two targets can be investigated simultaneously, thus more complex selectivity problems may be analyzed. Moreover, multiple structures for each target may be employed; this is a means of incorporating information about the experimentally determined conformational flexibility of the protein which goes beyond the side chain flexibility included with the MOVE directive in GRID.

Consequently, the applications highlighted above show that the methodology is not only able to identify regions important for selectivity and affinity, but also allows the classification of whole target families. A potential application of target family landscapes could be the identification of related proteins for a particular target that should be preferentially assayed for selectivity or an early assessment of potential problems in the design of selective ligands.

A similar concept to classify protein binding sites has been employed in the CavBase [61] or the SIFt [62] approach, where the binding pockets of proteins are described by a three-dimensional collection of interaction centers which are encoded into a fingerprint. These fingerprints can then be used to judge the similarity of binding sites. An important advantage of these approaches is that they are alignment–free, i.e. no previous superimposition of the protein sites is necessary. This facilitates the comparison for even distant proteins. However, the positions of the interaction centers are based on heuristic rules, and no quantitative calculation of interaction energies, as in the GRID-MIFs, is employed. At present it is not clear if the mapping of the binding site is accurate enough for ligand selectivity considerations.

Despite its successes, it is important to keep in mind the following limitations and simplifications of the methodology described in this contribution:

1. The method can be applied only to targets for which three-dimensional structures are available. An even more restrictive requisite is that the targets should be similar enough to permit a rational superimposition of the binding sites. However, usually only in these cases the design of selective compounds is a real problem which needs computational help.
2. Often, all the water is removed from the target protein structures prior to analysis. However, water molecules may play an important role in the enzyme and may be considered constitutive of the protein structure. In such cases, the final results and the success of the method may depend on keeping certain water molecules in the protein structure. The decision, which water molecules should be kept, depends mainly on external hints, such as a high occupancy and a low temperature factor in the crystallographic refinement or previous knowledge about its function in the protein.
3. Only enthalpy is considered, but entropy is also known to be an important determinant of protein–ligand interactions.
4. Probably the most serious problem is the limited consideration of protein flexibility. In many problems, differential flexibility of similar proteins causes large experimental affinity differences of a ligand that cannot be explained by looking at static protein structures from e.g. an X-ray experiment. Some conformational freedom can be incorporated by considering side chain flexibility via the MOVE directive in GRID; the possibility to use several structures for a given protein in the CPCA methodology is another means to allow for different conformations of the targets. However, it is clear that these approximations cannot fully replace more sophisticated methods to deal with protein flexibility [63].

Nevertheless, despite these limitations, the GRID/PCA and GRID/CPCA methodology has proved useful for extracting relevant information from three-dimensional structures. the GRID/CPCA approach is an especially efficient and reliable tool for the comparison of structurally related proteins. The method provides a large amount of information that may be exploited for ligand design, predicting the outcome of protein point mutations, and for the design of enzymes having tailored activities and selectivities.

References

1 C. J. Manly, S. Louise-May, J. D. Hammer, The impact of informatics and computational chemistry on synthesis and screening, *Drug Discovery Today* **2001**, *6*, 1101–1110.
2 *Modern Methods of Drug Discovery*, A. Hillisch, R. Hilgenfeld (Eds.) Springer Verlag, Berlin, **2003**.
3 W. L. Jorgensen, The Many Roles of Computation in Drug Discovery, *Science* **2004**, *303*, 1813–1818.
4 H. Matter, W. Schwab, Affinity and selectivity of matrix metalloproteinase inhibitors: A chemometrical study from the perspective of ligands and proteins, *J. Med. Chem.* **1999**, *42*, 4506–4523.
5 G. Cruciani, P. J. Goodford, A search for specificity in DNA-drug interactions, *J. Mol. Graphics* **1994**, *12*, 116–129.
6 M. Pastor, G. Cruciani, A novel strategy for improving ligand selectivity in receptor-based drug design, *J. Med. Chem.* **1995**, *38*, 4637–4647.
7 E. Filipponi, V. Cecchetti, O. Tabarrini, D. Bonelli, A. Fravolini, Chemometric rationalization of the structural and physicochemical basis for selective cyclooxygenase-2 inhibition: Toward more specific ligands, *J. Comput.-Aided Mol. Des.* **2000**, *14*, 277–291.
8 M. A. Kastenholz, M. Pastor, G. Cruciani, E. E. J. Haaksma, T. Fox, GRID/CPCA: A new computational tool to design selective ligands, *J. Med. Chem.* **2000**, *43*, 3033–3044.
9 M. Ridderström, I. Zamora, O. Fjellström, T. B. Andersson, Analysis of selective regions in the active sites of human cytochromes P450, 2C8, 2C9, 2C18, and 2C19 homology models using GRID/CPCA, *J. Med. Chem.* **2001**, *44*, 4072–4081.
10 T. Naumann, H. Matter, Structural classification of protein kinases using 3D molecular interaction field analysis of their ligand binding sites: Target family landscapes, *J. Med. Chem.* **2002**, *45*, 2366–2378.
11 G. E. Terp, G. Cruciani, I. T. Christensen, F. S. Jørgensen, Structural differences of matrix metalloproteinases with potential implications for inhibitor selectivity examined by the GRID/CPCA approach, *J. Med. Chem.* **2002**, *45*, 2675–2684.
12 P. Braiuca, C. Ebert, L. Fischer, L. Gardossi, P. Linda, A homology model of penicillin acylase from Alcaligenes faecalis and in silico evaluation of its selectivity, *ChemBioChem* **2003**, *4*, 615–622.
13 H. Ji, R. B. Silverman, H. Li, M. Flinspach, T. L. Poulos, Computer Modeling of Selective Regions in the Active Site of Nitric Oxide Synthases: Implication for the Design of Isoform-Selective Inhibitors, *J. Med. Chem.* **2003**, *46*, 5700–5711.
14 M. Kurz, V. Brachvogel, H. Matter, S. Stengelin, H. Thüring, W. Kramer, Insights into the bile acid transportation system: The human ileal lipid-binding protein-cholyltaurine complex and its comparison with homologous structures, *Proteins: Struct., Funct. Genet.* **2003**, *50*, 312–328.
15 E. Myshkin, B. Wang, Chemometrical classification of ephrin ligands and Eph kinases using GRID/CPCA approach, *J. Chem. Inf. Comput. Sci.* **2003**, *43*, 1004–1010.
16 L. Afzelius, F. Raubacher, A. Karlén, F. S. Jørgensen, T. B. Andersson, C. M.

Masimirembwa, I. Zamora, Structural analysis of CYP2C9 and CYP2C5 and an evaluation of commonly used molecular modeling techniques, *Drug Metab. Dispos.* **2004**, *32*, 1218–1229.

17 P. Braiuca, G. Cruciani, C. Ebert, L. Gardossi, P. Linda, An innovative application of the "flexible" GRID/PCA computational method: Study of differences in selectivity between PGAs from Escherichia coli and a Providentia rettgeri mutant, *Biotechnol. Prog.* **2004**, *20*, 1025–1031.

18 B. Pirard, Peroxisome proliferator-activated receptors target family landscape: A chemometrical approach to ligand selectivity based on protein binding site analysis, *J. Comput.-Aided Mol. Des.* **2003**, *17*, 785–796.

19 GRID, Molecular Discovery Ltd., 215 Marsh Road, 1st Floor, HA5 5NE, Pinner, Middlesex, UK, www.moldiscovery.com

20 D. N. A. Boobbyer, P. J. Goodford, P. M. McWhinnie, R. C. Wade, New hydrogen-bond potentials for use in determining energetically favorable binding sites on molecules of known structure, *J. Med. Chem.* **1989**, *32*, 1083–1094.

21 P. J. Goodford, A computational procedure for determining energetically favorable binding sites on biologically important macromolecules, *J. Med. Chem.* **1985**, *28*, 849–857.

22 R. C. Wade, K. J. Clark, P. J. Goodford, Further development of hydrogen bond functions for use in determining energetically favorable binding sites on molecules of known structure. 1. Ligand probe groups with the ability to form two hydrogen bonds, *J. Med. Chem.* **1993**, *36*, 140–147.

23 R. C. Wade, P. J. Goodford, Further development of hydrogen bond functions for use in determining energetically favorable binding sites on molecules of known structure. 2. Ligand probe groups with the ability to form more than two hydrogen bonds, *J. Med. Chem.* **1993**, *36*, 148–156.

24 P. Goodford, Multivariate Characterization of Molecules for QSAR Analysis, *J. Chemom.* **1996**, *10*, 110–117.

25 GOLPE4.5, Multivariate Infometric Analysis S.r.l., Viale dei Castagni, 16, I-06143 Perugia, Italy.

26 M. Baroni, G. Constantino, G. Cruciani, D. Riganelli, R. Valigi, S. Clementi, Generating Optimal Linear PLS Estimations (GOLPE): An Advanced Chemometric Tool for Handling 3D-QSAR Problems, *Quant. Struct.-Act. Relat.* **1993**, *12*, 9–20.

27 M. D. Miller, S. K. Kearsley, D. J. Underwood, R. P. Sheridan, FLOG: a system to select 'quasi-flexible' ligands complementary to a receptor of known three-dimensional structure, *J. Comput.-Aided Mol. Des.* **1994**, *8*, 153–174.

28 R. P. Sheridan, M. K. Holloway, G. McGaughey, R. T. Mosley, S. B. Singh, A simple method for visualizing the differences between related receptor sites, *J. Mol. Graphics Model.* **2002**, *21*, 217–225.

29 S. Wold, K. Esbensen, P. Geladi, Principal component analysis, *Chemom. Intell. Lab. Syst.* **1987**, *2*, 37–52.

30 L. Eriksson, E. Johansson, N. Kettaneh-Wold, S.Wold, Multi- and Megavariate Data Analysis, 2001. Umetrics AB.

31 T. Fox, presentation at the International Workshop 'New Approaches in Drug Design & Discovery', Schloss Rauischolzhausen, March 19–22, 2001.

32 J. A. Westerhuis, T. Kourti, J. F. Macgregor, Analysis of multiblock and hierarchical PCA and PLS models, *J. Chemom.* **1998**, *12*, 301–321.

33 S. Wold, N. Kettaneh, K. Tjessem, Hierarchical multiblock PLS and PC models for easier model interpretation and as an alternative to variable selection, *J. Chemom.* **1996**, *10*, 463–482.

34 A. K. Smilde, J. A. Westerhuis, S. de Jong, A framework for sequential multiblock component methods, *J. Chemom.* **2003**, *17*, 323–337.

35 T. D. Warner, J. A. Mitchell, Cyclooxygenases: new forms, new inhibitors, and lessons from the clinic, *FASEB J.* **2004**, *18*, 790–804.

36 D. Banner, J. Ackermann, A. Gast, K. Gubernator, P. Hadvary, K. Hilpert, L. Labler, E. Myshkin, G. Schmid, T. B. Tschopp, H. Van de, Waterbeemd, B. Wirz, Serine Proteases: 3D Struc-

tures, Mechanisms of Action, in *Perspectives in Medicinal Chemistry*, B. Testa (Ed.), VCH, Weinheim, pp. 27–43, **1993**.

37 M. T. Stubbs, W. Bode, A player of many parts: The spotlight falls on thrombin's structure, *Thrombosis Res.* **1993**, *69*, 1–58.

38 T. Steinmetzer, J. Stürzebecher, Progress in the development of synthetic thrombin inhibitors as new orally active anticoagulants, *Curr. Med. Chem.* **2004**, *11*, 2297–2321.

39 F. Al Obeidi, J. A. Ostrem, Factor Xa inhibitors by classical and combinatorial chemistry, *Drug Discovery Today* **1998**, *3*, 223–231.

40 S. Ekins, M. J. De Groot, J. P. Jones, Pharmacophore and three-dimensional quantitative structure activity relationship methods for modeling cytochrome p450 active sites, *Drug Metab. Dispos.* **2001**, *29*, 936–944.

41 L. Afzelius, C. M. Masimirembwa, A. Karlén, T. B. Anderson, I. Zamora, Discriminant and quantitative PLS analysis of competitive CYP2C9 inhibitors versus non-inhibitors using alignment independent GRIND descriptors, *J. Comput.-Aided Mol. Des.* **2002**, *16*, 443–458.

42 S. Ekins, J. Berbaum, R. K. Harrison, Generation and validation of rapid computational filters for CYP2D6 and CYP3A4, *Drug Metab. Dispos.* **2003**, *31*, 1077–1080.

43 D. Korolev, E. Kirillov, T. Nikolskaya, K. V. Balakin, Y. Nikolsky, Y. A. Ivanenkov, N. P. Savchuk, A. A. Ivashchenko, Modeling of human cytochrome P450-mediated drug metabolism using unsupervised machine learning approach, *J. Med. Chem.* **2003**, *43*, 3631–3643.

44 R. G. Susnow, S. L. Dixon, Use of robust classification techniques for the prediction of human cytochrome P450 2D6 inhibition, *J. Chem. Inf. Comput. Sci.* **2003**, *43*, 1308–1315.

45 C. W. Locuson II, J. P. Jones, D. A. Rock, Quantitative binding models for CYP2C9 based on benzbromarone analogues, *Biochemistry* **2004**, *43*, 6948–6958.

46 Y.-T. Chang, N. S. Gray, G. R. Rosania, D. P. Sutherlin, S. Kwon, T. C. Norman, R. Sarohia, M. Leost, L. Meijer, P. G. Schultz, Synthesis and application of functionally diverse 2,6,9-trisubstituted purine libraries as CDK inhibitors, *Chem. Biol.* **1999**, *6*, 361–375.

47 H. D. Foda, S. Zucker, Matrix metalloproteinases in cancer invasion, metastasis and angiogenesis, *Drug Discovery Today* **2001**, *6*, 478–482.

48 J. W. Skiles, N. C. Gonnella, A. Y. Jeng, The Design, Structure, and Therapeutic Application of Matrix Metalloprotease Inhibitors, *Curr. Med. Chem.* **2001**, *8*, 425–474.

49 B. C. Finzel, E. T. Baldwin, J. Bryant, G. F. Hess, J. W. Wilks, C. M. Trepod, J. E. Mott, V. P. Marshall, G. L. Petzold, R. A. Poorman, T. J. O'Sullivan, H. J. Schostarez, M. A. Mitchell, Structural characterizations of nonpeptidic thiadiazole inhibitors of matrix metalloproteinases reveal the basis for stromelysin selectivity, *Protein Sci.* **1998**, *7*, 2118–2126.

50 J. Schröder, A. Henke, H. Wenzel, H. Brandstetter, H. G. Stammler, A. Stammler, W. D. Pfeiffer, H. Tschesche, Structure-based design and synthesis of potent matrix metalloproteinase inhibitors derived from a 6H-1,3,4-thiadiazine scaffold, *J. Med. Chem.* **2001**, *44*, 3231–3243.

51 V. Lukacova, Y. Zhang, M. Mackov, P. Baricic, S. Raha, J. A. Calvo, S. Balaz, Similarity of Binding Sites of Human Matrix Metalloproteinases, *J. Biol. Chem.* **2004**, *279*, 14194–14200.

52 T. S. Rush III, R. Powers, The application of X-ray, NMR, and molecular modeling in the design of MMP inhibitors, *Curr. Top. Med. Chem.* **2004**, *4*, 1311–1327.

53 H. M. Berman, J. Westbrook, Z. Feng, G. Gilliland, T. N. Bhat, H. Weissig, P. E. Bourne, H. M. Berman, I. N. Shindyalov, The Protein Data Bank, *Nucleic Acids Res.* **2000**, *28*, 235–242.

54 D. P. Becker, G. DeCrescenzo, J. Freskos, D. P. Getman, S. L. Hockerman, M. Li, P. Mehta, G. E. Munie, C. Swearingen, α-alkyl-α-amino-β-sulphone hydroxamates as potent MMP inhibitors that spare MMP-1, *Bioorg. Med. Chem. Lett.* **2001**, *11*, 2723–2725.

55 A.-M. Chollet, T. Le Diguarher, N. Kucharczyk, A. Loynel, M. Bertrand, G. Tucker, N. Guilbaud, M. Burbridge, P. Pastoureau, A. Fradin, M. Sabatini, J.-L. Fauchere, P. Casara, Solid-phase synthesis of α-substituted 3-bisarylthio N-hydroxy propionamides as specific MMP Inhibitors, *Bioorg. Med. Chem.* **2002**, *10*, 531–544.

56 G. E. Terp, I. T. Christensen, F. S. Jørgensen, Structural differences of matrix metalloproteinases. Homology modeling and energy minimization of enzyme-substrate complexes, *J. Biomol. Struct. Dynam.* **2000**, *17*, 933–946.

57 B. Lovejoy, A. R. Welch, S. Carr, C. Luong, C. Broka, R. T. Hendricks, J. A. Campbell, K. A. M. Walker, R. Martin, H. Van Wart, M. F. Browner, Crystal structures of MMP-1 and -13 reveal the structural basis for selectivity of collagenase inhibitors, *Nature Struct. Biol.* **1999**, *6*, 217–221.

58 W. K. Alderton, C. E. Cooper, R. G. Knowles, Nitric oxide synthases: Structure, function and inhibition, *Biochem. J.* **2001**, *357*, 593–615.

59 H. Matter, P. Kotsonis, Biology and chemistry of the inhibition of nitric oxide synthases by pteridine-derivatives as therapeutic agents, *Med. Res. Rev.* **2004**, *24*, 662–684.

60 K. Kullander, R. Klein, Mechanisms and functions of Eph and ephrin signalling, *Nature Rev. Mol. Cell Biol.* **2002**, *3*, 475–486.

61 S. Schmitt, D. Kuhn, G. Klebe, A new method to detect related function among proteins independent of sequence and fold homology, *J. Mol. Biol.* **2002**, *323*, 387–406.

62 Z. Deng, C. Chuaqui, J. Singh, Structural Interaction Fingerprint (SIFt): A Novel Method for Analyzing Three-Dimensional Protein-Ligand Binding Interactions, *J. Med. Chem.* **2004**, *47*, 337–344.

63 H. A. Carlson, Protein flexibility and drug design: How to hit a moving target, *Curr. Opin. Chem. Biol.* **2002**, *6*, 447–452.

64 B. Pirard, Computational methods for the identification and optimisation of high quality leads, *Combinatorial Chem. High Throughput Screening* **2004**, *7*, 271–280.

65 K. Kamata, H. Kawamoto, T. Honma, T. Iwama, S.-H. Kim, Structural basis for chemical inhibition of human blood coagulation factor Xa, *Proc. Natl. Acad. Sci. U. S. A.* **1998**, 6630–6635.

66 V. Dhanaraj, Q. Z. Ye, L. L. Johnson, D. J. Hupe, D. F. Ortwine, J. Dunbar, J. R. Rubin, A. Pavlovsky, C. Humblet, T. L. Blundell, X-ray structure of a hydroxamate inhibitor complex of stromelysin catalytic domain and its comparison with members of the zinc metalloproteinase superfamily, *Structure (London)* **1996**, *4*, 375–386.

67 C. K. Esser, R. L. Bugianesi, C. G. Caldwell, K. T. Chapman, P. L. Durette, N. N. Girotra, I. E. Kopka, T. J. Lanza, D. A. Levorse, M. MacCoss, K. A. Owens, M. M. Ponpipom, J. P. Simeone, R. K. Harrison, L. Niedzwiecki, J. W. Becker, A. I. Marcy, M. G. Axel, A. J. Christen, J. McDonnell, V. L. Moore, J. M. Olszewski, C. Saphos, D. M. Visco, F. Shen, A. Colletti, P. A. Krieter, W. K. Hagmann, Inhibition of stromelysin-1 (MMP-3) by P1′-biphenylylethyl carboxyalkyl dipeptides, *J. Med. Chem.* **1997**, *40*, 1026–1040.

4
FLAP: 4-Point Pharmacophore Fingerprints from GRID

Francesca Perruccio, Jonathan S. Mason, Simone Sciabola, and Massimo Baroni

Abstract

FLAP (fingerprints for ligands and proteins) is a software developed at the University of Perugia (Gabriele Cruciani and Massimo Baroni, Molecular Discovery, Italy) in collaboration with Pfizer (Sandwich UK, Jonathan Mason and Francesca Perruccio) able to describe small molecules and protein structures in terms of 4- or 3-point pharmacophore fingerprints. The molecular interaction fields (MIF) calculated in GRID [1, 2], representing the interactions between probes and small molecules or defined regions of protein structures, contain relevant information on which kind of critical interactions a ligand may have with a receptor, or, in the case of proteins, which possible sites of interaction are present in a selected area of the macromolecular structure.

GRID associates specific atom types to chemical features of a ligand: these selected atom types can be used within FLAP to build all the possible 3- or 4-pharmacophores of the investigated small molecule. A similar approach can be applied to protein studies: FLAP can build the 3- or 4-point pharmacophores present in the protein active site using site points. Site points are calculated from MIFs and they indicate favorable interactions between given probes (miming specific chemical groups) and the investigated protein region. FLAP presents several applications: it can be used as a docking tool, for ligand based virtual screening (LBVS) and structure based virtual screening (SBVS), to calculate descriptors for chemometric analysis and to investigate protein similarity. In this chapter we will present a general overview of the FLAP software and case studies for the various applications of the approach.

4.1
Introduction

4.1.1
Pharmacophores and Pharmacophore Fingerprints

Pharmacophores are a key concept in drug design that are commonly defined as an arrangement of molecular features or fragments forming a necessary, but not necessarily sufficient, condition for biological activity (or of features required for binding) [3, 4]. The history of pharmacophores has recently been reviewed by van Drie [5] and their use described by Martin and others [6–11]. A three-dimensional (3D) pharmacophore is defined by a critical geometric arrangement of such features. The use of pharmacophores derived from ligands is well established, with many methods available for their perception, the concept of pharmacophore mapping being to discover the common 3D patterns present in diverse molecules that act at the same site (e.g. target enzyme or receptor, or a potential "anti-target" such as the cytochrome P450 metabolising enzymes or HERG receptor). These patterns can be defined by distances between "pharmacophoric" features (e.g. atoms, functional groups or groups of atoms) with a particular property such as hydrogen bond donors and acceptors, acidic and basic groups, and lipophilic/hydrophobic groups. Pharmacophores have been widely used as inputs for 3D database searching [7, 12, 13] to generate new leads and for automated 3D design and QSAR. Their application has been further expanded by the concept of pharmacophore "fingerprints" (see below), that represent a more systematic view of the potential pharmacophores a molecule can exhibit. 3D pharmacophores have been used as a diversity and similarity method for the design of combinatorial libraries [14–17] as well as for virtual screening (using both single defined pharmacophores and fingerprints of potential pharmacophores) [11, 18, 19]. The ability to generate complementary pharmacophores to a protein binding site gives powerful methods that provide a common reference framework for the analysis of both ligands and their binding sites. Mason and Cheney [14–16] have used them to compare serine protease binding sites, through a GRID analysis, with automated generation of the pharmacophore fingerprint from the complementarity site points, but with only a semi-automated generation of the site points and without using the shape of the site. For docking, the pharmacophores could be used individually as 3D database search queries, with the site as an added constraint, and the use of site pharmacophores was automated in the "DiR" (design in receptor) approach [15, 16, 20], but this software developed with Chemical Design is no longer available. There is thus a need for automated methods to generate and use complementary pharmacophores of protein binding sites together with ligand-based pharmacophores; this need led to the development of FLAP that provides additional capabilities, such as protein similarity studies.

Molecular similarity and diversity methods typically represent molecules by a vector of real-valued properties (molecular weight, log$P$ etc.) or binary values (0 for absence, 1 for presence of a substructure feature for example) in a bit-string or binary fingerprint, optionally including a count of the number of times the feature is exhibited. The term "fingerprint" or "key" or "signature" is generally used to refer to an encoding of features/characteristics a molecule exhibits (e.g., substructures, topological or 2–4 point pharmacophoric feature combinations) as a string of bits (indicating the presence or not of a particular characteristic), with an optional count. A wide variety of one-dimensional and two-dimensional methods have been used, that require knowledge of the "flat" or 2D structure which represents the bonds between the atoms, together with, more recently, 3D properties (e.g. pharmacophoric fingerprints) that require knowledge of the 3D conformational space available to a molecule. A 3D pharmacophoric fingerprint marks the presence or absence of potential pharmacophores (combinations of different features and distances between them, often for 3- or 4-points, i.e. triplets/triangles or quartets/tetrahedra) within a molecule. 3D pharmacophore fingerprints can also be calculated for the target protein binding sites, being derived from site points complementary to the functional groups in the protein backbone and side chains, thus bridging the ligand-based and protein structure-based universes.

The representation of a set of active compounds by a single or small set of pharmacophores that is necessary for that activity is a well established concept, and remains an excellent model for lead optimization. The ability to readily identify active compounds that contain a different core structure from the compounds used to generate the model ("lead-hopping") is an important advantage over structure-focused methods: pharmacophores have the ability to divorce the 3D structural requirements for biological activity from the 2D chemical make-up of a ligand [19]. This success and the importance of the pharmacophore hypothesis in understanding the interaction of a ligand with a protein target led to the use of 3D pharmacophores as a molecular descriptor for similarity and diversity related tasks [14–16]. The descriptor thus generated can identify in a systematic way, within the conformational sampling constraints, all the potential pharmacophores that a molecule could exhibit, and when extended to complementary site points all the pharmacophores of a perfectly complementary molecule. By generating the descriptors in a common frame of reference, ligand–ligand, ligand–receptor and receptor–receptor comparisons are all possible, enabling additional capabilities, including selectivity analysis. Distances between pairs of features (2-point), triplet (3-point) and quartet (4-point) pharmacophore representations have been extensively used, with a variety of features sampled at each point and inter-feature distances considered in a discrete set of ranges ("bins").

Using 4-point pharmacophores enables chirality to be handled and adds some elements of volume/shape linked to electronic properties, increasing separation in similarity and diversity studies. There is a large increase in the number of potential pharmacophores that need to be considered. For example, using six possible feature types for each point, and 10 distance ranges (bins) for each feature–feature distance, the number of potential pharmacophores increases from 33 000 for 3-point pharmacophores to 9.7 million for 4-point pharmacophores [14–16]. Reducing the number of distance bins to seven reduces these numbers to 9000 and 2.3 million respectively. The granularity of conformational sampling, generally performed by torsional sampling of rotatable bonds, affects the useful resolution that can be used, as defined by the number and size of the distance bins. With protein site-derived pharmacophores, binding site flexibility can be addressed by generating fingerprints for several different conformations and optionally combining the fingerprints.

4.1.2
FLAP

FLAP is an approach and software that is specifically designed for:
- fast quantification on properties and shape complementarities between ligands and receptors;
- extraction of chemical pattern from 3D molecular interaction field maps;
- obtaining useful 3D-descriptors for optimizing pharmacodynamic properties in lead optimization;
- structure based drug design;
- selectivity analysis in proteins or receptors;
- 3D pharmacophoric properties calculation to bias combinatorial libraries;
- fast generation of lattice independent molecular descriptors for quantitative structure–property relationships;
- *in silico* ADME and DMPK predictions;
- ADME database analyses and filters determination for early phase drug discovery;
- working with small, medium and large molecules.

FLAP can be used for automatic generation of site points for docking, automatic generation of 3D fingerprints descriptors for ligands and proteins ready for chemometric analyses and lead optimization. After a brief review of the theory underpinning the FLAP software, this chapter will illustrate some of its applications and case studies.

4.2
FLAP Theory

3D pharmacophores in FLAP consist of triplets or quartets of distances between chemical features. As we mentioned before, FLAP is a computational procedure able to explore the 3D-pharmacophore space of small molecules and protein structures. All the potential 3- and 4-point 3D pharmacophores expressed by ligands and/or receptors are calculated taking conformational flexibility and molecular or receptor shape into account. With 4-point pharmacophores chirality is evaluated. Starting from GRID force field parametrization, FLAP provides a common frame of reference to allow ligand–ligand, ligand–protein or protein–protein comparison. Molecular and receptor shape are precisely evaluated "on-the-fly" and compared only when required.

For a small molecule as well as for a macromolecule the features are automatically identified. Then all the accessible geometries for all the combinations of four features are calculated. In a small molecule a pharmacophore can be defined by the atoms (described by corresponding atom types), which may have critical interactions with a receptor. In a macromolecule a pharmacophore can be defined as a combined set of all site points located in the macromolecule active site. Site points are favorable places for ligand atoms on the corresponding molecular interaction field maps. Site points are a key concept within FLAP and they correspond to interaction energy points showing best interaction energy (local minima). Site point positions, calculated for example in the active site of a given protein, define locations at which ligand atoms might be able to make favorable strong nonbonded interactions. Thus they define pharmacophoric features in proteins with a common frame of reference with pharmacophoric features in ligands.

Site points are automatically selected by FLAP but users can inspect and modify the proposed selection according to externally available information. For instance, FLAP may suggest site points on the protein surface or in locations not so important for selective binding. However, these positions may be inspected, deleted and/or modified by the user. The next step will be an automatic selection of site points based on certain criteria: this approach is, however, work in progress but it will have the advantage of making the procedure user independent and so reproducible (in the actual situation different users might select different site points and therefore the resulting pharmacophore fingerprint will be different). Once site points are stored, they are used to define the 3- or 4-point pharmacophore features inside the protein, producing millions of potential combinations of pharmacophore feature locations.

As stated before, in the FLAP software the pharmacophores for a ligand are defined in terms of atom types, which can interact with the receptor. The types of interaction are categorized as:

- hydrogen bond donors (N1 type in GRID)
- hydrogen bond acceptors (O type in GRID)
- positive charge centers (N+ type in GRID)
- negative charge centers (O– type in GRID)
- hydrophobic centers (DRY probe in GRID)
- hydrogen bond donor–acceptor centers (OH, O1, N1: or N2: type in GRID)
- shape (H type in GRID)

This approach allows the pharmacophore to be defined as 3- or 4-centers, thus forming a triangle or a tetrahedron. The shape probe is optional, but when selected, it allows a precise depiction of molecular shape in a protein cavity, or around ligand molecule(s).

These six types of interaction described above are automatically identified for each ligand molecule; this means that each ligand atom will be associated to the corresponding atom type in GRID. However, the user can assign ligand atoms to these six categories through a customisable parametrization database. Hydrophobic atoms are automatically identified by looking for atom charges, bond polarities and donor and acceptor properties. For a protein, site points are identified via energetic sampling of the putative active site using the six GRID probes reported above. For example, a site point with assigned hydrogen bond donor feature (N1 GRID probe) will be likely to be placed at a hydrogen bonding distance from a protein's carbonyl group (carbonyl group which should be however sterically accessible to the N1 GRID probe). The combined set of all site points represents a theoretical molecule that binds to all possible positions inside or in the surface of the protein cavity. Potential pharmacophores are generated from these site points in the same way as for a normal ligand. A pharmacophore key is thus generated that indicates the presence or absence of all the theoretically possible combinations of features and distances (potential pharmacophores); an additional chirality indicator is then added for 4-point pharmacophores. The GRID-probe interaction energy evaluated in the site-point locations is also recorded in the fingerprint. This allows better description of the 3D-pharmacophores. The shape of the ligands or proteins under investigation, as well as the flexibility of the molecules, are also taken into consideration during the calculation of pharmacophore fingerprints. In the next sections of this chapter we illustrate the different applications for the approaches available in the FLAP software.

4.3
Docking

The program FLAP fits ligand molecules into a set of GRID MIFs of a protein structure. Thus FLAP can be used as docking program, which uses all the GRID force fields options and capabilities.

The input target structure (protein) is investigated using the "GREATER" [21] interface. FLAP is then able to carry out the various steps needed to obtain one or more docked positions of the ligand into the target in an automatic way. The FLAP program requires two input files: the target protein in the "kout" [21] format, and the ligand molecule(s) to dock into the protein. Optionally the user may provide the location of the docking by using a simple grid cage in ASCII format. GRID MIFs are generated for the probes that are going to best simulate the ligand interactions. The GRID maps can be further elaborated and used as input for the docking process. Finally FLAP will generate the results of the docking saved in a single file and in some individual files for graphical analysis.

This information can be of particular value for ligand design selectivity studies (anti-receptor studies), because it might be possible to make a small structural alteration of the ligand(s) in order to tune desired binding modes at the expense of others. The application of computational methods to study the formation of intermolecular complexes has been the subject of intensive research during the last decade, indicating their importance to drug design projects [22]. It is widely accepted that drug activity is obtained through the molecular binding of one molecule (the ligand) to the pocket of another, usually larger, molecule (the receptor), which is commonly a protein. Assuming the receptor structure is available, a primary challenge in lead discovery and optimisation is to predict both ligand orientation and binding affinity. The computational process of searching for a ligand that is able to fit both geometrically and energetically the binding site of a protein is called molecular docking [23].

The number of algorithms available to assess and rationalise molecular docking studies is large and ever increasing. Many algorithms share common methodologies with novel extensions, and the diversity in both their complexity and computational speed provides a plethora of techniques to deal with modern structure-based drug design problems [24]. Due to the increase in computer power and algorithm performance, it is now possible to dock thousands to millions of ligands on a time scale which is useful to the pharmaceutical industry [25].

4.3.1
GLUE: A New Docking Program Based on Pharmacophores

GLUE [26] is a new docking program aimed at detecting favorable modes of a ligand with respect to the protein active site using all the options and capabilities of the GRID force field [1, 2]. The protein cavity is mapped using several GRID runs (Fig. 4.1(a–c)): a set of different probes is used to mimic each chemical group carried out by the ligand and the resulting maps are encoded into compact files, which store the local energy minima (Fig. 4.1(d)).

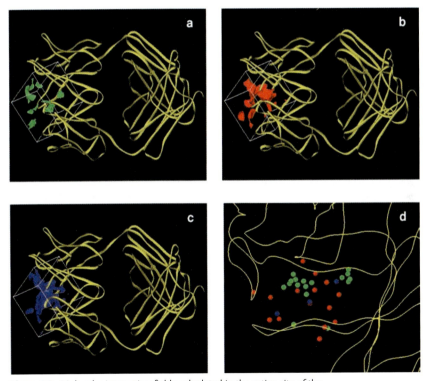

Figure 4.1. Molecular interaction fields calculated in the active site of the protein structure with the DRY probe (a), O probe (b) and N1 probe (c). Site points selected from the corresponding MIFs (d).

The energy minima are exhaustively combined into 3D pharmacophores consisting of quartets of distances between chemical features. With 4-point pharmacophores chirality is evaluated with a significant increase in the amount of information on fundamental requirements for ligand–receptor recognition. For a (macro)molecule the features are automatically identified. Then all the accessible geometries for all the combinations of four features are calculated and stored in a fingerprint of the binding site (Fig. 4.2 (a)). Afterwards, an iterative procedure identifies all the ways in which four atoms of the ligand could bind to the target, by pairing every atom to the nearest MIF used. Hydrophobic and polar atoms of the ligand for which several conformers are quickly produced are fitted over their corresponding energy minima, giving rise to sometimes millions of ligand orientations, which are temporarily stored (Fig. 4.2(b)).

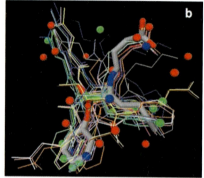

Figure 4.2. (a) All the possible pharmacophore built with site points within the active site of the protein structure. (b) Fit of each conformer of the ligands under investigation over their corresponding energy minima (site points).

Then, many orientations are quickly eliminated due to redundancy and steric hindrance constraints. Redundancy occurs whenever two or more orientations are close enough to each other, i.e. the RMSD calculated over their 3D structures is lower than 2.0 Å: therefore they are grouped by a clustering process and only one orientation will be the candidate in order to represent the entire group. Conversely, steric hindrance (Fig. 4.3) occurs whenever part of the ligand clashes into the binding site: if possible the clashing part is accommodated along the site, otherwise the orientation is excluded.

Indeed, this refinement allows only reliable orientations to be processed in the next step: each orientation is optimized within the cavity by means of successive torsions and translations. These are driven by the ligand–target interaction energy computed by the GRID force field: each small movement is followed by an energy reassessment according to the GRID standard equation ($E_{GRID} = E_{LJ} + E_{EL} + E_{HB} + E_{ENTROPY}$) applied over the whole ligand and active site (Fig. 4.4).

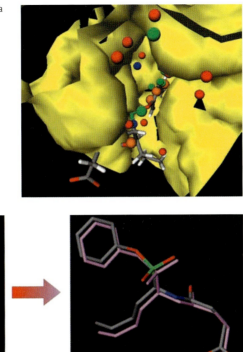

Figure 4.3. Steric hindrance can act as a filter for the many solutions found for each ligand when "docked" in the protein active site.

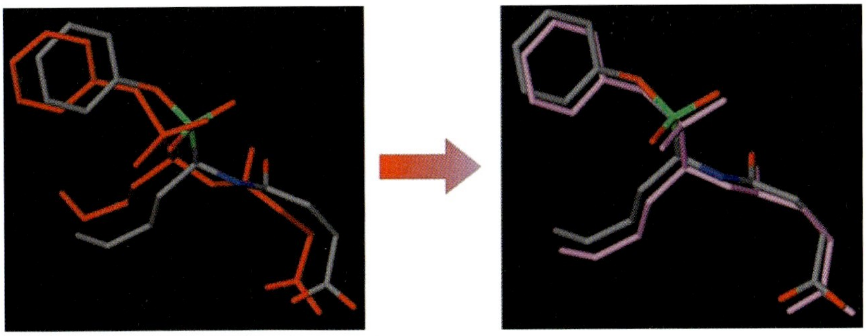

Figure 4.4. Each pose for a given ligand (red structure) is then minimized inside the protein active site (purple structure) using the GRID force field. In atom-type color is reported the X-ray crystallographic structure as a comparison.

The optimized orientations represent possible binding modes of the ligand within the site. The interaction energy between the entire ligand and the protein binding site is calculating by using the GLUE equation, which provides an energy scoring function (E_{GLUE}) composed of the following contributions: $E_{GLUE} = E_{SR} + E_{ES} + E_{RHB} + E_{DRY}$, where E_{SR} = steric repulsion energy, E_{ES} = electrostatic energy, E_{RHB} = hydrogen bonding charge reinforcement, E_{DRY} = hydrophobic energy.

The final output of the docking procedure is a set of solutions ranked according to the corresponding scoring function values, each defined by the 3D coordinates of its atoms and expressed as a PDB file.

4.3.2
Case Study

The docking procedure GLUE has been evaluated using a dataset of 230 different protein–ligand X-ray structures extracted from the Protein Data Bank at Brookhaven, according to the following criteria: (i) varied crystallographic resolution of chosen target, (ii) wide spectrum of receptor families, (iii) metal presence in the binding pocket, (iv) varied flexibility of receptor-bound ligands and (v) activities of bound ligands varying from the low micromolar to the nanomolar range. In addition, analysis of several descriptors (MW, NROT, HBA, HBD, MlogP and PSA) has been carried out, showing that the chosen ligands cover a broad spectrum of physicochemical properties, most of them within the realms of what can be considered as drug likeness [27].

Docking experiments were carried out successfully, with 60% of the studied ligand–protein complexes predicted with high accuracy, RMSD value within 2.00 Å, when the best ranked solution was considered. Also, keeping the same RMSD cut-off of 2.00 Å, it is worth mentioning that taking the first three solutions proposed by GLUE is sufficient to find reliable binding modes of the ligand within the binding site, with only 15% uncertainty.

Using a comparative study on the docking programs DOCK, FlexX and GOLD, performed by Paul and Rognan [28], we could compare 83 ligand–protein complexes out of the whole dataset of 230 complexes.

Defining as "best pose" a well-docked solution with the value of its RMSD to the X-ray 3D structure lower than 2.00 Å, the best pose obtained by GLUE was among its three first solutions for 69 out of 83 cases (83%), whereas DOCK, FlexX and GOLD obtained 18 (22%), 20 (24%), 20 (24%) respectively (Fig. 4.5).

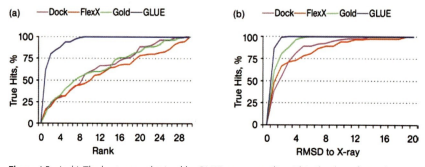

Figure 4.5. (a, b) The best pose obtained by GLUE appears to be within the three first solutions.

4.4
Structure Based Virtual Screening (SBVS)

The FLAP program can identify the pharmacophores that are in common between a ligand and a putative active site "on-the-fly". First, the protein pharmacophores are generated and recorded together with the shape of the cavity.

As stated before, the procedure to generate the pharmacophore for the protein includes first "sampling" a grid around a region of the protein (the putative active site, the coordinates of which have been selected by the user). The probes used for the calculations of the molecular interaction fields within the protein active site can be selected by default (from hydrogen bond donors (N1 type in GRID), hydrogen bond acceptors (O type in GRID), positive charge centers (N+ type in GRID), negative charge centers (O- type in GRID), hydrophobic centers (DRY probe in

GRID), hydrogen bond donor-acceptor centers (OH or O1 type in GRID) and shape (H type in GRID)). There is always the opportunity for the user to customize the selection of the probes, especially in the case where a known interaction between the target and a given is of particular interest.

From the molecular interaction fields so calculated, the points of minimum energy (site points) are consequently extracted and stored to generate all the possible pharmacophores within the protein active site. At this stage of the procedure for structure based virtual screening, the user can easily modify the number of site points to be stored, removing those which appear to be located out of the active site (region of interest) or in "inaccessible" narrow ramifications of the protein cavity. In the same way, in the case of known conserved interactions between a target and its ligands, more site points can be added to stress a particular position of interest (such as a hydrogen bond acceptor or donor atom, or a hydrophobic region). All the potential pharmacophores of the protein active site are calculated on the basis of the stored site points (modified or not).

Then the pharmacophores of the ligands to be screened are generated (each atom of the each ligand is classified as a GRID probe, such as hydrogen bond donors (N1 type in GRID), hydrogen bond acceptors (O type in GRID), positive charge centers (N+ type in GRID), negative charge centers (O– type in GRID), hydrophobic centers (DRY probe in GRID), hydrogen bond donor-acceptor centers (OH or O1 type in GRID) and shape (H type in GRID)). The generation of the pharmacophore for the ligands under investigation is performed using conformational sampling methods (random or systematic). For flexible ligands a conformational sampling is indeed needed. The method used is based on an on-the-fly generation of conformers done at search time: a quick evaluation of each conformation is performed based on an internal steric contact check to reject poor or invalid ones. The method selects automatically the rotamers strategically located in the ligand in such a way that their modifications produce the maximum variation of the molecular atom positions. Once the rotamers have been selected, a random perturbation generates a population of possible rotamer solutions. Alternatively to random generation, the user can select a systematic search method. In the latter, customizable angular steps and steric bump factors can be selected to tune the number of solutions. Moreover, a systematic selection of the systematic search solutions can be applied in order to reduce further the final number of rotamers. Conversely to many other pharmacophoric methods, which append the fingerprint for each of the conformers in a unique resulting fingerprint, FLAP produces a single fingerprint for each of the molecule conformations. For each conformation of each ligand under investigation, protein–ligand matches between all the possible pharmacophores of the putative active site of the protein and the pharmacophores for each ligand conformation are then calculated.

A unique integrated feature in FLAP may be appreciated when a ligand is processed together with receptor information. In such a case the receptor shape can be used to "bias" (filter) the generation of conformers of the ligand. Thus, conformers are generated not only to populate the conformational space, but with the intention of matching the ligand pharmacophore with the shape and the chemical features of the protein cavity. The FLAP program will have now identified all the ways in which up to four atoms of a ligand could make polar interactions with the target. These can be of the order of thousands. Shape indeed seems to be a fundamental characteristic for addressing target–ligand selectivity. The process consists of matching of the ligand pharmacophoric features into the protein pharmacophoric features. The matches are accepted only when they show shape complementarities and feature complementarities. The resultant matches are thus strongly biased by protein–ligand shape similarity. Finally, the ligands (in the new docked coordinates) are written out to a file, together with the number of matches and other similarity indicators.

The structure based virtual screening process includes also the use of some keywords. With these keywords FLAP filters out matches and keeps them only if they make sense in terms of binding site shape. FLAP can also allow additional binding site volume (cavity expansion, useful when the protein structure under investigation is an homology model) and with the use of regions (definition of a sphere within each pharmacophore needs to have at least one point) or selection of a probe (enforcing a particular feature to be present in the calculated pharmacophores) certain constraints can also be added.

4.5
Ligand Based Virtual Screening (LBVS)

In the case of ligand based virtual screening, ligands can be compared to each other similarly to the comparison between ligands and a protein structure.

FLAP computes the ligand pharmacophores and it can identify the pharmacophores that are in common between a ligand template and other ligands under investigation. As for the protein–ligand case, ligand–ligand complementarity may be generated using conformational sampling biased by shape complementary and feature complementarities with one or more template molecules. The shape can be defined around a unique template molecule, or around a combination of template molecules. The resultant matches are then written out to a file, together with the number of matches and other similarity indicators.

If the target of the ligands under investigation is known, another possible approach is to compare ligands using the shape of the protein as a shape constraint and features in the protein cavity as additional constraints. As in the case of structure based virtual screening, keywords are used such as regions to define a sphere within which each pharmacophore needs to have at least one point, and the selection of a particular probe.

Ligand based virtual screening has been performed with FLAP on an in-house project at Pfizer, Sandwich Laboratories. For reasons of confidentiality we cannot disclose details of the structures for the project. Both active ligands and receptor structures were available.

Seven different pharmacophores were built in FLAP representing different series of ligands active against the same target. The virtual screening was carried out with the seven pharmacophores one at a time to rank a given library. The results from the seven runs of ligand based virtual screening were then merged. In one case the shape of the ligands was used (FLAP LB, see legend on Fig. 4.6) as a constraint and in another case the shape of the receptor (FLAP SB, see legend on Fig. 4.6). Exactly the same procedure was applied in Catalyst [29] for comparison.

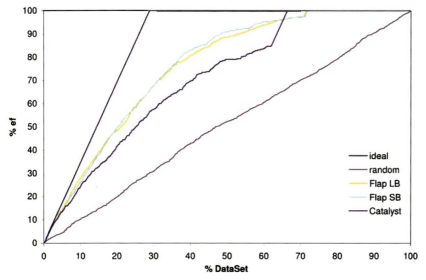

Figure 4.6. Enrichment plot using FLAP with shape constraint of the ligands (FLAP LB), with shape constraint of the receptor (FLAP SB) and comparison with Catalyst.

4.6
Protein Similarity

Proteins can also be compared, with or without using information from ligands that bind to them. From the protein site points FLAP identifies the pharmacophores that are in common on-the-fly. Then complementarities between the proteins are evaluated maximizing the site-point features and the shape complementary between protein cavities. The resultant matches are then written out to a file.

An earlier study by Mason and Cheney [14–16] with serine proteases used site-derived fingerprints to quantify the range of different pharmacophoric shapes complementary to the target protein binding sites, and illustrated the large differences in 3D pharmacophoric fingerprints between related targets. These can be exploited for selectivity, whereas the common pharmacophores could represent common binding motifs. The 4-point pharmacophore fingerprints were generated from atoms added in the most favorable interaction regions from GRID favorable energy contours from five pharmacophoric probes: H-bond donor (NH of amide);

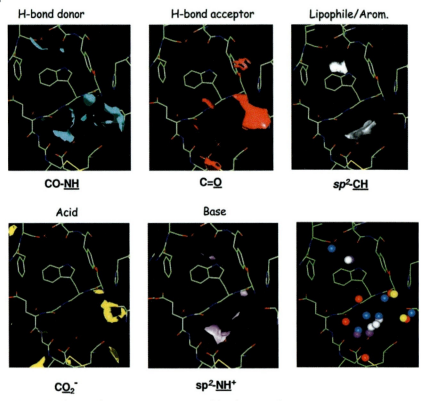

Figure 4.7. The complementary site points used for pharmacophore fingerprint calculations (lower right), together with the energetically favorable contours from 5 GRID probes on a Factor Xa binding site.

H-bond acceptor (C=O), acid (CO_2^-), base (NH^+) and lipophilic (aromatic CH). Figure 4.7 illustrates the contours and the atoms which were added (with associated pharmacophore features) for the Factor Xa serine protease active site.

The ensemble of atoms added to the "hotspots" was treated as a hypothetical molecule that interacts at all favorable positions in the binding site, and the pharmacophore fingerprint was calculated and analysed from this in the same way as for ligands [14]. For example, the Factor Xa and Thrombin serine protease active sites had 13 complementary site points added, leading to fingerprints of 2103 and 2063 4-point pharmacophore shapes respectively, with 234 in common. The third serine protease, Trypsin, which has a less defined S4 pocket, had only 11 significant complementary site points, leading to a fingerprint of 1233 pharmacophore shapes, of which 243 were in common with Factor Xa, and 120 in common for all three serine proteases. Using 3-point fingerprints the numbers for Factor Xa, Thrombin and Trypsin are 491, 430 and 350 respectively, with overlaps of 202 between factor Xa and Thrombin, and a common 131 pharmacophores for all three proteins. There are clearly less "unique" pharmacophores for each protein using the 3-point pharmacophores with a higher proportion of the total in common; the common pharmacophores provide a set of useful common binding motifs that can be used to drive docking studies. Ensembles of pharmacophores can thus be identified that can be used both to differentiate the sites (selectivity) and to identify common features.

Comparison of these protein derived pharmacophore fingerprints with known ligands, using 4-point fingerprints, shows that they can be used to search for novel ligands within a database and that they are specific enough to capture ligand selectivity between similar proteins such as these three serine proteases [14,16]. A thrombin inhibitor (NAPAP, 6nM) showed most overlap (352) with thrombin using 4-point pharmacophores (210 and 82 for Factor Xa and Trypsin respectively), whereas with 3-point fingerprints selectivity was not captured and there was more overlap with Factor Xa (64) than with Thrombin or Trypsin (64 and 31 respectively). Using a "decoy" molecule, a fibrinogen receptor antagonist that contained the benzamide serine protease S1 pocket binding motif, the 4-point pharmacophore fingerprint comparisons clearly indicated a lack of complementarity (4, 2 and 0 common pharmacophores for Thrombin, Factor Xa and Trypsin respectively), whereas with 3-point pharmacophore fingerprint comparisons failed to differentiate this compound, with as many common pharmacophores found (60, 57 and 48 for Thrombin, Factor Xa and Trypsin respectively) as with the thrombin inhibitor. The comparisons possible in FLAP (see SBVS section) enable binding site characteristics such as shape to be retained when comparing proteins and protein to ligands, greatly enhancing the signal, providing a needed capability.

4.7
TOPP (Triplets of Pharmacophoric Points)

In this section a case study (cytochrome P450 metabolic stability) introduces a further application of 3D fingerprint descriptors for ligands and proteins for chemometric analyses. Metabolic stability is used in order to describe the rate and the extent to which a molecule is metabolized. It usually refers to the susceptibility of compounds to undergo biotransformation. This is normally an issue in selecting and/or designing drugs with favorable pharmacokinetics properties. The ability to evaluate the metabolic stability of compounds in the very early stages of drug discovery improves the chance of selecting a molecule with good *in vivo* activity. In this context our intent was to build an *in silico* model able to discriminate between metabolically stable or unstable compounds relative to cytochrome CYP2D6. The computational tool we used to build this model is TOPP (Triplets Of Pharmacophoric Points), a new *in silico* QSAR approach able to use 3-point pharmacophores as 3D descriptors and GOLPE [30] as the tool to perform multivariate statistical analysis.

The base theory underpinning TOPP is similar to the FLAP approach, previously described in this chapter. First, atoms in the molecules are classified by the GRID force field parametrization. In this way, atoms are described according to their charge and hydrogen bonding properties as DRY (hydrophobic), DONN (HBD hydrogen bond donor interactions), ACPT (HBA hydrogen bond acceptor interactions) and DNAC (both HBD or HBA interactions). Once the classification of the atoms in each molecule under investigation has been performed, an iterative procedure is able to generate all possible combinations of three points and four different atom types (DRY, DONN, ACPT and DNAC) encoding them in two possible ways. One of the two possible approaches is to run TOPP in order to store only the presence or the absence of a 3-point pharmacophore combination. The other approach is to run TOPP in order to count the number of times that a combination is present in the molecule (Fig. 4.8).

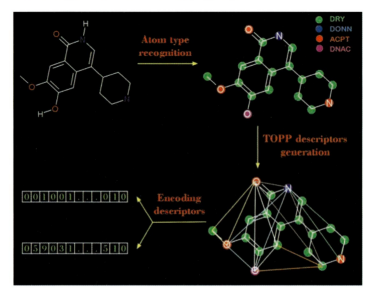

Figure 4.8. Flowchart of TOPP methodology.

The kinds of calculations described above are done for all the molecules under investigation and then all the data (combinations of 3-point pharmacophores) are stored in an X-matrix of descriptors suitable to be submitted for statistical analysis. In theory, every kind of statistical analysis and regression tool could be applied, however in this study we decided to focus on the linear regression model using principal component analysis (PCA) and partial least squares (PLS) (Fig. 4.9). PCA and PLS actually work very well in all those cases in which there are data with strongly collinear, noisy and numerous X-variables (Fig. 4.9).

Applying this procedure to investigation of the metabolic stability of CYP2D6, we were able to find a model to correctly classify metabolically stable and unstable compounds. This model was trained using a set of 129 compounds from the BioPrint [31] database. Drug-likeness and solubility properties were used as primary filter in order to eliminate unattractive compounds and all those compounds classified as not soluble, which are always classified as metabolically stable. The data-

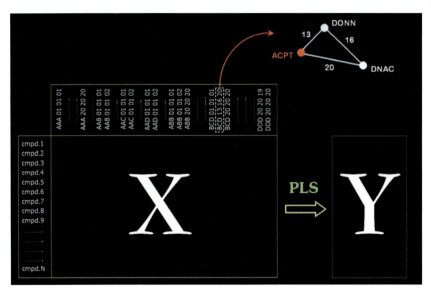

Figure 4.9. Multivariate statistical analysis and regression tools can be applied to the TOPP descriptors.

set was divided into two classes: unstable class (% metabolic stability < 60; 52 compounds) and stable class (% metabolic stability ≥ 60; 77 compounds). The molecules, modelled in their cationic form, were converted into 3D structures using in-house software. From the 3D structures, molecular descriptors were calculated using the TOPP program. The descriptors were further correlated to the experimental metabolic stability classes by a partial least squares discriminate analysis and three significant latent variables were extracted from the PLS model with cross validation. The score plot of the first two principal components shows the compounds color-coded according to their metabolic stability (red points represent stable compounds whereas blue points indicate unstable compounds). So far, the validation of the model with external datasets has been performed to predict the yes or no response for a set of 265 stable BioPrint compounds. The results indicate that 215 out of 265 compounds are well predicted as stable compounds (Fig. 4.10).

Using this model as initial filter in the very early stages of drug discovery, for example in library design and virtual screening, could be very useful in order to decrease the large amount of compounds to be tested *in vitro* or *in vivo*.

100 | 4 FLAP: 4-Point Pharmacophore Fingerprints from GRID

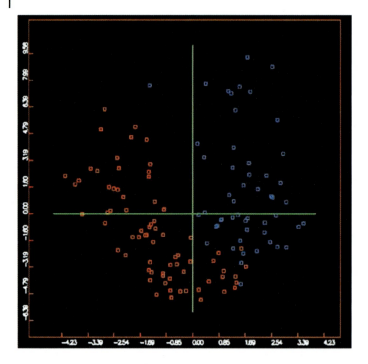

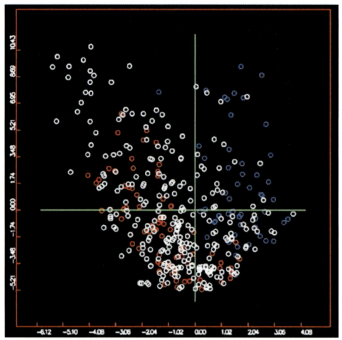

Figure 4.10. PLS score plot separating the stable compounds (red circles) from unstable compounds (blue circles). Prediction (white circles) of the yes or no response for a set of 265 stable BioPrint compounds.

4.8 Conclusions

The FLAP program represents a promising approach to gaining information from the molecular interaction field calculated by the GRID software within a region of a protein structure and from the atom classification in GRID probes for atoms in a ligand molecule. The key feature of this software is the transformation of a molecular field describing an interaction between a particular probe and a region of interest of a protein structure (such as the active site) into site points, which in turn describe the location of the most favorable interaction between the given probe and the protein structures. The site points so calculated are used to build all 3- or 4-point pharmacophores present in the protein region under investigation and these pharmacophores are encoded in a fingerprint.

In the same way each atom of the ligand, once classified in the corresponding GRID probes describing features such as hydrophobicity, hydrogen bond donor or/and acceptor capabilities and charge. These are equivalent to the protein site points, which are used to build the 4- and 3-point pharmacophores present in the ligand. These pharmacophores are encoded in the fingerprint the ligand under investigation.

By generating 4- or 3-point pharmacophore fingerprints for proteins and ligands, FLAP is able to perform comparison between protein and ligand pharmacophore fingerprints, between ligand pharmacophore fingerprints and between protein pharmacophore fingerprints. This kind of approach can be exploited very straightforwardly in structure based virtual screening and docking, ligand based virtual screening and protein similarity studies.

A key enhancement is that the flexibility and shape of the ligand or/and of the active site of the protein are taken into consideration. Constraints can be set by the user as well as other keywords able to describe particular features of the protein active site or within the ligand molecules.

The calculation of the pharmacophore fingerprints is fast and a reasonably large number of molecules can be handled. The speed of these calculations permits the user to readily add to the overall procedure more information about the target or active compounds as probes in the ligand based virtual screening.

Finally, the possibility to study with linear regression models using principal component analysis (PCA) and partial least squares (PLS) regression analysis pharmacophores as descriptors for the corresponding molecules represents an interesting and novel approach in QSAR.

Acknowledgments

We would like to thank Prof Gabriele Cruciani (University of Perugia, Italy) for his contribution and support to the FLAP project and Massimo Baroni (Molecular Discovery, Ltd) for developing the FLAP code. Particular thanks to Simone Sciabola for the work done on testing and validation of the software in the docking approach and to Dr Jonathan Mason for constructive discussions.

References

1. Goodford, P. J. *J. Chemometrics* **1996**, *28*, 107–117.
2. Carosati, E., Sciabola, S., Cruciani, G., *J. Med. Chem.* **2004**, *47 (21)*, 5114–5125.
3. Gund, P. Three-dimensional pharmacophoric pattern searching, in *Progress in Molecular and Subcellular Biology*, Springer-Verlag, Berlin, **1977**, Vol. 5, pp. 117–143.
4. Marshall, G.R. Binding-Site Modeling of Unknown Receptors, in *3D QSAR in Drug Design*, ESCOM, Leiden, **1993**, pp. 80–116.
5. Van Drie, John H. Pharmacophore discovery: A critical review, Comput. Med. Chem. Drug Discovery **2004**, 437–460. CODEN: 69FIPX CAN 141:306806 AN 2004:371615 CAPLUS.
6. Ghose, A. K., Wendoloski, J. J. *Perspect. Drug Discovery Des.* **1998**, *9/10/11*, 253–271.
7. Milne, G. W. A., Nicklaus, M. C., Wang, S. *SAR QSAR Environ. Res.* **1998**, *9 (1–2)*, 23–38.
8. Van Drie, J. H., Nugent, R. A. *SAR QSAR Environ. Res.* **1998**, *9 (1–2)*, 1–21.
9. Martin, Y. C. Pharmacophore Mapping, in *Design of Bioactive Molecules*, Martin, Y. C., Willett, P. (eds.), **1998**, pp. 121–148.
10. Bures, M. G. Recent Techniques and Applications in Pharmacophore Mapping, in *Practical Applications of Computer-Aided Drug Design*, Charifson, P. S.(ed.), Dekker, New York, **1997**, pp. 39–72.
11. Mason J. S., Good A. C., Martin E. J. *Curr. Pharm. Des.* **2001**, *7*, 567.
12. Good A. C., Mason J. S. Three-dimensional structure database search, in *Reviews in Computational Chemistry*, VCH, New York, **1995**, Vol. 7, pp. 67–117.
13. Warr W. A., Willett P. The Principles and Practice of Three-Dimensional Database searching, in *Design of Bioactive Molecules*, American Chemical Society, Washington D.C., **1998**, pp. 73–95.
14. Mason J. S., Morize I., Menard P. R., Cheney D. L., Hulme C. R., Labaudiniere R. F. *J. Med. Chem.* **1999**, *42*, 3251.
15. Mason J. S., Cheney D. L. *Proc. Pac. Symp. Biocomput.* **1999**, *4*, 456.
16. Mason J. S., Cheney D. L. *Proc. Pac. Symp. Biocomput.* **2000**, *5*, 576.
17. Mason J. S., Beno B. R. *J. Mol. Graphics Modell.* **2000**, *18*, 438.
18. Mason J. S., Pickett S. D. Combinatorial Library Design, Molecular Similarity and Diversity Applications, in *Burger's Med. Chem. and Drug Discov.*, 6th edn, Wiley, NY, **2003**, Vol. 1.
19. Good, A. C., Mason, J. S.; Pickett, Stephen D. Pharmacophore pattern application in virtual screening, library design and QSAR, in *Methods Princ. Med. Chem.* **2000**, *10* (Virtual Screening for Bioactive Molecules), 131–159.
20. Murray C. M., Cato S. J., *J. Chem. Inf. Comput. Sci.* **1999**, *39*, 46.
21. GRID version 22, Molecular Discovery Ltd. (www.moldiscovery.com).
22. Blaney, J. M., Dixon, J. S. *Perspect. Drug Discov.* **1993**, *1*, 301–319.
23. Lybrand, T. P. *Curr. Opin. Struct. Biol.* **1995**, *5*, 224–228.
24. Kuntz, I. D. *Science* **1992**, *257*, 1078–1082.
25. Abagyan, R., Totrov, M. *Curr. Opin. Chem. Biol.* **2001**, *5*, 375–382.
26. Sciabola, S., Baroni, M., Carosati, E., Cruciani, G. Recent improvements in the GRID force field. 1. The docking procedure GLUE, poster presented at the 15th Eur. Symp. QSAR & Molecular Modelling, Istanbul, Turkey, **2004**.
27. Pickett, S. D., McLay, I. M., Clark, D. E., *J. Chem. Inf. Comput. Sci.* **2000**, *2*, 263–272.
28. Paul, N., Rognan, D., ConsDock: A New Program for the Consensus Analysis of Protein–Ligand Interactions, *Proteins* **2002**, *47*, 521–533.
29. Catalyst (Accelrys Inc., San Diego CA).
30. GOLPE version 4.5.12, M.I.A.
31. Krejsa, C. M., Horvath, D., Rogalski, S. L., Penzotti, J. E., Mao, B., Barbosa, F., J. C. *Curr. Opin. Drug Discov. Dev.* **2003**, *6 (4)*, 470–480.

5
The Complexity of Molecular Interaction: Molecular Shape Fingerprints by the PathFinder Approach

Iain McLay, Mike Hann, Emanuele Carosati, Gabriele Cruciani, and Massimo Baroni

5.1
Introduction

It is well known that molecular shape plays a key role in ligand–receptor binding, since molecular recognition is largely mediated by shape, and similar biological activity often reflects similar molecular shape. In fact, although the right matching of pharmacophore features is required for a small molecule to bind to its target, the establishment of surface-to-surface contact between ligand and target along the surface of the small molecule is also important.

Several computational methods, ranging from docking to virtual screening and molecular superposition/alignment, make use of various shape descriptions. Shape similarity is the foundation of many ligand-based methods, which seek compounds with structure similar to known actives, and shape-complementarity is also the basis of many receptor-based designs, where the goal is to identify compounds with high complementarity in shape to a given receptor.

Clearly a method for describing a shape in simple numerical terms would assist such work greatly. Indeed, it can be said that an efficient way of describing the shape of any kind of molecule is nowadays central to drug discovery.

The difficulty of encoding the shape into numbers increases with the shape complexity. Comparison of the shapes of two objects, or two molecules, is intuitive to the human brain, but the task becomes far from trivial when the complexity of the problem increases, i.e. when comparing and classifying several compounds or sorting a set of compounds according to their molecular shape.

The molecular interaction fields (MIF) obtainable by GRID [1] may be used to define the solvent accessible surface, which resembles the molecular shape. However, MIFs are descriptors that depend on the 3D-location, and usually several thousand are required to describe a shape. In this chapter we present a novel procedure, called PathFinder, which encodes MIF into a compact alignment-free description of molecular shape.

5.2
Background

A chemist will usually consider the structure of a drug molecule as a 2D representation (e.g. Fig. 5.1(a)), whilst for a computer it may be necessary to encode the structure as a molecular graph (e.g. Fig. 5.1(b)). Molecular graphs are suitable for similarity searching and substructure searching. However, the receptor, or other drug target, does not recognise the drug through 2D or graph representations. The drug interacts with the target as a 3D object with appropriate 3D molecular complementarity. In a simple way it can be considered that during the binding process the receptor senses the drug and recognises the complementarity.

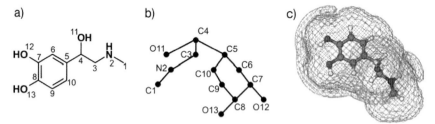

Figure 5.1. A molecule of adrenaline is presented as it is drawn by chemists (a), and as it is encoded by computational methods into molecular graphs (b) and molecular interaction fields (c).

GRID-derived molecular fields (MIF) are designed to explore numerically the way a receptor could "feel" a drug molecule. The GRID force field calculates an interaction potential between probe and molecule by assessing many different nonbonding interactions: hydrogen bonding, charge–dipole, dipole–dipole, Lennard-Jones and electrostatic. The shape description for a molecule is certainly embedded within the MIFs with positive/negative values representing repulsive or attractive interactions and the presence/absence of molecular fields (MIF) representing occupied or unoccupied volumes.

The GRID-derived MIFs have been processed mathematically in several different ways to provide descriptors which are of great use both for QSAR and for general ADME prediction. For QSAR the GRID/GOLPE procedure has been developed, in which the intermolecular comparison is performed by first aligning by either their fields or the molecules themselves. Accordingly, although the molecular shape is of great relevance in such analyses it is strictly dependent upon the alignment-superposition of the compounds. The alignment problem has been addressed to some extent with the GRIND procedure, which extracts characteristic features of the molecular fields. However, although available, the GRIND description of shape is not comprehensive. The well-known ADME tool VolSurf uses another method to convert GRID fields into simple descriptors. However, very few VolSurf descriptors are qualitatively related to molecular shape.

The absence of a precise shape description, based on the GRID molecular interaction fields, prompted the development of a new procedure, called PathFinder, that is described here.

5.3 The PathFinder Approach

The PathFinder procedure is aimed at describing the shape of objects starting from the MIF surface representation. Although the method presented in the following is general and could be applied to any kind of surface, we will refer to the isopotential surface obtained by the program GRID, when using the water probe (OH2) and setting the energy level to $+0.2\,\text{kcal mol}^{-1}$. In this way a molecular surface is obtained which resembles the solvent-accessible molecular surface.

A subset of points, uniformly distributed on the molecular surface, is selected from this surface. The default value is usually set to 100, which is sufficient to describe drug-like molecules and receptor cavities. However, it is possible to select a lower number of surface nodes if thought appropriate for the situation.

Each selected node is paired to all the remaining nodes of the molecular surface, one by one, and the shortest connection "walking on the molecular surface" between two nodes is computed. This is implemented through the graph theory which is explained briefly here: (i) the molecular surface is encoded into a weighted and undirected graph in which each single point from the surface represents a node and the nodes are connected by arcs; (ii) the weight for each arc subtended by adjacent nodes is the euclidean distance between the two points, calculated from the cartesian coordinates; (iii) only arcs lying on the molecular surface are utilised. Summarising, the molecular surface is encoded into a graph; its nodes are connected to each other, when adjacent, through surface-lying paths. Indeed, the surface computationally resembles a grid in which all nodes can be reached from all other nodes.

The walk of minimum weight is sought by applying the Dijkstra algorithm [2]. The corresponding path is approximated by a set of segments: the minimum path is the weight of the entire walk, representing the sum of the corresponding weights of edges composing the walk. Therefore, two numerical values are finally related to each node pair: their corresponding euclidean distance and the minimum path.

5.3.1 Paths from Positive MIF

The minimum path has been defined as the shortest connection between two grid nodes obtained by "walking on the molecular surface". When coupled with the corresponding euclidean distance it provides a novel descriptor derived from each node pair.

Searching the minimum paths over the entire set of node pairs yields a huge amount of path–distance couples, stored into a path–distance matrix. Its elements count the frequency of the corresponding path–distance pairs: therefore, the matrix expresses the probability factor for the existence of two points on the molecular surface at a specific distance and path.

Consequently, elements of the matrix corresponding to nonexisting path–distance pairs contain zero frequency value, whereas all nonzero values indicate the molecular elongation, size, and variegation of the surface.

The frequency of path–distance pairs for a single molecule can be viewed by means of three-dimensional plots in which the frequency distribution (z axis) is reported versus distances (x axis) and the path–distance (y axis), see Fig. 5.2. Molecular shape peculiarities are condensed in the frequency distribution graph in which low path–distance values (path ≈ distance), represented by points close to the distance axis of the plot in Fig. 5.2, characterize more planar surfaces, whereas high path–distance values (path > distance) characterize more wrinkled surfaces.

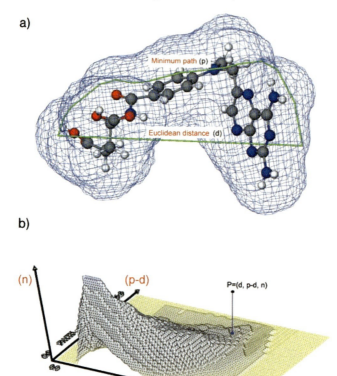

Figure 5.2. GRID positive MIF (a) and path–distance frequency distribution (b) for the methotrexate. Two nodes are used as an example: their euclidean distance and minimum path are highlighted with green lines (a) and the point representing that node pair is indicated by P (b).

The minimum paths provide a molecular shape description, coded as a two-way matrix table. The matrix may then be unfolded into a single array, which describes the molecular shape, as illustrated in Fig. 5.3.

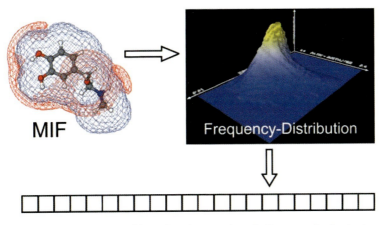

Figure 5.3. Schematic view of the PathFinder procedure: the frequency distribution is extracted from GRID MIF, and is consequently encoded into a molecular fingerprint.

5.3.2
Paths from Negative MIF

It has been shown how shape-related molecular description can be extracted starting from the GRID positive MIF. Similarly, the nodes of attractive interaction between the probe and the molecule (negative MIF) can be handled. The red isocontour surfaces in Fig. 5.4 represent the interaction of methotrexate with the water probe. In this case the application of the procedure described above will highlight the relative position on the molecular surface of the chemical moiety interacting with the water probe. The representation in Fig. 5.4(b) is not a strict shape function but more a new pharmacophoric representation of the molecule.

The main differences from the previously presented method (positive MIF), are:

1. The nodes are selected on the basis of their different energy values. Then, only relevant nodes, with the most attractive energy, are selected; that is up to five per region with the default setting.
2. Only nodes from different chemical groups are connected. Although the minimum paths are calculated by connecting only nodes from negative MIF, the walk still lies on the molecular surface (blue surface on Fig. 5.4) as it does for positive MIF.
3. The GRID energy value of every node is combined with the frequency distribution. In fact, the GRID energies from the two nodes are multiplied, and the product contributes to the probability factor.

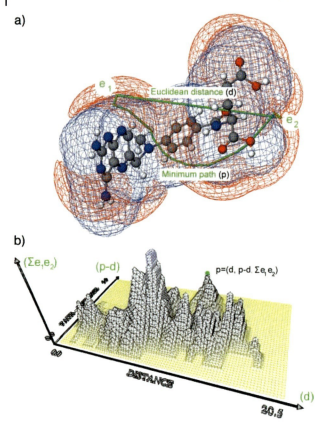

Figure 5.4. GRID negative MIF – red colored surface – (a) and frequency distribution (b) are reported for the methotrexate. Two nodes are used as examples: their euclidean distance and minimum path are highlighted with green lines (a), whereas the point responsible for that node pair is indicated by P (b). Each point in (b) is characterized by three coordinates: the distance, the path–distance value, and the sum of the GRID energy products.

The negative-to-negative minimum path can be applied to the MIF obtained with all GRID probes. A common procedure makes use of water (OH2) and hydrophobic (DRY) probes. Accordingly, a further description OH2–DRY is given by connecting combined node pairs, one node from water negative MIF, the other from the hydrophobic negative MIF. However, other probes could be used, such as the combination of N1 (hydrogen bonding donor) and O (acceptor). In this way, a molecular description of molecular functionality is obtained.

The resulting matrix is then unfolded into a one-dimensional vector, which can be merged with the shape description, and is suitable for multivariate statistics analysis such as principal component analysis (PCA) and partial least squares (PLS).

5.4
Examples

PathFinder is a novel procedure which numerically describes molecules on the basis of 3D shape and functionality. Since shape is a fundamental property in many fields, applications range from 3D QSAR to metabolism, as well as from molecular diversity and/or similarity issues, such as clustering, classification or database searching of drug-like molecules, to end up with shape analysis and functional complementarity in the macromolecules recognition. A key peculiarity of the PathFinder method is that the shape profile can be derived not only for ligands but also for enzymes and receptor sites. Thus, as well as ligand–ligand and target–target similarity studies it is also possible to perform mixed target–ligand comparisons.

5.4.1
3D-QSAR

The use of the PathFinder descriptor for QSAR is exemplified here through application to two datasets taken from the literature: (i) 31 steroids utilized in the original CoMFA study [3], since considered a benchmark for evaluating QSAR methods [4]; (ii) 55 inhibitors of HIV-1 reverse transcriptase proposed by Chan and coworkers [5], interestingly this series could only be successfully correlated by means of a structure-based approach [6].

For the first dataset, testosterone-binding globulin (TBG) and/or corticosteroid-binding globulin (CBG) binding affinities were available. The GRID probes OH2 and DRY were used with both positive and negative MIF being encoded for the OH2 probe whilst the DRY contributed with its negative MIF to the "self" paths and to the so-called Mix description. 1136 active variables were obtained and reported in the loading plots, Fig. 5.6, according to their type.

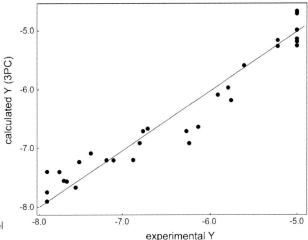

Figure 5.5. Calculated versus experimental values for the PLS model with 3 latent variables.

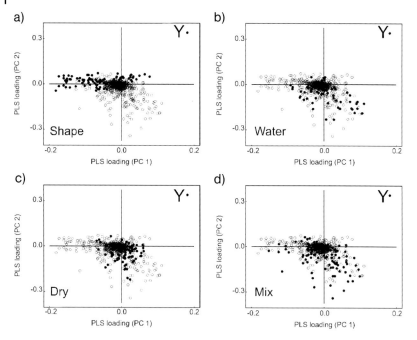

Figure 5.6. Loading plots for the PLS model. Each block of variables (black-colored) is highlighted in the corresponding plot. The shape is the most important block of variables, inversely correlated with the activity.

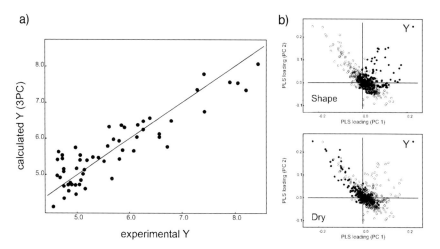

Figure 5.7. (a) Calculated versus experimental values for the PLS model with 3 latent variables. (b) Loading plots for the model. The upper plot highlights high collinearity of the shape descriptors with activity; the lower plot shows hydrophobic regions with orthogonal behavior with activity.

The PathFinder descriptors were pretreated using block-unweighted (BUW) scales, as available in the Golpe program [7], using the following BUW scales: 0.50 for shape, 2.00 for Water and Dry, 1.50 for the combined Water–Dry analysis. The impact of shape was reduced, whereas the impact of the Water, Dry and Mix was reinforced. Then the data were correlated to the activity by means of PLS analysis. The PLS method condensed the overall information into two smaller matrices, which can be visualized by means of the score plot (which shows the pattern of the compounds) and the loading plot (which shows the pattern of the descriptors). The optimal model was obtained with three components, exhibiting a significant statistical quality, as evinced by good $R^2 = 0.94$ and $Q^2 = 0.71$ values. No superposition was required.

By screening corporate databases Chan and co-workers [5] identified 6-arylthio-2-aminobenzonitriles with micromolar antiviral activity against HIV-1; modification to 6-arylsulfinyl- and 6-arylsulfonyl-2-aminobenzonitriles yielded classes with nanomolar activity. These data have recently been used [6] for a comparison of ligand- and structure-based GRID-derived procedures. Herein the same set is used to test the procedure.

Again, PathFinder descriptors were generated as previously described and pretreated using block-unweighted (BUW) scales, using the following BUW scales: 1.50 for shape, 1.00 for Water and Dry, 0.75 for the combined Water–Dry analysis.

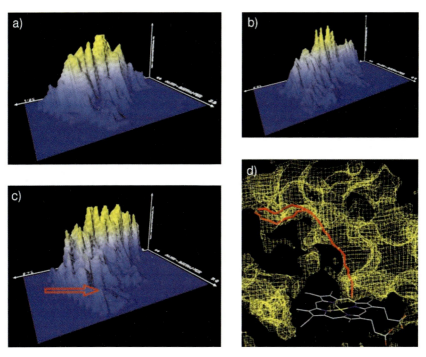

Figure 5.8. Frequency distribution plots for the cytochromes 2D6 (a), 2C9 (b), and 3A4 (c). For 3A4 the paths responsible for the peak are indicated with the red arrow in (c) and reported, in red, in (d).

Here, the impact of shape was reinforced whereas the impact of the combined was reduced.

PLS analysis resulted in a three-component model that revealed good fitting ($R^2 = 0.81$) and internal validation ($Q^2 = 0.54$) for HIV1-RT inhibition.

5.4.2
CYP Comparison

The rate and the site of metabolism for xenobiotics are due to a complex mixture of recognition, shape and chemical reactivity. The human cytochrome P450 shape in proximity to the reactive heme plays a fundamental role in molecular recognition and orientation. Thus, the CYP cavity shape modulates the likelihood of a compound reacting with the enzyme, since it has to enter into the cavity, reach the reactive site (the heme), adopt a stable orientation to allow the reaction to occur and subsequently exit from the cavity. Different cytochromes show different cavity shape, and *in silico* prediction cannot neglect their key role.

The PathFinder approach was used to compare the CYP cavities of CYP2C9 [8], 2D6 [9], and 3A4 [10], the most important human cytochrome enzymes. CYP2C9 and 3A4 were available as protein crystal structures whilst an homology model was used for 2D6. Frequency distribution plots (Fig. 5.8) were obtained from non-superposed CYP structures, selecting the iron in the heme moiety as a root departure path.

Using the iron anchor point produces a cavity description which is relative to the reactive site position. Such a description is different from the simple distance coordinate used before, because the site complexity here refers to a precise spatial position (the reactive center). Figure 5.8 shows that CYP2C9 has a smaller cavity than the other two CYPs and that CYP3A4 show the most complex shape within the series.

These qualitative statements are due to a simple graphical analysis of the cavity frequency distribution plot compared in Fig. 5.8. However, each cavity can be inspected and compared in detail. For example, the peak indicated by the arrow in Fig. 5.8 for 3A4 corresponds to the path–distance pairs reported in red color in Fig. 5.8(d). They end up far away from the heme in a subpocket region generated by the residues Leu 211 and Tyr 307. This subpocket is not present in the other CYPs and can be involved in a selective recognition of the substrate molecule.

The reported procedure can be used to map the entire active site of an enzyme, or, if linked with appropriate statistical analysis, the selective regions in a protein family.

5.4.3
Target–Ligand Complexes

Two diverse target–ligand complexes were taken as examples. The shape frequency distributions obtained from positive MIF for the complexes 1acl–DME [11] and 1xli–GLT [12] are compared, as shown in Fig. 5.9. Both target (1acl) and ligand

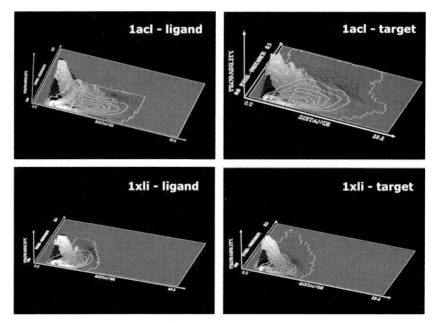

Figure 5.9. PathFinder description of target and ligand are reported for complexes 1acl–DME and 1xli–GLT.

(DME) of the former complex are quite elongated and mostly hydrophobic. Conversely, both target (1xli) and ligand (GLT) of the latter are more compact and hydrophilic. In addition, the pair 1acl–DME has greater dimensions, both surface and volume, compared to the 1xli–GLT complex. All these considerations can be assessed quantitatively by the use of the PathFinder descriptors derived from the positive MIF. Protein 1acl in Fig. 5.9(a) shows larger distance values when compared with 1xli in Fig. 5.9(b).

The two complexes may also be compared by looking at the path–distance axis: protein 1acl is characterized by higher values. This means that the variegated and elongated protein surface allows the recognition of the branched ligand DME and both variegated surfaces of target and ligand fit together. It is noteworthy how the shape frequency distribution of protein 1acl resembles that of ligand DME, whereas the shape frequency distribution of 1xli resembles that of ligand GLT.

The analysis can be extended to the search for complementarity using negative MIF from different GRID probes. The complementarity of the hydrogen bonding pattern using GRID probes O for the ligand and N1 for the protein or vice versa, while the complementarity between hydrophobic regions can be addressed simply using the probe DRY.

Figure 5.10 shows the distribution graphs of protein 1xli (a) and the ligand GLT (b), respectively. MIF were produced using probe O for the protein, and the N1 probe for the ligand. The protein frequency distribution is usually more populated than the corresponding ligand frequencies. In fact, protein cavities are larger than ligand volumes. Some paths, two in the case of the ligand, and six in the case of the target, refer to the same chemical entities, such as two hydrophilic regions in the protein active site, where hydrophilic hydroxy groups of the ligand are placed. The correspondence between the paths in the protein and in the ligand is reported in Fig. 5.10 (c) and (d). The path in (c), obtained with the probe O on the protein corresponds to the paths in (d) obtained with the probe N1 on the ligand.

Similarly, Fig. 5.11 shows the distribution graphs of protein 1acl (a) and the ligand DME (b), respectively. MIF were produced using the probe DRY for both the ligand and the protein. Here again, some paths are highlighted. These paths link some chemical entities in the ligand which have corresponding entities in the protein active site, in this case hydrophobic regions in the protein active site with corresponding hydrophobic ligand groups, the path obtained on the ligand corresponds to the paths obtained on the protein.

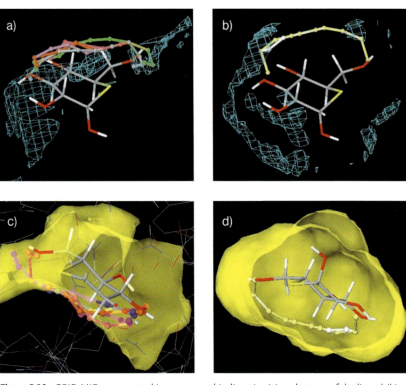

Figure 5.10. GRID MIF, represented in cyan, were obtained on the target (1xli) with the O probe (a) and on the ligand (GLT) with the N1 probe (b). A few paths are highlighted, connecting two hydrophilic MIF of the binding site (a) and atoms of the ligand (b). From a diverse orientation it is more evident how the paths lie on the target (c) and ligand (d) molecular surfaces, represented in yellow.

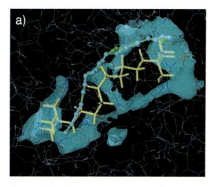 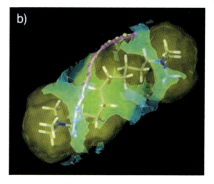

Figure 5.11. GRID MIF, represented in cyan, were obtained on the protein target 1acl (a) and on the ligand DME (b) with the DRY probe. A few paths are highlighted which connect two hydrophobic MIF of the binding site (a) and atoms of the ligand (b). The ligand molecular surface is represented in yellow, whereas it was omitted for the protein for clarity.

5.5 Conclusions

We have presented a new procedure, called PathFinder, aimed at encoding the GRID molecular interaction fields into invariant shape-descriptors, suitable for similarity and complementarity issues. Shape similarity is the underlying foundation of ligand-based methods while shape complementarity is the basis of many receptor-based designs.

Since PathFinder works in path/distance space, the frame of reference for every molecule is internal and, therefore, no pairwise alignment is necessary when molecules are compared. PathFinder at the same time incorporates information on both overall shape (long distances) and local topology (shorter distances).

The new PathFinder procedure has recently been successfully applied to quinolones [13] and to the NAD within its site of the protein L-aspartate oxidase [14]. It shows a number of uniquely promising attributes: the new descriptors are alignment-independent, highly relevant for describing the pharmacological properties of the compounds, well suited for describing the macromolecules, ready to compare macrostructures with their potential ligands without the need for virtual docking. Moreover, path–distance descriptors are well suited for 3D quantitative structure–metabolism relationships in which substrates can be structurally so different as to make the relative superposition impossible using standard techniques.

References

1 (a) P. J. Goodford, *J. Med. Chem.* **1985**, *28*, 849; (b) E. Carosati, S. Sciabola, G. Cruciani, *J. Med. Chem.* **2004**, *47*, 5114.

2 E. Dijkstra, W. *Numer. Math. 1*, **1959**, 269.

3 R. D. Cramer III; D. E. Patterson, J. D. Bunce, *J. Am. Chem. Soc.* **1988**, *110*, 5959.

4 A. C. Good, S.-S. So, W. G. Richards, *J. Med. Chem.* **1993**, *36*, 433.

5 J. H. Chan, J. S. Hong, R. N. Hunter III, G. F. Orr, J. R. Cowan, D. B. Sherman, S. M. Sparks, B. E. Reitter, C. W. Andrews III, R. J. Hazen, M. St. Clair, L. R. Boone, R. G. Ferris, K. L. Creech, G. B. Roberts, S. A. Short, K. Weaver, R. J. Ott, J. Ren, A. Hopkins, D. I. Stuart, D. K. Stammers, *J. Med. Chem.* **2001**, *44*, 1866.

6 S. Sciabola, E. Carosati, M. Baroni, R. Mannhold, *J. Med. Chem.* **2005**, *48*, 3756.

7 Golpe program is available at http://www.miasrl.com

8 M. R. Wester, J. K. Yano, G. A. Schoch, C. Yang, K. J. Griffin, C. D. Stout, E. F. Johnson, *J. Biol. Chem.* **2004**, *279*, 35630.

9 F. DeRienzo, F. Fanelli, M. C. Menziani, P. G. De Benedetti, *J. Comput.-Aided Mol. Des.* **2000**, *14*, 93.

10 J. K. Yano, M. R. Wester, G. A. Schoch, K. J. Griffin, C. D. Stout, E. F. Johnson, *J. Biol. Chem.* **2004**, *279*, 38091.

11 M. Harel, I. Schalk, L. Ehret-Sabatier, F. Bouet, M. Goeldner, C. Hirth, P. H. Axelsen, I. Silman, J. L. Sussman, *Proc. Natl. Acad. Sci. U S A*, **1993**, *90*, 9031.

12 C. A. Collyer, K. Henrick, D. M. Blow, *J. Mol. Biol.* **1990**, *212*, 211.

13 G. Cianchetta, R. Mannhold, G. Cruciani, M. Baroni, V. Cecchetti, *J. Med. Chem.* **2004**, *47*, 3193.

14 G. Cruciani, E. Carosati, S. Clementi, in *The Practice of Medicinal Chemistry*, **2003**, Elsevier, Amsterdam, p. 405.

6
Alignment-independent Descriptors from Molecular Interaction Fields

Manuel Pastor

6.1
Introduction

6.1.1
The Need for MIF-derived Alignment-independent Descriptors

Molecules are complex entities that can be numerically described in many different ways. The molecular interaction fields (MIF) represent a very particular molecular property: the ability of a molecule to establish energetically favorable interactions with other molecules. For this reason, MIF have been widely used in the field of drug discovery and development, since most biological properties of a compound depend on its ability to bind different kinds of biomolecules, such as transporters, receptors and enzymes.

There are different ways of applying MIF to the study of the ligand–receptor interaction. MIF can be computed in the receptor binding site in order to find out structural and physicochemical characteristics of potentially binding compounds, as was originally described in [1]. It is also possible to proceed the other way around and compute MIF in one or many small compounds in order to characterize them according to their potential to act as ligands, binding a certain receptor. When used in this manner, the MIF can be seen as computationally obtained descriptor variables ("molecular descriptors"), which represent properties of the molecules, much like the calculated $\log P$ represents the molecule's lipophilicity.

In general, the molecular descriptors allow one to obtain a mathematical representation of the molecules and open the possibility of applying mathematical and statistical methods to answer many useful questions. For example: how similar are two different molecules? The simple observation of their structure might provide some answer, but this will always be subjective. However, if we describe the molecules using some descriptor variables, the values assigned to both compounds can be compared and an objective similarity index can be provided, although it can be argued that the value of such an index is relative, and depends on the appropriate choice of the variables representing the molecules. For example, the results will be different depending on whether we use the lipophilicity or the

Molecular Interaction Fields. Edited by G. Cruciani
Copyright © 2006 WILEY-VCH Verlag GmbH & Co. KGaA, Weinheim
ISBN: 3-527-31087-8

molecular weight or both. Moreover, no single choice can be considered correct, and the suitability of the molecular descriptors depends on the problem to be addressed. As a general rule, the most suitable variables describe properties of the molecules which are strongly related to the problem being studied and which are said to be "relevant" to the problem. In this sense, MIF are often a good choice for describing small molecules in areas related to drug discovery, since the MIF represent well the ability of these molecules to act as ligands and bind other molecules, and this binding is the first step in many chain events responsible for their biological properties (for example, interaction with a receptor with respect to their pharmacodynamic properties). For this reason, MIF-based molecular descriptors have been extensively applied in different 3D QSAR methods, like CoMFA, COMSIA, GRID/GOLPE, etc. For a review of such applications see [2].

Unfortunately, even if we can admit that MIF make good, highly relevant molecular descriptors, it must be borne in mind that every single MIF usually contains several hundred or thousand variables, each representing the ability of the molecule described to interact favorably with a certain kind of chemical group at a certain point in space. This particular organization of the data leads to several practical problems. The first is related to the implicit association of every variable to a certain spatial coordinate. Let us imagine that we want to evaluate the similarity of two molecules using MIF. The first step would be to compute a MIF for both and then to compare their values for every single variable, however two values can only be compared when the variable represents the same information and this means that in both MIF the variable must represent the same position in space. In practice, this only happens when both molecules are perfectly aligned and placed in the precise orientation in which they would bind a certain biomolecule. In the rare event that the structures of the ligand–receptor complexes are known for both compounds, the alignment of the ligands can be obtained by superimposing equivalent receptor atoms. If the structure of the complexes is not available but the structure of the receptor is known, then it is possible to attempt the docking of the structures into the receptor binding site using manual or automatic methods. If the structure of the receptor is not available, as is often the case, one can try to align the molecules using all the information at hand, which could be some pharmacophoric hypothesis, common structural motifs in the ligands, etc. However, apart from the first situation, in which the structure of the complex was experimentally solved, every alignment operation introduces a certain amount of subjectivity. The less information available the worse the alignment and, in the limit, when the structure of the receptor is unknown and the structural similarity of the compounds to be aligned is low, obtaining a suitable alignment becomes nearly impossible. Moreover, the alignment is a time-consuming step, difficult to perform using automatic procedures and often constituting the bottleneck in the whole computational study. Computational methods used to perform the automatic alignment on the basis of the ligand structures have been reviewed recently [3]. However, even if some of these methods produce reasonable results, they are far from providing a definitive answer to the problem. The problem of the alignment should not be underestimated and it has been recognised for a long time as the key step of any 3D QSAR study [4]. Since the alignment of the molecules is the first step in the generation of their descriptors, any mistake will affect the rest of the study. No data treatment can minimize the effect of mistakes or inaccuracies at this step, which will lead to unpredictable results.

Apart from the alignment problem, the use of raw MIF variables as molecular descriptors has some practical problems, many of which are a consequence of the fact that in the MIF the information is scattered in many different variables. In any MIF, a lot of variables represent empty regions of space containing very little or no information at all. The total number of variables is always large and often rises to some hundred thousands or even more! This huge number of variables prevents the use of simpler regression analysis methods, like multiple linear regression, and requires the use of multivariant and megavariant methods like partial least squares (PLS) regression or principal component analysis (PCA). Moreover, the MIF are so huge that they are difficult to store, transmit or handle.

For all the above-mentioned reasons we decided to develop new molecular descriptors, obtained from the MIF, but transformed in order to condense the more relevant information into fewer variables and to solve the main drawbacks reported above. It is important to stress that such transformation will inevitably be associated with some loss of information and will introduce some bias, implicit in the assumptions on the basis of which we will choose the "most relevant information". In our opinion, this is a price worth paying in order to obtain more useful descriptors. Finally, the result of these efforts were the VolSurf descriptors and the GRid INdependent Descriptors (GRIND), each one representing a different method of condensing the MIF information and biased towards a certain application. The VolSurf molecular descriptors [5] are mainly aimed at representing MIF information relevant to the description of the phamacokinetic and physicochemical features of the compounds. The GRIND [6] are more oriented towards representing the ability of a small compound to bind biomolecules (for example, receptors) and therefore are better suited to represent pharmacodynamic properties. This chapter will focus on a thorough discussion of the second type of descriptors, the GRIND, as a representative example of the alignment-independent molecular descriptors that can be obtained from the MIF.

6.1.2
GRIND Applications

The GRIND were published in the year 2000, and the first version of the software used for generating and manipulating these descriptors (ALMOND) [7], was available in the same year. Since then, more than 20 applications of GRIND in the most diverse fields have been published (see Table 6.1).

Apart from their application in practical problems, the GRIND have been the subject of some theoretical papers, dealing for instance [28] with their information content, performing comparisons with other descriptors and discussing their suitability in drug discovery.

Table 6.1 Published applications of the GRIND (at January 2005).

	References
3D-QSAR	8–16
3D-QSPR	17–24
Molecular similarity	25, 26
Binding site characterization	27

Since the first publication of the method, much experience has been gained and the method has been extended and improved. This chapter discusses for the first time the details of the method, the problems more commonly found in its practical applications and the approach for the interpretation of GRIND-derived 3D QSAR models.

6.2
GRIND

6.2.1
The Basic Idea

The GRIND method was developed with the aim of extracting the most relevant information from a MIF and compressing it into a handful of variables. The requirements were challenging:
- The resulting variables should not depend on the position or the orientation of the molecule, thus making the molecular descriptors alignment-independent.
- The variables should be chemically meaningful and interpretable from a chemical point of view.
- A few variables should condense the most important information.
- They should be suitable for analysis using standard chemometric methods like PCA and PLS.
- It should be possible to compute them in a fast and automatic way.

The first constraint, the fact that the resulting variables should not depend on the spatial position nor orientation of the molecule, suggested the use of a geometrical description based on "internal coordinates", not making use of any external reference system. The basic idea was to recognize in the MIF a number of highly relevant regions and to describe their spatial distribution on the basis of their mutual distances. Therefore, for a certain compound, we obtain a vector of values, each one representing the presence or not in the MIF of a couple of nodes separated by a certain distance and each one belonging to a different "highly relevant region". The value will be zero when the MIF contains no such couple of nodes at the given distance, and positive and the product of the MIF values at these positions, when it does. This method allows one to condense into the variables two kinds of information: the presence or not in the MIF of a couple of nodes separated by a certain distance and the overall intensity of the MIP at both ends of the distance. The variables so obtained are therefore richer in information and more suitable for analysis using chemometric methods like PCA and PLS.

In greater detail, the procedure for obtaining the GRIND involves three steps, as shown in Fig. 6.1.

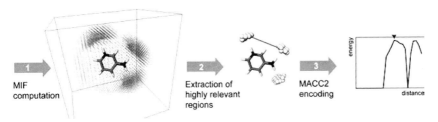

Figure 6.1. Steps involved in the computation of the grid-independent descriptors.

6.2.1.1 Computation of MIF

In most cases, there is no a priori knowledge about which chemical groups are in the binding site and are therefore relevant for the description. Situations in which the structures of the binding site are known are an exception. However, in most cases it can be assumed that the most important interactions with the ligand are due to hydrophobic, hydrogen bond acceptor and hydrogen bond donor groups found at the receptor-binding site. In the original GRIND formulation, the MIF were computed with the program GRID and, by default, the GRIND method suggests computing three MIF, with the following GRID probes: DRY, characterizing hydrophobic interactions; O (an sp2 oxygen) representing hydrogen bond acceptor groups; N (a planar, amide-like nitrogen) representing hydrogen bond donor groups. It should be noted that this is only a default choice, sensible in the absence of further information, but in the presence of any knowledge hinting at other kinds of interactions it would be wiser to reconsider this choice, removing some of the probes or adding some others, for example probes representing charged groups when there is suspicion of the presence of ionised residues at the binding site.

It should not be forgotten that the MIF is continuous and that programs like GRID provide only an approximation consisting of a discrete sampling of the MIF at certain positions, represented by the nodes of a regularly spaced three-dimensional grid. The smaller the grid spacing, the better and more accurate is the sampling, even if the time of computation and the storage space impose practical limits to this sampling. In the context of GRIND, it is important to start from an accurate representation of the MIF, because the descriptors may be affected by the field discretization errors. Even if the descriptors are insensitive to the position of the molecules in space, if the situation of the molecule within the sampling grid

6.2.1.2 Extraction of Highly Relevant Regions

In the context of GRIND methodology, we intend by "highly relevant regions", areas of the MIF where the ligand can establish strongly favorable interaction and therefore the position of binding site groups which can putatively bind the compound. This definition of "highly relevant" clearly orients the application of the GRIND towards the description of ligand–receptor binding, making the resulting variable more focused and less general.

From a practical point of view, selecting such highly relevant regions is not an easy task. Confronted with a 3D isovolume representation of a MIF, a trained human expert usually has no difficulty in recognizing potential regions where the molecule could establish strong interactions. These are characterised by having negative energies and large absolute values. However, computational algorithms for extracting these regions must pay attention to other aspects as well as the field values. For example, if the molecule has a region with very strong negative values, computational algorithms could extract only this region, neglecting the rest of the molecule (Fig. 6.2(a)). However, certain substituents can produce several favorable regions, not too intense, and a selection based only on the field values runs the risk of extracting an arbitrary sample that does not represent the true potential of the molecule for interacting with the receptor (Fig. 6.2(b)).

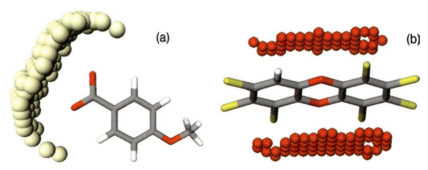

Figure 6.2. Problems that might arise in the selection of highly relevant regions when it is based only on the MIF values: (a) charged groups, like this COO⁻ group, concentrate on themselves all the selected nodes, neglecting the effect of other groups producing weaker interactions, (b) probes that do not establish strong interactions with the target compounds, like this O probe, select a sample of scattered nodes.

In GRIND, the selection is performed using a balanced combination of two different criteria. On the one hand, the values of the field in the selected regions should be negative (favorable) and intense. On the other hand, the distance between the regions should be as large as possible, in order to prevent excessive concentration on a few areas. In principle, the relative weight given to both criteria is equal, but it can be finely tuned to provide a better selection in certain situations. For example, when the compounds simultaneously contain groups producing very intense MIF interactions and weak MIF interactions, a higher weight on the distance criterion might prevent situations like the one represented in Fig. 6.2(a), in which all the nodes are concentrated in one of the regions. These criteria were used to compute a scoring function, which was used by an interchange optimisation algorithm [29]. This algorithm works by selecting randomly a certain number of MIF coordinates and moves coordinates in and out of the selection until the value of the scoring function is maximized. Finally, the algorithm yields a certain number of grid nodes maximizing the above-mentioned criteria, and these are identified with the "highly relevant regions" thereafter. Figure 6.3 shows the aspect of the selected regions.

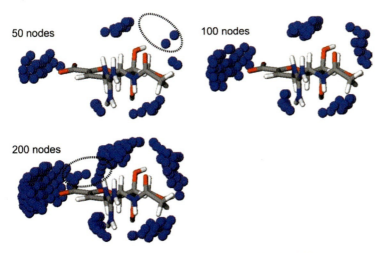

Figure 6.3. Aspect of the highly relevant regions extracted by the GRIND method with different settings: 50 nodes, 100 nodes and 200 nodes. The encircled areas highlight regions which have been misrepresented due to the selection of an incorrect number of nodes. See text for details.

This step is one of the more critical in the GRIND procedure for several reasons. By selecting only a few of the grid positions (nodes), all the rest are removed and no longer used. Much of the information of the MIF is thrown out at this stage, remarkably all the positive part of the MIF. Therefore, any mistake in the selection of the regions, either conceptual or computational, has a large impact on the results. For this reason, this is the only part of the method which often requires some human intervention in order to adjust two algorithm parameters: the relative weight of the field in the scoring function and the number of selected nodes. The first parameter has already been discussed. The second parameter, the number of nodes, was fixed by default at one hundred and in most drug-like compounds this number is adequate to obtain a suitable representation of the relevant regions. However, when the compounds are large, or they contain many substituents producing large favorable regions, this number of nodes is not enough to build a complete map of these regions and should be increased. On the other hand, if the number of nodes is set to too high a value the algorithm will include in these nodes positions of the MIF representing weak, nonspecific interactions, and the resulting map will again not be adequate. Figure 6.3 shows the result of selecting 50, 100 and 200 nodes on the MIF obtained with a N1 probe on the structure of Zanamivir (Relenza®). If can be clearly seen how the 50 nodes fail to represent the regions produced by the hydroxy group substituent, while in the 200 nodes representation the algorithm saturates some regions with nodes, thus failing to extract the most relevant regions.

In our experience, the best way to tune these parameters is to carry out the procedure using default parameters and then inspect the nodes selected in order to detect misrepresented regions. This visual inspection can be carried out on a small series containing only a few representative structures if the original series contains many compounds. Once the parameters have been adjusted, the procedure can be run on the whole series. It is not advisable to tune-up the parameters using as a guide statistical parameters like the r^2 and q^2 of the PLS models obtained, since this can produce the opposite effect and lead to artifacts. Certainly, the method described is not perfect and is at present being reviewed in order to develop a better selection procedure that might work without human supervision.

6.2.1.3 MACC2 Encoding

Once the MIF is reduced to a set of nodes representing the most relevant regions, the next step is to describe the spatial position of these nodes without using their absolute coordinates nor any external reference system. The solution adopted in GRIND is based on describing the node–node distances and the MIF energies represented by such nodes. In order to do this, the node–node distances were first converted to a discrete set of distance ranges or distance "bins", then every couple of selected nodes was analysed in turn, measuring their mutual distance and assigning them to a certain distance bin. At the end of the analysis, every bin is represented by the couple of nodes for which the product of their MIF is higher, thus representing the more favorable energy at both ends of the line linking the nodes. The result of this analysis is an array of values, one for each distance bin, containing 0 if no couple of nodes was found and an energy score representing the largest product of MIF, if one or more couples was found. The vector contains "energy scores" and not plainly "energy products" because, in order to equalize the scale of the correlograms obtained with different GRID probes, the values were scaled by dividing the MIF products by the maximum MIF value obtained for this probe in a series of drug-like compounds: the full Maybridge HTS database of September 2002 (see manual of [7] for further details). The effect of this scaling is to obtain values approximately in the range 0.0 to 1.0. Other scaling schemes can be tried, but the application of autoscaling is strongly discouraged, because the scale of the variables obtained from the same MIF is informative and should not be removed.

These results can be seen, from a mathematical point of view, as a set of n quantitative-continuous variables, each representing a range of distances. It is possible to represent these values in a graphic like the one represented in Fig. 6.4, which is called a "correlogram", in which the distances are represented on the horizontal axis and the scaled product of energies on the vertical axis.

The first characteristic of this geometry representation that must be emphasized is that it is completely alignment independent, since the values assigned to every variable depend only on the mutual distance of the nodes and not on the position of the nodes in space. Therefore, if we carry out this analysis for a series of molecules not aligned in space, the variables obtained for every compound do nevertheless have the same meaning and the values obtained can be combined to build a consistent descriptor matrix, without the need to align or otherwise superimpose their structures.

One of the critical steps for carrying out the encoding is the definition of the distance ranges. Every range defines one descriptor variable and only one couple of nodes will be extracted to represent this range. Therefore, the ranges should be chosen carefully: selecting too many will produce a lot of empty distances and selecting too few will condense many different distances into a single descriptor variable, thus losing information. It should also be borne in mind that the nodes are regularly spaced in a 3D grid and therefore not all distances are defined, especially for small distances. The default range width was set to 0.8 times the grid spacing, and this setting was found to work well in most cases.

As mentioned above, the descriptor variables obtained take values that represent two different things: the presence or absence of a couple of nodes at a certain distance and the combined intensity of the MIF at both ends. If we assume that these MIF values represent adequately the ability of the molecules to bind a certain receptor, then the molecular descriptors obtained by the encoding should preserve the original information, since they are simply scaled MIF products, and can therefore be expected to be relevant for representing the binding properties of the compounds. It should also be noticed that the GRIND produce a "fuzzy" encoding of the structural features since the presence or absence of a certain group is not reflected in a single descriptor variable but in many of them, often grouped around a peak in the correlogram. In this respect, the GRIND differs from other molecular descriptors and, in particular, from the structural fingerprints.

So far we have mentioned how to encode a single MIF in order to obtain a single correlogram. In order to encode all the MIF computed for a structure, the same procedure is repeated for the k MIF computed. First the encoding is used to represent couples of nodes belonging to the same MIF, obtaining k so-called "auto-correlograms". Then, another set of correlograms is computed, this time representing distances between couples of nodes in which one of the members of the couple belongs to MIF_i and the other to MIF_j, the so-called "cross-correlograms". The total number of cross-correlograms can be calculated as the number of picking MIF pairs from the set of k MIF, using the formula of combinations:

$$_2C_k = \binom{k}{2} = \frac{k}{(k-2)!2!} \tag{1}$$

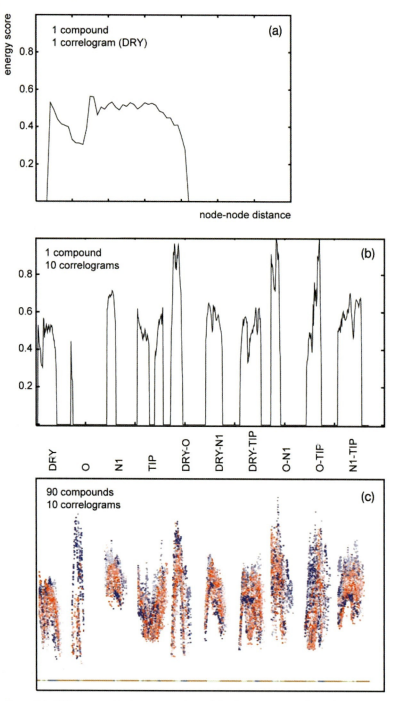

Figure 6.4. Different graphical representations of the GRIND-generated correlograms.

Each such correlogram can be seen as a different block of descriptor variables characterising different binding abilities of the compounds. Usually, the whole correlogram set is used to characterize the compounds and, accordingly, they are represented side by side graphically (see Fig. 6.4). The total number of variables generated by the procedure depends on the size of the largest compound, which defines the largest node–node distance found in the series. Typically, drug-like compounds use 50–80 variables per correlogram and the total number of variables obtained for a three-field description oscillates between 300 and 800.

As mentioned above, the values assigned to the descriptor variables for every compound were obtained from a single couple of nodes. By using appropriate software, the identity and therefore the 3D coordinates of these two nodes, for every compound, can be stored during the encoding and then represented together with the compound structure, in order to identify the structural correspondence of the descriptors. This is extremely useful for the model interpretation since it allows one to represent the results of the analysis on top of the structures of the compounds, in a language which can be understood by any chemist. Indeed, this is one of the strong points of the descriptors: even if they can be computed without aligning the structures, it is possible to revert the transform and obtain a representation of every variable, for every compound, indicating clearly which couple of nodes is represented by this variable.

However, the practical application of the encoding procedure produced some unexpected problems. One of the nicest features of program GRID is its ability to consider the conformational freedom of some polar hydrogen atoms in the computations and obtain MIF reflecting the most favorable interaction produced with the most favorable of such conformations at each grid node. This is clearly seen in the MIF map produced by a hydroxy group, in which the favorable regions of interaction with a polar probe have a crown shape, produced by the rotation of the hydrogen and/or oxygen lone pair. However, it is obvious that two opposite nodes of this crown cannot coexist, since they represent two different conformations of the hydroxy group. Unfortunately, the GRIND encoding procedure often considered couples of nodes produced by the same atom in different conformations, thus generating descriptors without physical meaning. In order to prevent this effect, it is advisable to generate the MIF using in GRID the ALM directive (this is the default in the latest ALMOND versions). The ALM directive adds some extra information to the MIF, labelling every single node with the identity of the atom contributing most to the energy of interaction. When this information is available in the MIF, the GRIND encoding removes from the list of couples those in which both nodes are generated by the same atom, thus preventing the generation of meaningless descriptors.

6.2.2
The Analysis of GRIND Variables

The way in which the GRIND were obtained produces large redundancies in the descriptor matrix. For example, the presence of a hydrogen bond donor substituent will probably be reflected in several consecutive variables of the hydrogen bond acceptor probe auto-correlogram as well as in many variables of other cross-correlograms. In order to overcome this problem it is convenient for the analysis of GRIND to use chemometric methods like principal component analysis (PCA) or partial least squares (PLS), in which the original variables are not used directly but are used to build a few "principal components" (PC) or "latent variables" (LV), obtained as a linear combination of the original ones and forming the basis of the subsequent analysis. Indeed, the GRIND were designed to be analysed using these powerful techniques and without their application the original descriptor matrix might not be adequate for direct analysis, mainly due to problems related to the internal correlation and information redundancy.

GRIND can be obtained for a series of compounds of very different size and can be used in any sort of computational study. However, we will mention here some typical ways in which they are often analysed:

1. GRIND have often been used to build QSAR or QSPR models correlating the structure of a series of compounds of small/medium size (typically from 10 to 500 compounds) with their experimentally measured biological properties. The final goal is to obtain a predictive model or rationalise the differences in the biological property or both. In these cases, PLS is the regression analysis tool of choice. A detailed description of the PLS technique is obviously beyond the scope of this chapter and can be found elsewhere [30]. It should nevertheless be mentioned that the general principles of PLS modelling must be applied: the number of LV to incorporate should be carefully assessed; small increases in q^2 do not justify the incorporation of new LV. Also, the TU scores plot (also called PLS plot) for every LV should be carefully inspected, in order to detect outliers and object clustering. If the series contain very dissimilar compounds or many correlograms, it is often advisable to apply a mild variables selection using one or two sequential runs of FFD variables selection [31]. Models with a q^2 over 0.80 can be considered of sufficient quality, however this figure should not be seen as a strict cut-off and in some series in which the Y values are expected to be only approximate (for example, when they are obtained from *in vivo* data, or the data was collected from multiple sources) smaller values could be acceptable and vice versa. Since GRIND provides only an approximate description of the most relevant regions, PLS models with extremely high values of r^2 are seldom obtained. Some of the fine effects are not well described and in most cases the r^2 are lower than those obtained with other 3D QSAR methods. Conversely, these "crude" descriptions are very robust and perform well in prediction, which is reflected in high q^2 values, often very similar to the r^2. The interpretability of the models is also

rather good, and is discussed in the next section. Several examples of QSAR/QSPR models using GRIND can be found in the references reported in Table 6.1.

2. In other cases, the GRIND were used to analyse larger series of compounds (of the order of 100 to 10 000) in order to obtain models describing some biological property. In most of these studies, the structural dissimilarity of the compound was large or very large and/or the quality of the experimentally measured biological property was poor (obtained from multiple sources, qualitative data, etc.). In these circumstances the goal is not so ambitious and does not aim to obtain a quantitative model: it is sufficient to obtain a discriminant model that performs well in prediction, and from the beginning the interpretation is assumed to be not affordable. GRIND are rather suitable descriptors in these situations because it is possible to compute them quickly and without supervision. A PCA exploratory analysis can reveal whether the descriptors are able to form clusters of compounds with similar properties. When this is so the biological measurement can be categorized (as active or inactive) and PLS-discriminant analysis (PLS-DA) can be used to obtain simple discriminant models on the basis of which the properties of new compounds may be predicted. Some published applications of these methods are listed in Table 6.1 [17, 23].

3. Sometimes, the study does not aim to obtain models of any sort, but to characterise the structure of the compounds in a biologically relevant way. This is the case for example in combinatorial library design, when the aim is to extract a set of compounds, bioisosteric to an active prototype, or alternatively, to obtain a series of compounds as dissimilar as possible. GRIND provide a description of the compounds based on relevant features and are therefore highly suitable for these applications. However, as mentioned above, GRIND are largely redundant and therefore the direct use of the variables should be avoided. For these applications, the best approach is to start carrying out a PCA and work on the PCA score spaces, using a reasonable number of principal components (PC). Selecting compounds in the PC space is not difficult and can be carried out with standard chemometric tools like simple Euclidean distances, k-means selection or other clustering techniques. It is worth mentioning that in such score spaces, every PC represents features or a combination of features representing the maximum structural variability in the series. Not surprisingly, the first PC nearly always represents the size of the compounds. The meaning of subsequent PCs changes from series to series, but could be investigated both by inspecting the PCA loadings for the particular PC and by inspecting structures of compounds situated in opposed positions of the PCA scores plot. An example of this kind of application can be found in [26].

6.3
How to Interpret a GRIND-based 3D QSAR Model

6.3.1
Overview

One of the greatest advantages of 3D QSAR methods is the possibility of showing the results in 3D graphics, in a format that can be understood by chemists and that can be of help in obtaining new ideas and in designing new compounds with enhanced properties. 3D QSAR models generated with GRIND can be represented in this way and can serve this purpose, but the peculiarities of the descriptors make the procedure a little more complex and require some more effort from the researcher.

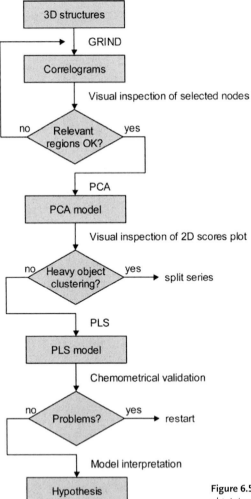

Figure 6.5. Proposed work flowchart for obtaining GRIND-based 3D QSAR models.

The suggested work flowchart to follow in GRIND-based 3D QSAR studies is shown in Fig. 6.5. The GRIND should be generated taking account of the principles mentioned above, adjusting parameters like the number of nodes. The descriptors should then be inspected by carrying out a PCA to help to detect unacceptable clustering of the series. Only then is it possible to obtain PLS models, which must be validated from a chemometric point of view taking account of the general principles mentioned above.

The next step is to attempt interpretation of the model, aiming to obtain a set of hypotheses rationalizing the major reasons that explain the differences in activity observed in the series. As stated before, the GRIND make possible a graphical representation, as lines linking couples of nodes, but it is important to notice that this representation does not represent a GRIND variable, but a single value of the variable: the particular instance of this distance in a particular compound. Therefore, since the compounds are usually not aligned, in another compound the corresponding couple of nodes used for the same variable might fall in a completely different region of the space. In this sense, the GRIND 3D representations are different from other 3D QSAR methods like CoMFA or GRID/GOLPE, in which the 3D coefficient plots, for example, are able to represent the contribution of the variables to the model.

For this reason, the recommended way to proceed is to identify single relevant variables from a PLS (or PCA) plot and then visualize these variables in 3D in some representative compounds, for example, highly active compounds and inactive compounds or compounds from different structural families. The goal is to inspect the node couples in these structures in order to identify the structural features that are contributing to the descriptors and to rationalize the reasons explaining the importance of these descriptors (see Fig. 6.6). This method of interpretation requires a clear understanding of the two alternative graphical representations of the GRIND: the correlograms and the 3D graphics representing node couples, which are described in detail in the next sections. The proposed method does not aim to provide a full understanding of the effect in the model of every single descriptor variable, because these contribute with different weights (often very small ones) to the PC or LV which are actually correlated with the activity. Therefore, our advice is to concentrate on the variables producing the higher and lower peaks in the PLS coefficients bar plot (see Fig. 6.6).

6.3.2
Interpreting Correlograms

The collection of auto- and cross-correlograms computed for a single molecule can be represented side by side in a graphic which resembles a spectrum, like those represented in Fig. 6.4 (b) and (c). In QSAR/QSPR applications, where the GRIND are computed for a series of compounds, it is often interesting to represent together the correlograms for all the compounds, coloring every compound in a spectrum scale representing the value of the biological property; for example, inactive compounds are colored blue and active compounds are colored red. In

132 | *6 Alignment-independent Descriptors from Molecular Interaction Fields*

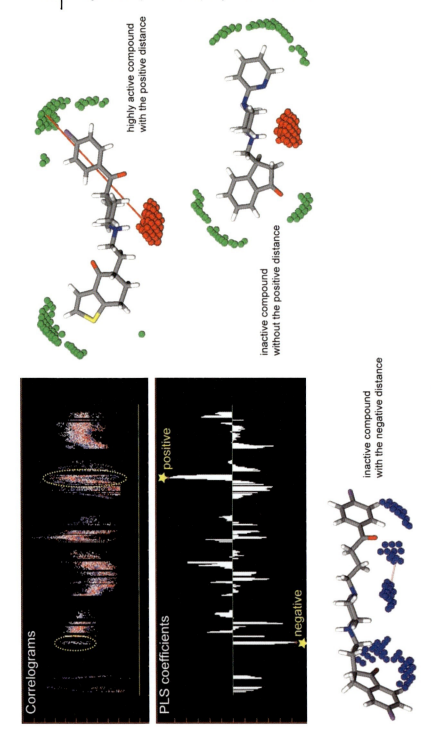

Figure 6.6 Graphics involved in the interpretation of GRIND-based 3D QSAR models. Variables with the highest and lowest PLS coefficients were investigated, first in a plot representing the correlograms obtained for all the compounds in the series where the colour represents the value of the Y variable (red for active compounds and blue for inactive). Then, these variables were represented in 3D for highly active and inactive compounds, in order to identify the structural and physicochemical features they represent.

these graphics, relevant descriptor variables can be easily identified, because red and blue points are not mixed up, indicating that active and inactive compounds consistently take either high or low values for this particular variable (see the areas enclosed by the yellow dashed line in Fig. 6.6).

After chemometrical analysis of the matrix, interesting variable information like the PCA loadings or the PLS coefficients can be represented in bar plots. An informative graphic can be obtained by situating a correlogram plot and a PLS coefficient plot on top of each other (Fig. 6.6). Variables with large positive contributions in the PLS coefficients (high positive variables) generally identify structural features present in highly active compounds and absent in low activity compounds. These can be seen in the correlogram plot as a band where the blue and red dots are not too mixed and where the red points are above the blue points (see the elongated area enclosed by the yellow dashed line on the right-hand side of the correlograms in Fig. 6.6). In some circumstances, the difference between active and inactive is due to the absence of a feature, for example, inactive compounds might lack a substituent and as a consequence, some descriptors in a certain range of distances take a value of zero. In other cases the difference is quantitative and reflects that active compounds have substituents able to establish stronger interactions with the receptor than the inactive compounds. Often, the differences in the values of the field nodes are actually reflecting differences in the position of certain groups, for example, inactive compounds might have lower values at certain distances because the center of the interaction region can be located in a "nonoptimal" position. Then, in inactive compounds the nodes located within the optimum distance range are nodes from the border of the region and therefore weaker than the nodes at the center. All these considerations are also applicable to variables with negative contributions in PLS models, simply by inverting the reference to highly active and low active compounds.

6.3.3
Interpreting Single Variables

Every GRIND variable represents both the presence and the intensity of a couple of nodes present at a certain distance. Using the appropriate software it is possible to visualize the couple of nodes which has been used to assign a value for a certain GRIND variable in a certain compound. Unfortunately, it is not possible to obtain a representation applicable to every compound in the series, as is the case in other 3D QSAR methods like CoMFA or GRID/GOLPE. From the point of view of interpretation this implies that in order to understand the meaning of a relevant variable, one should visualize couples of nodes involved in several structures, and at least in structures with extreme differences in structure and biological activity. In other words, the process of interpreting a variable requires one to visualize the nodes which were used to compute values in the most active compounds, in the less active and in at least a representative of every structural family. The goal of this inspection is to unveil the structural features which differ in active/inactive compounds. The difference could be qualitative, representing the presence/

absence of the structural feature, or quantitative, due to a difference in the intensity of the interaction. Additionally, it is also important to inspect different structural families of compounds in order to detect the consistency of the interactions and any potential artifacts in the model. In particular, if the series contains different families of compounds that exhibit different overall biological activity, then any QSAR model will identify as detrimental for the activity any descriptor which identifies a structure as belonging to the less active structural family. Models built using GRIND are also susceptible to this problem and the only solution is appropriate series design and careful model interpretation. This problem is so common and so important that it is discussed further in Section 6.3.4.

At the end of the process, the significant variables that are identified represent couples of MIF nodes which are related to the biological properties of the molecule and these can be transformed into a set of hypotheses with potential usefulness for the design of new compounds or the rationalization of the results. These MIF nodes correspond with the spatial position where a chemical probe can establish a favorable interaction with our compound, and therefore represent chemical groups of the binding site and not chemical groups of the ligands. The distance which separates these positions can be easily computed from the sequential identifier of the variable, by multiplying this value by the grid spacing and the width of the ranges in grid units (the value of the "smoothing window" in ALMOND). For example, if the variable 23 of the N auto-correlogram is found to be important, this means that there are two hydrogen donor groups in the binding site, separated by approximately $23 \times 0.5 \times 0.8 = 9.2$ Å (assuming that the GRIND were generated using the default values of 0.5 Å of grid spacing and a smoothing window of 0.8 grid units). This distance represents the spacing between the atomic centers and is only approximate, because the GRIND do not represent single distances but distance ranges, which in this case have a width of 0.4 Å.

6.3.4
GRIND-based 3D QSAR Models are not Pharmacophores

A pharmacophore can be described as an ensemble of distances and angles between atoms or molecular features which constitutes an essential requirement of the molecular structures for exerting a particular biological effect. The methods for identifying pharmacophores were reviewed recently [32]. The GRIND share with the concept of pharmacophore the description in terms of distances, even if in GRIND the distances are between MIF nodes and therefore more related to the distances in the receptor than to the distances in the ligand. In any case, the results of a 3D QSAR model interpretation are a number of highly relevant variables, representing distances which can be shown in 3D and which look very much like a pharmacophore. However, in the vast majority of cases these representations are not pharmacophores, for reasons that are worth discussing here.

First, as was discussed above, the distances represented by GRIND refer to atoms in the binding site and not to atoms in the ligands, as is the case in most pharmacophores. This makes these distances unsuitable for carrying out searches in 3D structural databases.

The results of any QSAR model (not only GRIND generated models) express the correlation between the differences in the structure of the compounds and their differences in biological properties. Structural features which are common to every compound in the series are simply not considered in the analysis. Any pharmacophoric interpretation of the results of a QSAR analysis will miss structural features shared by all the compounds. This fact is not trivial, because most series are constituted mainly of compounds with a certain degree of affinity for the receptor and this reveals that most of them have relevant common features. Therefore, as stated previously, none of these features will be present in the results of the models and no "pharmacophore" derived from this analysis will incorporate them. Moreover, the QSAR model results also inform us of structural features that are detrimental for the activity, while the pharmacophores are often focused only on the structural features that are needed to obtain active compounds.

Another reason to avoid a pharmacophoric interpretation of the 3D QSAR/GRIND results is the presence of correlation effects. Often, the compounds in the series studied belong to different structural families, each one characterised by a number of structural features, like the presence of certain groups, rings or chains. The problem is that all these features are present or absent together. For this reason, the results of the QSAR analysis will assign the same importance to all these features and if only one of them is relevant for the activity, no model would be able to distinguish this single feature from the others, since all are present or absent simultaneously in the tested compounds. However, if these results are extrapolated to external compounds, in which maybe only some of the features are present, the predictions will be wrong.

For all the above reasons, the results of the QSAR models obtained with GRIND should not be considered as pharmacophores. A correct and sensible interpretation of these results would be extremely useful, but its over-interpretation can be misleading and produce unrealistic expectations. It should also be stressed that the above-mentioned considerations are applicable to most QSAR and 3D QSAR results and are not a problem strictly linked to GRIND.

6.4
GRIND Limitations and Problems

6.4.1
GRIND and the Ligand Conformations

The GRIND were designed to be invariant to the position of the ligand in space but not to conformational changes in the ligands. The problem of the ligand flexibility and how to choose the appropriate ligand conformation is extremely difficult. Most 3D QSAR methods published so far are sensitive to the conformation of the ligands, with the exception of only a few methods [33], specifically designed to address this problem.

It is also true that not all molecular descriptors exhibit the same sensitivity to conformational changes. Descriptions like the MIF, based on the absolute 3D coordinates of every atom in the structure, are very sensitive even to minute conformational changes. The transformation applied to MIF in descriptors like Vol-Surf and GRIND make the resulting molecular descriptors much more independent of conformation, in particular for small conformational changes. In the case of GRIND, this is easy to understand: a small displacement in space of a single polar group can change the values of energies assigned to hundreds of MIF values, but in the GRIND correlogram we will only appreciate a small shift in a single peak. The effect of conformational changes on the results of GRIND-derived 3D QSAR models have been explored in a series of butyrophenones with activity as 5-HT_{2A} antagonists [34]. In this example, multiple QSAR models were obtained, using randomly selected 3D structures from a pool of alternative conformations, obtained by 2D to 3D conversion programs as structures with nearly equivalent conformational energies. The results of this study showed that the statistical parameters of the analysis exhibit moderate variation (r^2 ranged from 0.69 to 0.77 and q^2_{LOO} from 0.44 to 0.56) but the main structural features related to the activity were identified in all instances and the predictions for external compounds oscillated by less than 0.5 logarithmic units.

The claim for a really alignment and conformational independent method is justified by the need for fast assessment methods, able to work without human supervision on a large series of candidate compounds. In these situations, a reasonable approach is to represent every compound by a single structure, representing the extended conformation or the lowest energy conformation, which can be obtained automatically from the 2D representation of the compounds using different computational tools. Unfortunately, there is nothing to justify the idea that the bioactive conformation will be similar to the lowest energy one and numerous exceptions can be found in the literature. Alternatively, every compound can be represented by several conformations, generated by systematic conformational analysis, the Monte Carlo method or some other methods. However, in these approaches there are also no clear criteria to decide which of the conformations should represent the compound. Choosing the one producing the best fit to our models is certainly tempting but there is a high risk of producing "self-consistent" results (the fitting justifies the conformer choice and the conformer choice justifies the fitting) and hence models with very low predictive power. It is also possible to average the description obtained for different conformers, even applying weighting schemes based on energetic criteria or others, but the resulting variables cannot be considered as a description of the bioactive conformation and are more the description of an artificial ensemble of accessible conformations. In the particular case of GRIND, these approaches tend to produce nearly flat correlograms for all the compounds and have no practical application. All in all, the problem of ligand conformation should be considered still open.

An interesting alternative to the above-mentioned approaches is the computation of GRIND starting from the special type of MIF generated by program GRID with different values of the directive MOVE. When this directive is set, the GRID program automatically recognises conformationally flexible groups in the target compounds and computes the MIF taking into consideration, from all the accessible structures in the conformational space, those producing the most favorable interaction. Therefore, the resulting MIF does not represent a single structure, but all the accessible conformations, and the energy values describe the most favorable energies between the probe and any of the considered conformations. As a consequence, GRIND computed from these MIF incorporate, indirectly, information about the flexibility of the compounds, thus inheriting the conformational flexibility described by the so-obtained MIF. Unfortunately, the flexible GRIND are not a general solution. In many cases, the correlograms obtained have broad, overlapping peaks, which do not describe the structure of the compounds with the required precision. However, some studies [19] demonstrated that this somewhat fuzzy description can be used to describe the affinity to cytochromes. This is not surprising, if we consider that the cytochromes are part of the natural detoxification mechanism of the body and have evolved to act on structurally diverse compounds. Therefore, the structural requirements for binding a certain cytochrome cannot be expected to be strict, as supported by the structural diversity of experimentally determined substrates, and depends more on the fulfilment of some generic structural and physicochemical conditions. In these cases, the description provided by the flexible GRIND can be used, as demonstrated by published models [19] providing reasonable predictions for the inhibition of a large series of extremely diverse compounds on cytochrome CYP2C9.

6.4.2
The Ambiguities

The GRIND method assumes that the couples of nodes selected to represent a given distance in different compounds do actually represent the same kind of potential interaction with the receptor. This is often not true, for two reasons, as represented in Fig. 6.7. In the first situation (Fig. 6.7 (a)) all the compounds in the series contain alternative sites representing exactly the same distance and from which the candidate couples can be selected. Since the criteria for selecting the nodes is based only on the value of the MIF energy product, different sites can be selected in different compounds. The result is that the 3D graphic shows a non-consistent representation of the sites and can look messy.

In the second possible scenario, represented in Fig. 6.7 (b), the alternative sites represented by the same GRIND variable are not present in all the compounds in the series. Actually, the same variable represents two different and unrelated positions of the compound that cannot be expected to produce the same interactions with the receptor. These variables can be considered to be the sum of two separate ones, each one representing separate pieces of information about the compound, but confounded into a single variable.

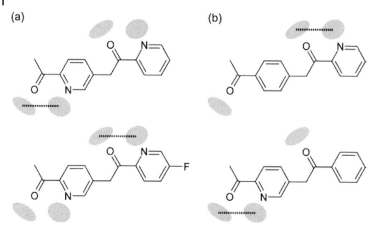

Figure 6.7. Two potential situations leading to ambiguous GRIND descriptors. See text for details.

The effect of these problems on the models is different depending on the type of situation as described in Fig. 6.7 (a) or (b). In the first situation, where the alternative sites are present in all the compounds, the effect of all the sites is correlated and the model is not altered. This situation is not too dangerous, since the problem is only related to the position of the nodes represented in the 3D graphics and affects only the interpretation. However, the second situation could be much more dangerous, in particular when the alternative sites have different effects on the biological properties of the compounds. In this situation, the variables are no longer consistent and represent different information in different compounds. Fortunately, the use of projection methods helps to minimise the negative impact of this problem: most MIF regions are represented in many GRIND descriptors and even if a couple of such regions are confounded in some variables, they will participate differently in some others of the same correlogram or of different correlograms. Since the projection methods make simultaneous use of all the variables, the regions which could be confounded for a single variable are never confounded for the ensemble of descriptors, since they will produce different patterns of distances. This again stresses the importance of using multivariate projection methods in association with GRIND and explains why the models behave so well both in fitting and in prediction, even in the presence of the above described ambiguities. The main problem in this case is that, again, the problem makes the interpretation harder, because the individual inspection of PLS coefficients will not easily unveil the true relationships.

6.4.3
Chirality

The GRIND descriptors are insensitive to the chirality of the structures. This has the undesirable side effect of providing exactly the same description for the two enantiomers associated with any chiral center. Diastereomers might, on the contrary, produce different correlograms, due to the presence of differences in the internal geometry.

6.5
Recent and Future Developments

6.5.1
Latest Developments

The best way to detect the weakness of any method or program is to use it in practice. After nearly five years of applying GRIND and ALMOND to a wide range of series, and collecting the suggestions and complaints of numerous users, we detected several aspects of the method which required improvement. We will mention here two of these, which were solved by applying extensions to the original GRIND method as well as other aspects which remain still open.

6.5.1.1 Shape Description

The GRIND provide a reasonable description of the spatial position of groups able to establish favorable interaction with the receptor, neglecting completely the effect of other regions, maybe unable to make strong interactions but important because for steric reasons they can prevent the binding. Therefore, some users asked us to incorporate into the GRIND some description of the molecular shape and, particularly, of those aspects of the shape which could be relevant to binding, like the existence of protrusions preventing ligand docking.

With this aim, we developed the TIP probe, a pseudo-MIF representing the local curvature of the molecular surface. The values provided by this probe have the ability to identify any sharp change in the local curvature, like those produced by substituents or present at the molecule ends (see Fig. 6.8), which are highlighted as a set of relevant nodes. Even if this description is not formally a MIF, the probe-like characterization allows a seamless integration of the description into the GRIND. A detailed description of the computations involved as well as some examples of the advantages of its incorporation was published recently [35].

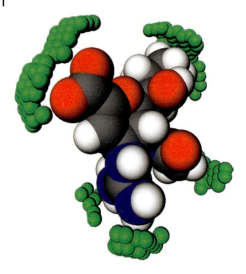

6.5.1.2 Anchor GRIND

GRIND provide a solution in the more general case, in which there is no a priori knowledge about how to superimpose the compounds. However, situations where there are some hints about common positions on the structure of the compounds are not rare. For example, many series contain compounds with a charged group which is known to interact with a charged residue of the binding site. In other cases, the reason for knowing this common position are not related to the receptor, but to the chemistry of the compounds, since all the structures share a common scaffold, connected with diverse substituents at a certain position. In these situations there is not enough information to attempt a full alignment of the compounds but the application of the GRIND method discards valuable information. The anchor-GRIND method [36] proposes an intermediate solution.

In the anchor-GRIND, the MIF are generated and processed as in the regular GRIND, but, at the encoding stage, the MACC2 transform is not applied between node couples, but between every node and a pre-defined anchor point. The resulting correlograms are therefore less sensitive to the above described ambiguities and the interpretation is much more simple, as demonstrated in some applications published recently [36]. The anchor-GRIND method has been implemented in the latest versions of ALMOND software [7].

6.5.2
The Future

The above two examples of GRIND extensions indicate that the GRIND are becoming more a family of descriptors than a closed methodology. There are aspects of the method which need improvement and some others which would open the possibility of applying the descriptors to other fields. With respect to the technical improvements, as was pointed out in this chapter, it is important to improve the way the original MIF is reduced to a set of nodes, reducing the need for user intervention at this step. There are also promising routes to explore in order to encode chirality into the GRIND.

6.6
Conclusions

MIF are a powerful tool for characterizing drug–receptor interactions. The GRIND described here represent an effort to overcome some of the major limitations of MIF for certain applications. However, condensing the MIF into a bunch of highly relevant variables has a price in terms of loss of information and interpretability of the results, as was thoroughly discussed here. In our opinion, for GRIND, the balance between benefits and risk is clearly positive. Many published applications in the last five years demonstrate the usefulness of the approach in very diverse fields. However, we believe that there is still much work to be done before we can extract from the MIF and the GRIND their full potential.

Acknowledgments

The author is grateful to John S. Masson, Iain McLay, Stephen Pickett, Sergio Clementi and Gabriele Cruciani, the other architects of the GRIND methodology. Both the GRIND method and the software ALMOND were developed with generous funding from Rhone-Poulenc Rorer Inc. and MIA Srl.

References

1 P. J. Goodford, A Computational-Procedure for Determining Energetically Favorable Binding-Sites on Biologically Important Macromolecules, *J. Med. Chem.* **1985**, *28 (7)*, 849–857.
2 H. Kubinyi, G. Folkers, Y. C. Martin (Eds.) *3D QSAR in Drug Design*, Kluwer/ESCOM, Dordrecht, *1998*, Vol. 2–3.
3 C. Lemmen, T. Lengauer, Computational methods for the structural alignment of molecules, *J. Comput.-Aided Mol. Des.* **2000**, *14 (3)*, 215–232.
4 G. Folkers, A. Merz, D. Rognan, CoMFA: Scope and Limitations, in *3D QSAR in Drug Design. Theory Methods and Applications*, H. Kubinyi (Ed.), ESCOM, Leiden, **1993**, pp. 583–618.
5 G. Cruciani, M. Pastor, W. Guba, VolSurf: a new tool for the pharmacokinetic optimization of lead compounds, *Eur. J. Pharm. Sci.* **2000**, *11*, S29–S39.
6 M. Pastor, G. Cruciani, I. McLay, S. Pickett, S. Clementi, GRid-INdependent descriptors (GRIND): A novel class of alignment-independent three-dimensional molecular descriptors, *J. Med. Chem.* **2000**, *43 (17)*, 3233–3243.
7 *ALMOND 3.3.0*, Molecular Discovery Ltd (UK) and Tripos (USA), 2004.
8 P. Prusis, M. Dambrova, V. Andrianov, E. Rozhkov, V. Semenikhina, I. Piskunova, E. Ongwae, T. Lundstedt, I. Kalvinsh, J. E. S. Wikberg, Synthesis and quantitative structure-activity relation-

ship of hydrazones of N-Amino-N'-hydroxyguanidine as electron acceptors for xanthine oxidase, *J. Med. Chem.* **2004**, *47*(12), 3105–3110.

9 M. L. C. Montanari, A. D. Andricopulo, C. A. Montanari, Calorimetry and structure–activity relationships for a series of antimicrobial hydrazides. *Thermochim. Acta* **2004**, *417 (2)*, 283–294.

10 C. Gnerre, U. Thull, P. Gaillard, P. A. Carrupt, B. Testa, E. Fernandes, F. Silva, M. Pinto, M. M. M. Pinto, J. L. Wolfender, K. Hostettmann, G. Cruciani, Natural and synthetic xanthones as monoamine oxidase inhibitors: Biological assay and 3D-QSAR. *Helv. Chim. Acta* **2001**, *84 (3)*, 552–570.

11 G. Cruciani, P. Benedetti, G. Caltabiano, D. F. Condorelli, C. G. Fortuna, G. Musumarra, Structure-based rationalization of antitumor drugs mechanism of action by a MIF approach, *Eur. J. Med. Chem.* **2004**, *39 (3)*, 281–289.

12 P. Cratteri, M. N. Romanelli, G. Cruciani, C. Bonaccini, F. Melani, GRIND-derived pharmacophore model for a series of alpha-tropanyl derivative ligands of the sigma-2 receptor, *J. Comput.-Aided Mol. Des.* **2004**, *18 (5)*, 361–374.

13 J. Brea, C. F. Masaguer, M. Villazon, M. I. Cadavid, E. Raviña, F. Fontaine, C. Dezi, M. Pastor, F. Sanz, M. I. Loza, Conformationally constrained butyrophenones as new pharmacological tools to study $5\text{-}HT_{2A}$ and $5\text{-}HT_{2C}$ receptor behaviours, *Eur. J. Med. Chem.* **2003**, *38 (4)*, 433–440.

14 P. Benedetti, R. Mannhold, G. Cruciani, M. Pastor, GBR compounds and mepyramines as cocaine abuse therapeutics: Chemometric studies on selectivity using grid independent descriptors (GRIND), *J. Med. Chem.* **2002**, *45 (8)*, 1577–1584.

15 P. Benedetti, R. Mannhold, G. Cruciani, G. Ottaviani, GRIND/ALMOND investigations on CysLT(1) receptor antagonists of the quinolinyl(bridged)aryl type. *Bioorg. Med. Chem.* **2004**, *12 (13)*, 3607–3617.

16 F. P. Ballistreri, V. Barresi, P. Benedetti, G. Caltabiano, C. G. Fortuna, M. L. Longo, G. Musumarra, Design, synthesis and in vitro antitumor activity of new trans 2- 2-(heteroaryl)vinyl -1,3-dimethylimidazolium iodides, *Bioorg. Med. Chem.* **2004**, *12 (7)*, 1689–1695.

17 L. Afzelius, C. M. Masimirembwa, A. Karlen, T. B. Andersson, I. Zamora, Discriminant and quantitative PLS analysis of competitive CYP2C9 inhibitors versus non-inhibitors using alignment independent GRIND descriptors, *J. Comput.-Aided Mol. Des.* **2002**, *16 (7)*, 443–458.

18 L. Afzelius, F. Raubacher, A. Karlen, F. S. Jorgensen, T. B. Andersson, C. M. Masimirembwa, I. Zamora, Structural analysis of CYP2C9 and CYP2C5 and an evaluation of commonly used molecular modeling techniques, *Drug Metab. Dispos.* **2004**, *32 (11)*, 1218–1229.

19 L. Afzelius, I. Zamora, C. M. Masimirembwa, A. Karlen, T. B. Andersson, S. Mecucci, M. Baroni, G. Cruciani, Conformer- and alignment-independent model for predicting structurally diverse competitive CYP2C9 inhibitors, *J. Med. Chem.* **2004**, *47 (4)*, 907–914.

20 L. Afzelius, I. Zamora, M. Ridderstrom, T. B. Andersson, A. Karlen, C. M. Masimirembwa, Competitive CYP2C9 inhibitors: enzyme inhibition studies, protein homology modeling, and three-dimensional quantitative structure-activity relationship analysis, *Mol. Pharmacol.* **2001**, *59 (4)*, 909–919.

21 G. Cianchetta, Y. Li, J. Kang, D. Rampe, A. Fravolini, G. Cruciani, R. Vaz, Predictive Models For hERG Potassium Channel Blockers, *Bioorg. Med. Chem. Lett.* **2005**, *15 (15)*, 3637–3642.

22 G. Cianchetta, R. Singleton, M. Zhang, M. Wildgoose, D. Giesing, A. Fravolini, G. Cruciani, R. Vaz, A pharmacophore hypothesis for P-Glycoprotein substrate recognition using GRIND based 3D-QSAR, *J. Med. Chem.* **2005**, *48 (8)*, 2927–2935.

23 P. Crivori, I. Zamora, B. Speed, C. Orrenius, I. Poggesi, Model based on GRID-derived descriptors for estimating CYP3A4 enzyme stability of potential drug candidates, *J. Comput.-Aided Mol. Des.* **2004**, *18 (3)*, 155–166.

24 I. Zamora, L. Afzelius, G. Cruciani, Predicting drug metabolism: A site of metabolism prediction tool applied to

the cytochrome P4502C9, *J. Med. Chem.* **2003**, *46 (12)*, 2313–2324.

25 G. Cruciani, M. Pastor, R. Mannhold, Suitability of molecular descriptors for database mining. A comparative analysis, *J. Med. Chem.* **2002**, *45*, 2685–2694.

26 F. Fontaine, M. Pastor, H. Gutierrez de Teran, J. J. Lozano, F. Sanz, Use of alignment-free molecular descriptors in diversity analysis and optimal sampling of molecular libraries, *Mol. Diversity* **2003**, *6 (2)*, 135–147.

27 H. Gutierrez-de-Teran, N. B. Centeno, M. Pastor, F. Sanz, Novel approaches for modeling of the A(1) adenosine receptor and its agonist binding site, *Proteins: Struct., Funct., Bioinform.* **2004**, *54 (4)*, 705–715.

28 T. I. Oprea, On the information content of 2D and 3D descriptors for QSAR. *J. Brazil. Chem. Soc.* **2002**, *13 (6)*, 811–815.

29 V. Fedorov, *Theory of Optimal Experiments*, Academic Press, New York, **1972**.

30 S. Wold, E. Johansson, M. Cocchi, PLS – Partial Least-Squares Projections to Latent Structures, in *3D QSAR in Drug Design. Theory Methods and Applications*, H. Kubinyi (Ed.), ESCOM, Leiden, **1993**.

31 M. Baroni, G. Costantino, G. Cruciani, D. Riganelli, R. Valigi, S. Clementi, Generating Optimal Linear PLS Estimations (GOLPE) – an Advanced Chemometric Tool for Handling 3D-QSAR Problems, *Quant. Struct.–Activity Relationships* **1993**, *12 (1)*, 9–20.

32 O. Dror, A. Shulman-Peleg, R. Nussinov, H. J. Wolfson, Predicting molecular interactions in silico: I. A guide to pharmacophore identification and its applications to drug design, *Curr. Med. Chem.* **2004**, *11 (1)*, 71–90.

33 A. J. Hopfinger, S. Wang, J. S. Tokarski, B. Q. Jin, M. Albuquerque, P. J. Madhav, C. Duraiswami, Construction of 3D-QSAR models using the 4D-QSAR analysis formalism, *J. Am. Chem. Soc.* **1997**, *119 (43)*, 10509–10524.

34 F. Fontaine, M. Pastor, F. Sanz, in *Utilidad potencial de los descriptores GRIND para la obtención de modelos QSAR-3D sin supervisión.*, XII Congreso Nacional de la Sociedad Española de Química Terapéutica., Sevilla, **2001**.

35 F. Fontaine, M. Pastor, F. Sanz, Incorporating molecular shape into the alignment-free GRid-INdependent Descriptors, *J. Med. Chem.* **2004**, *47 (11)*, 2805–2815.

36 F. Fontaine, M. Pastor, I. Zamora, F. Sanz, Anchor-GRIND: Filling the gap between standard 3D QSAR and the GRid-INdependent Descriptors, *J. Med. Chem.* **2005**, *48*, 2687–2694.

7
3D-QSAR Using the GRID/GOLPE Approach
Wolfgang Sippl

7.1
Introduction

Understanding the structural properties and features affecting the biological activity of a drug molecule is important in the ongoing process of drug design. After the initial characterization of a drug by physical means, additional quantitative information can be obtained by using quantitative structure–activity relationships (QSAR) [1]. Historically, the primary objective of QSAR was the understanding of which properties are important for the specific biological activity of a series of compounds. However, the main objective in today's drug design is the prediction of novel unknown compounds on the basis of previously synthesized molecules. The strategies that can be applied for this purpose can be separated into two major categories – the indirect ligand-based and the direct receptor-based approaches. The common aim of both strategies is to understand structure–activity relationships and to employ this knowledge in order to propose novel compounds with enhanced activity and selectivity profile for a specific therapeutic target. The ligand-based methods include the various QSAR and 3D-QSAR methods [2]. 3D-QSAR methods, with particularly widespread successes in analyzing structure–activity data, are the comparative molecular field analysis (CoMFA) [3] and the GRID/GOLPE approach [4]. These methods are based entirely on experimental structure–activity relationships for enzyme inhibitors or receptor ligands. For the direct receptor-based methods, which include molecular docking and molecular dynamics simulations, the 3D-structure of a target enzyme or even a receptor–ligand complex is required with atomistic resolution. The structures are generally determined by either X-ray crystallography, NMR spectroscopy or protein homology model building [5].

3D-QSAR methods, like the GRID/GOLPE method, are nowadays used widely in drug design, since they are computationally not demanding and afford fast generation of QSARs from which the biological activity of newly synthesized molecules can be predicted. The basic assumption in GRID/GOLPE is that a suitable sampling of the molecular interaction fields around a set of aligned molecules might provide all the information necessary for understanding their biological

Molecular Interaction Fields. Edited by G. Cruciani
Copyright © 2006 WILEY-VCH Verlag GmbH & Co. KGaA, Weinheim
ISBN: 3-527-31087-8

activities [2]. The suitable sampling is achieved by calculating interaction energies between each molecule and an appropriate probe at regularly spaced grid points surrounding the molecules. The resulting energies derived from simple potential functions can then be contoured to give a quantitative spatial description of molecular properties. If correlated with biological activity, 3D-fields can be generated, which describe the contribution of a region of interest surrounding the ligands to the target properties. However there is one main difficulty in the application of 3D-QSAR methods: for a correct model, a spatial orientation of the ligands towards one another has to be found, which is representative for the relative differences in the binding geometry at the protein binding site. The success of a molecular field analysis is therefore determined by the quality of the choice of the ligand superimposition [6–9]. In most cases, the first step in a 3D-QSAR study is the generation of a reliable pharmacophore model. Many alignment strategies have been reported and compared for this purpose (a detailed comparison of different methods can be found in [10]). Depending on the molecular flexibility and the structural diversity of the investigated compounds this task of unique pharmacophore generation becomes less feasible. Despite the difficulties concerning the molecular alignment many successful 3D-QSAR studies applying the GRID/GOLPE approach have been reported in the last few years [11–18].

Structure-based methods nowadays are able to calculate fairly accurately the position and orientation of a potential ligand in a receptor binding site. This has been demonstrated by various docking studies, described in the literature [19–23]. The docking methods yield important information concerning the spatial orientation of the ligands in the binding site and also towards other ligands binding to the same target. The major problem of today's docking programs is the inability to evaluate binding free energies correctly in order to rank different ligand–receptor complexes. Since docking programs generate a huge amount of possible ligand–receptor complexes, it is impossible to determine a priori which ligand conformation represents the bioactive one. The problem in predicting affinity has generated considerable interest in developing methods to calculate ligand affinity reliably for a widely diverse series of molecules binding to the same target protein of known structure [23–28]. For the calculation of ligand–receptor interaction energies, most approaches rely on molecular mechanics force fields that represent van der Waals and Coulombic interactions on the basis of empirical potentials. Other approaches use simpler scoring functions rather than calculating the affinity by molecular mechanics equations (for a detailed review see [27]). These methods commonly use available experimental data to obtain parameters for some relatively simple functions that allow fast estimation of the binding energy. The estimated binding energies or scores are widely used to discriminate between active and inactive ligands, for example in virtual database screening, but are mostly not accurate enough for 3D-QSAR analysis. The main problem in affinity prediction is that the underlying molecular interactions are highly complex and various terms should be taken into account to quantify the free energy of the interaction process. Only rigorous methods, such as free energy perturbation or thermodynamic integration are able to predict correctly the binding affinity. While these methods clearly have the potential of providing accurate evaluation of relative binding free energies, they are very expensive in a computational sense [29, 30].

Regarding the strengths of both approaches, the docking programs using protein information and the 3D-QSAR methods to develop predictive models for related molecules, prompted us and others to combine both in an automated unbiased procedure [11–15, 31–39]. In this context, the three-dimensional structure of a target protein, along with a docking protocol is used to guide alignment selection for comparative molecular field analysis [9]. This approach allows the generation of a kind of target-specific scoring method considering all the structure–activity data known for a distinct ligand data set.

In this chapter the application of the GRID/GOLPE method [3] in combination with a receptor-based alignment strategy is reported. The comprehensive utility of this approach is exemplified by recent molecular modeling studies on different classes of ligands from our laboratory. Special emphasis will be placed on a detailed description of the combined receptor/ligand-based approach and the successful application of this procedure for the design of novel drug molecules.

7.2
3D-QSAR Using the GRID/GOLPE Approach

3D-QSAR has been synonymous for many years with CoMFA [40], which was the first method to implement in a QSAR method the concept that the biological activity of a ligand can be predicted by its three-dimensional structure and that any binding between a protein and a ligand is the product of noncovalent reversible interactions. The idea behind any molecular field analysis is that the three-dimensional properties of a molecule can be fully described by embedding the molecule in a grid and calculating the interaction energies between the ligand and a probe atom at any node of the grid. In the GRID/GOLPE approach the interaction with the probe atom is described by GRID potentials [41]. GRID includes a wider variety of probes than other approaches so that more different types of interactions can be modeled. Another advantage of using GRID is the fact, that the underlying potentials have been carefully developed on the basis of experimentally determined protein-ligand complexes [41]. GRID has demonstrated its performance in various drug discovery projects and is one of the most successful modeling programs. For a detailed description the reader is referred to Chapter 1.

If we want to carry out a 3D-QSAR analysis, all training set ligands have to be aligned correctly. The relative spatial orientation within the grid plays a crucial role in any 3D-QSAR analysis [7]. Generating 3D conformations and alignment for compounds used in 3D-QSAR analysis is a difficult and time-consuming process, especially when the compounds are very flexible and large in size. Moreover, there is a risk of introducing user subjectivity in manual alignment. Thus, an automated generation of alignment for the 3D-QSAR model could be beneficial. When the alignment problem is solved, a descriptor matrix, whose rows represents the ligands and whose columns contain the GRID interaction energies, is generated. To analyze such a highly intercorrelated matrix, powerful statistical tools such as the PLS (partial least squares) method are needed [42–45].

In spite of using PLS, spurious results can still occur due to the noise hidden in the obtained matrix. The GOLPE (which stands for generating optimal linear PLS estimation) approach was developed to identify which variables are meaningful for the prediction of the biological activity and to remove those with no predictivity [43]. Within this approach, fractional factorial design (FFD) is initially applied to test multiple combinations of variables [45]. For each combination, a PLS model is generated and only variables which significantly increase the predictivity are considered. Variables are then classified considering their contribution to predictivity. A further advance in GOLPE is the implementation of the smart region definition (SRD) procedure [44]. This is aimed at selecting the cluster of variables, rather than the single variable mainly responsible for activity. The SRD technique seems less prone to change correlation than any single variable selection, and improves the interpretability of the models [11].

One of the biggest advantages of 3D-QSAR over classical QSAR is the graphical interpretability of the statistical results. Equation coefficients can be visualized in the region around the ligands. Upon visual inspection, regions of space contributing most to the activity can be easily recognized. The interpretation of the graphical results allows one both to check the reliability of the models easily and to design modified compounds with improved activity or selectivity. In this respect 3D-QSAR methods like CoMFA and GRID/GOLPE have proven to be very useful [2].

The final part of a 3D-QSAR analysis is the model validation, when the predictive power of the model and hence its ability to reproduce biological activities of novel compounds is established. Most of the 3D-QSAR modeling methods implement the leave-one-out (LOO) cross-validation procedure. The output of this procedure are the cross-validated q^2 and the standard deviation of error prediction (SDEP), which are commonly regarded as the ultimate criteria of both the robustness and the predictive ability of a model. The simplest cross-validation method is LOO, where one object at time is removed and predicted. A more robust and reliable method is the leave-several-out cross-validation. For example, in the leave-20%-out cross-validation five groups of approximately the same size are generated. Thus, 80% of the compounds are randomly selected for generation of the model, which is then used to predict the remaining compounds. This operation must be repeated a large number of times in order to obtain reliable statistical results. The leave-20%-out or also the more demanding leave-50%-out cross-validation results are much better indicators of the robustness and the predictive ability of a 3D-QSAR model than the usually used LOO procedure [46, 47]. LOO often yields too optimistic models, which fail when predicting real test set molecules.

Despite the known limitations of the LOO procedure, it is still uncommon to test 3D-QSAR models for their ability to correctly predict the biological activities of compounds not included in the training set. However, many authors claim that their models, showing high LOO q^2 values, have high predictive ability in the absence of external validation (for a detailed discussion on this problem see [48, 49]). In contrast with such expectations, it has been shown by several studies that a correlation between the LOO cross-validated q^2 value for the training set and the correlation coefficient r^2 between the predicted and observed activities for the test set, does not exist [48–50]. Therefore, it is highly recommended to use external test sets to further validate a generated 3D-QSAR model [48].

The last decade has shown that 3D-QSAR methods are able to provide much useful information which has helped in the understanding of structure–activity relationships, in proposals of chemical modifications to enhance biological activity, and in the activity prediction of unknown compounds. However, the application of the molecular field analysis requires a lot of a priori knowledge or at least a synergetic interaction with other molecular modeling methods to generate a reliable ligand alignment. Two application examples, where we have applied the GRID/GOLPE procedure in combination with a receptor-based alignment strategy to develop predictive 3D-QSAR models, are reported below.

7.3 GRID/GOLPE Application Examples

7.3.1 Estrogen Receptor Ligands

Intensive research has revealed that there is a plethora of xenoestrogens in our environment, i.e., both man-made and natural molecules that have been shown to bind to the estrogen receptor as either agonists or antagonists. Serious concern has recently arisen about the adverse effects of chemical compounds possessing estrogenic activity on humans and other species, but it is practically impossible to perform thorough toxicological tests on all of the more than 87 000 xenoestrogens that may ultimately need to be evaluated. Thus there is an obvious need to develop alternative methods to predict the estrogenic activity of molecules with sufficient accuracy. These methods should ultimately facilitate the rapid screening of untested xenoestrogens, particularly in order to distinguish which molecules should have the highest priority for entry into expensive and stressful testing on animals. In this context, computational methods such as QSAR seem attractive.

The effect of xenoestrogens are mediated by an intracellular estrogen receptor (ER), which belongs to the steroid/thyroid nuclear hormone superfamily [51]. The biological action of estradiol, the most active endogenous estrogen, and xenoestrogens and their primary interactions with the receptor protein have been topics of much interest over the years [52, 53]. Over the last years a large amount of structure–activity information for modified estrogens and nonsteroidal ER ligands has been reported in the literature [54]. Several models of the ER ligand pharmacophore have been published in the last few years [54–60]. They are merely based on the comparison of rigid or semi-rigid steroid molecules. For the present study structure–activity data for a series of 30 structurally diverse ER ligands were considered [60]. The molecular structures of these molecules are represented in Table 7.1.

Table 7.1 Structure of ER ligands of the training set.

No		No		No	
1		11		21	
2		12		22	
3		13		23	
4		14		24	
5		15		25	
6		16		26	
7		17		27	

Table 7.1 Continued.

No	No	No
8	18	28
9	19	29
10	20	30

The three-dimensional structure of the ER was resolved by Brzozowski et al. [61]. Four additional X-ray structures of the receptor liganded with different molecules – estradiol, diethylstilbestrol, a nonsteroidal stilben derivative, and two antagonists raloxifen and 4-hydroxytamoxifen – were also solved [61, 62]. The availability of several structurally diverse structures bound to the active site of the receptor provided important experimental information detailing the molecular alignment of the studied molecules.

The crystal structures of the human estrogen α receptor ligand binding domain complexed with estradiol and diethylstilbestrol (PDB code: 1ERE and 2ERD) were taken from the Protein Databank and were overlaid using the backbone atoms (Fig. 7.1). The ER shows a nearly identical three-dimensional structure in these X-ray structures. The only major conformational differences are the orientation of two sidechains in the binding pocket - His524 and Met421. As the analysis of the crystal structures shows, estradiol and diethylstilbestrol bind similarly to the receptor. Both ligands make hydrogen bonds with Glu353 and His524 and interact in addition with hydrophobic residues. In order to determine the bioactive conformation of all studied ligands, we carried out a molecular docking study. Auto-Dock, which has been shown to accurately reproduce experimentally observed binding modes [20, 63] was used for this purpose (the program is described in detail elsewhere [20]). AutoDock uses a simulated annealing procedure to explore the binding possibilities of a ligand in a binding pocket [64]. The interaction energy of ligand and protein is evaluated using atom affinity potentials calculated on a grid similar to GRID [41]. The obtained docking complexes were then refined and the interaction energies were calculated using the YETI force field [65, 66]. This force field uses more sophisticated energy potentials to calculate the binding energy compared to the more simple functions implemented in docking programs, such as AutoDock.

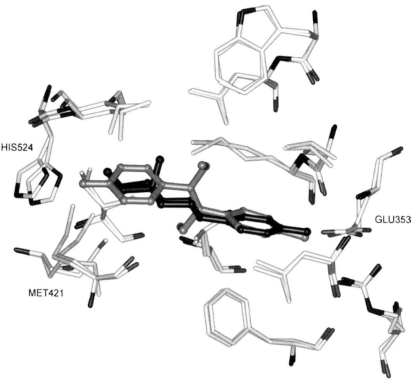

Figure 7.1. Comparison of the X-ray structures of the estrogen receptor bound to estradiol (dark-gray) and diethylstilbestrol (gray). The two crystal structures were overlaid using the backbone atoms. Only the amino acid residues in proximity to the binding pocket are shown for clarity.

Estradiol and diethylstilbestrol were taken as positive controls to test the performance of AutoDock. AutoDock was successful in reproducing the experimentally found binding position for estradiol and diethylstilbestrol (one may speak of reproduction if the root mean square deviation (RMSD) is below 2 Å [11]). The RMSD value between the observed and calculated position was 0.21 Å for estradiol and 0.37 Å for diethylstilbestrol. Subsequently the ligand–receptor complexes for the resulting 28 ligands were computed by applying the same docking protocol. For each ligand the complex which showed the lowest interaction energy applying the YETI force field was selected. Figure 7.2 shows the superimposition of all 30 ligands within the binding pocket.

To further validate the results obtained by the automated docking procedure, the receptor binding pocket was analyzed using program GRID [41]. The hydroxy,

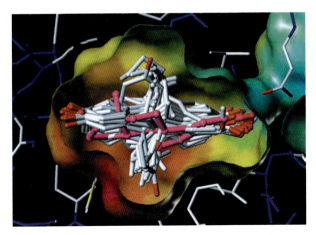

Figure 7.2. Alignment of all training set ligands obtained by the docking (estradiol is colored magenta). The solvent accessible surface area is colored according to the electrostatic potential (red: positive, blue: negative).

methyl and hydrophobic DRY probes were used in order to identify and visualize the main interactions between the protein and the ligands. The two main regions of interaction for the hydroxy probe correspond very well with the position of the aliphatic and aromatic hydroxy groups of the ligands. In an effort to study the areas of hydrophobic interaction the methyl and the hydrophobic DRY probe were tested inside the binding pocket. A good agreement between the molecular shape of diethylstilbestrol and estradiol and the interaction field of the methyl probe was observed. The interaction field obtained with a methyl probe indicates mainly the van der Waals interactions and corresponds to the location and size of the nonpolar parts of the ligands. The most negative areas obtained with the hydrophobic DRY probe are placed below and above the planar ring systems of the ligands and correspond to the positions of the alkyl groups of diethylstilbestrol and the other ligands. In general, visual inspection showed that the GRID interaction fields are in excellent agreement with the experimentally determined positions of the corresponding parts of the molecules. Thus, the GRID fields are perfectly suited to verify the results obtained by automatic docking programs like AutoDock.

Since we refined all protein–ligand complexes using the YETI force field we were also interested to see whether the interaction energies could be correlated to the observed biological activities. The interaction energies were calculated using the YETI force field, AM1 partial charges and a distance dependent dielectric function. A correlation coefficient of $r^2 = 0.541$ and a LOO cross-validated coefficient of $q^2 = 0.490$ were obtained for the 30 investigated ligands. Trends between experimental and calculated can be distinguished, but the standard deviation of 0.83 indicates the limitations of the method. Since solvation or entropic effects were not considered in the calculation of the binding energy, it is not surprising that the correlation was moderate.

With respect to the receptor-based methods, the 3D-QSAR approach has the advantage of dealing only with the differences in affinity of a special series of compounds. In this case the interaction energy of each ligand is not important, because some of the terms describing this energy (desolvation processes, entropic terms) take approximately the same value for every ligand. Since only differences in the binding affinity are regarded, these terms are not considered. Therefore we used the comparative field analysis to develop a quantitative structure–activity relationship for the investigated ligands.

The superimposition of the ligands derived from the molecular docking was taken as structural alignment for a comparative molecular field analysis. The interaction energies between the ligands and a water probe were calculated using the GRID program employing a grid spacing of 1 Å. The GRID calculations gave 11 350 variables for each compound. Many of the variables derived from the GRID analysis did not contribute to the correlation between the chemical structure and the biological activity and could be considered as noise, which decreased the quality of the model [43]. To obtain a robust QSAR model, the irrelevant variables were removed using the GOLPE program. From the active variables that were automatically selected by GOLPE, an initial pretreatment decreased the number of variables to 8740. The D-optimal preselection procedure allowed the selection of the most informative variables correlated with the biological activity from an initial PLS model. This procedure reduced the number of variables from 8740 to 1105. The SRD algorithm was performed with the aim being to select and group the regions of variables of highest importance for the model. These groups were then evaluated by fractional factorial design. This algorithm allowed the extraction of the most relevant variables by building a large number of reduced models similar to the complete model. The SRD variable preselection decreased the number of variables to 1105 without reducing the quality of the model, and after FFD, 493 variables were selected, resulting in a significant improvement of the quality of the model. It was concluded that many of the variables contributed to noise and not to the robustness of the predictive model.

To form the basis for a predictive statistical model the PLS method was used to analyze the 30 compounds. The analysis based on the receptor-based alignment yielded a correlation coefficient with a cross-validated q^2 of 0.900 using four principal components (the model generation and cross-validation was repeated 100 times). The conventional r^2 of this analysis is 0.992. This means, that the model explains approximately 99% of the variance in ligand binding of the investigated compounds. The model also expresses good predictive ability, indicated by the high correlation coefficient of $q^2 = 0.820$ obtained by using the leave-50%-out cross-validation procedure.

The comparison of the GRID/GOLPE results with a ligand-based CoMFA model, (q^2_{LOO} value of 0.796) [60], indicated that the ligand alignment constructed on the basis of the receptor structure supplies a better explanation of the biological activities. This is also indicated by smaller deviations of the calculated from the experimental values in the receptor-based model [35].

7.3 GRID/GOLPE Application Examples | 155

To further validate the GRID/GOLPE model it was necessary to test its predictivity for external test sets. We carefully checked the literature for further ER data sets, which were experimentally tested in the same assay under the same conditions [67–70]. In order to test the general predictivity 36 structurally diverse ER ligands were selected from the mentioned studies (examples are shown in Fig. 7.3, the whole data set is described in detail in [36]).

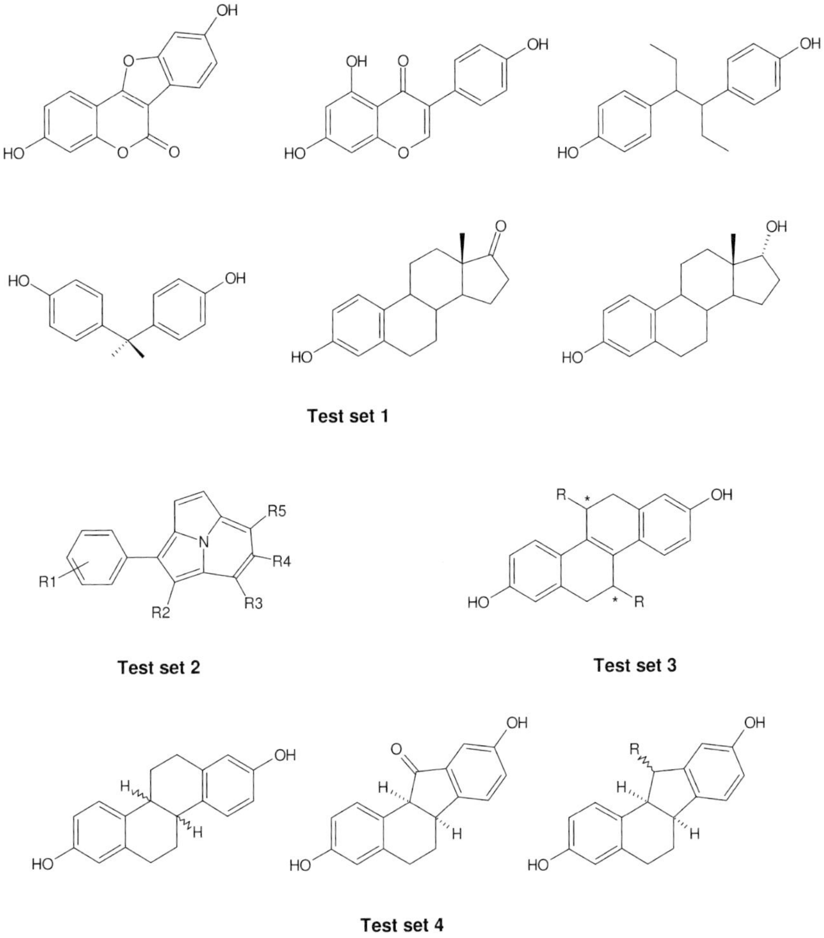

Figure 7.3. Examples of ER test set compounds.

All test set ligands were docked and scored as described in the methods section. For three compounds AutoDock was not able to find any low energy conformation. A visual inspection of the binding site revealed that these compounds can only bind in a low energy conformation if several amino acids change their side-chain orientation. In the present version of AutoDock receptor flexibility cannot be considered. The remaining compounds were successfully docked by the Auto-

Dock program. The ligand alignment, as obtained by the docking procedure, is in agreement with that of the training set molecules. All molecules form a hydrogen bond to Glu353/Arg394, indicated by a similar position of the phenolic ring systems in the particular alignment. A second hydrogen bond to the imidazole of His524 is formed by the potent ligands. The derived alignment indicates further that the alkyl substituents of the ligands occupy similar regions in space corresponding to two hydrophobic cavities located on both sides of the planar ring systems of the cocrystallized ligands.

The conformations, as obtained by the docking procedure, were extracted from the binding site and were used for the prediction of binding affinity applying the developed GRID/GOLPE PLS model. The external prediction yielded a predictive r^2 value of 0.656 and a SDEP value of 0.531 for the 33 test set compounds reflecting the good predictivity of the model. The average SDEP value of 0.531 for the external prediction is, as expected, larger than for the internal validation (0.345), but accurate enough to use the model for the prediction of structurally diverse compounds. The SDEP values for the individual test sets were quite similar, not depending on the composition of the individual test sets (Tables 7.2 and 7.3). Larger deviations were observed for molecules possessing flexible substituents. The flexibility of the substituents leads to energetically very similar conformations which made it difficult to select the correct ligand conformation for the prediction.

Table 7.2. Predictive r^2 values for the ER test sets.

Test set (n)	Receptor-based model	Ligand-based model	Interaction-energy model
Test set 1 (6)	0.778	−0.550	−1.513
Test set 2 (14)	0.675	0.338	0.227
Test set 3 (7)	0.595	0.195	−1.220
Test set 4 (6)	0.508	0.615	−1.112
All test sets (33)	0.656	0.203	−0.487

Table 7.3. External SDEP values obtained for the ER different test sets.

Test set (n)	Receptor-based model	Ligand-based model	Interaction-energy model
Test set 1 (6)	0.430	1.386	1.395
Test set 2 (14)	0.549	0.945	0.889
Test set 3 (7)	0.555	0.780	1.294
Test set 4 (6)	0.552	0.486	0.982
All test sets (33)	0.531	0.949	1.104

In conclusion, by applying the GRID/GOLPE procedure in combination with a receptor-base alignment a predictive and robust model was obtained. The quality of the 3D-QSAR model was demonstrated by the accurate prediction of the structurally diverse test set compounds. All other prediction methods that we applied to the same training and test set (ligand-based 3D-QSAR models, interaction energy-based model, scoring methods) showed much lower accuracy or were unable to correctly predict the test set ligands (Tables 7.3 and 7.4). A detailed comparison of the individual prediction methods is beyond the scope of this chapter and the reader is referred to the literature [36].

Table 7.4. Designed compounds predicted using the GRID/GOLPE model.

Cpd.	Structure	Observed[a]	Predicted[b]	Predicted[c]
4g		8.00	7.00	7.20
4h		7.41	7.62	7.66
4i		7.66	7.48	7.56
6g		7.24	6.90	6.77
6h		7.24	7.05	7.11
6i		7.27	7.25	7.2
6j		7.14	6.88	6.92

a Inhibitory activity measured on the AChE of Torpedo californica [12].
b Predicted activity using the water probe model.
c Predicted activity using the methyl probe model.

7.3.2
Acetylcholinesterase Inhibitors

In the second example the application of the described combined approach to a series of aminopyridazine acetylcholinesterase (AChE) inhibitors is reported [12]. According to the cholinergic hypothesis, memory impairments in patients with Alzheimer's disease result from a deficit of cholinergic functions in the brain [71]. One promising strategy to overcome this deficit is the inhibition of the AChE, the enzyme responsible for the hydrolysis of acetylcholine. The cholinergic hypothesis has led to first approaches for drug treatment of Alzheimer patients. In connection with this, different central acting inhibitors for AChE have been developed and introduced into therapy. Recent studies have shown that AChE inhibitors, especially mixed-type inhibitors, which interact with the peripheral site of the enzyme, could, in addition to AChE inhibition, act as potential inhibitors of the formation of βA4-amyloid protein [72]. Thus, centrally active AChE inhibitors, represent a promising approach to the treatment of AD. In this context we focussed on the search for novel potent and selective AChE inhibitors that interact with the peripheral site of the enzyme [73]. The chemical structures of known AChE inhibitors are diverse, ranging from quaternary compounds like decamethonium or edrophonium, to natural products like huperzine, up to the potent benzylpiperidines like the marketed drug donepezil [71]. The starting point of our AChE project was the finding, that the antidepressant minaprine showed weak inhibition of AChE [74]. Since minaprine has a unique structure among the known AChE inhibitors, it was taken as a promising lead compound. When the project was started four X-ray structures of AChE complexed with different inhibitors had been solved [75–77] whereas no information about the binding of minaprine and related inhibitors was available. As in the case of the estrogen receptor ligands we decided to carry out docking studies to determine the exact position of the inhibitors within the binding pocket.

A detailed inspection of the four AChE-inhibitor X-ray structures, obtained from the Protein Databank (PDB code: 1acl bound to decamethonium, 2ack bound to edrophonium, 1acj bound to tacrine and 1vot bound to huperzine) yielded relevant information concerning the orientation of the inhibitors within the binding pocket. AChE shows a nearly identical three-dimensional structure in all known X-ray structures. The active site is located 20 Å from the protein surface at the bottom of a deep and narrow gorge. The only major conformational difference between the four complexes is the orientation of Phe330, a residue located in the middle of the gorge (Fig. 7.4). Depending on the cocrystallized inhibitor this aromatic residue adopts a different conformation. However the positions of the four inhibitors in the binding pocket are quite different, indicating that more than one clearly defined binding region exists.

In the next step we inspected the interaction possibilities within the binding pocket by applying program GRID. The calculated GRID contour maps, obtained with a variety of different probes, were then viewed superimposed on the crystal structure of AChE. GRID fields are very helpful tools to detect the most favorable

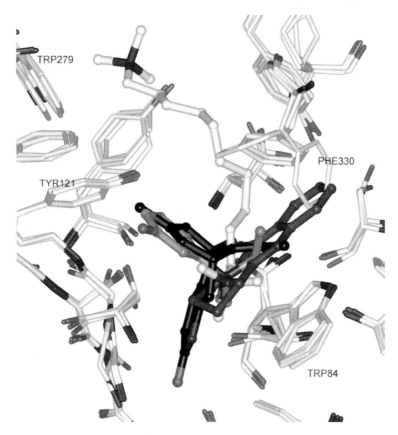

Figure 7.4. Superimposition of the investigated crystal structures of AChE in complex with huperzine (black), tacrine (dark-grey), edrophonium (grey) and decamethonium (light-grey). Only the amino acid residues close to the binding site are displayed.

interactions for a distinct functional group within a given binding pocket. We observed a nice agreement of the location and size of the GRID fields with the corresponding functional groups of the cocrystallized inhibitors. As an example, the results obtained with a trimethylammonium probe are shown in Fig. 7.5. GRID detected a favorable interaction field for the trimethylammonium probe 4 Å above the indole ring of Trp84. It agrees perfectly with the position found for the quaternary group of decamethonium (as well as edrophonium) in the corresponding X-ray structure (Fig. 7.5). Similar agreements were observed when applying other probes like a carbonyl, methyl or the hydrophobic DRY probe.

As in the previous example, the known crystal structures of AchE–inhibitor complexes were taken as positive control to test the usability of AutoDock. The same standard docking parameters were taken as for the analysis of the ER ligands [12]. Due to the flexibility and the few polar functional groups of the inhi-

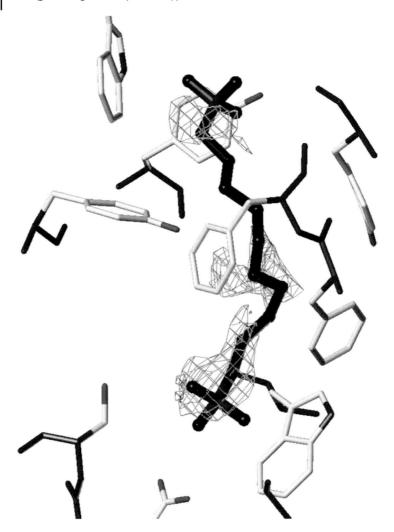

Figure 7.5. Favorable regions of interaction between the trimethylammonium probe and the uncomplexed active site (contour level −9.5 kcal mol^{-1}). Decamethonium is displayed for comparison.

bitors several favorable docking poses were obtained for each compound. In a subsequent step the top ranked poses were compared with the GRID interaction fields calculated with the trimethylammonium, methyl, carbonyl, amide and DRY probes. Taking the derived interaction fields as filters along with the complexes generated by the AutoDock procedure, we were able to select the cocrystallized conformation of all inhibitors. The closest agreement was observed when the protein structure extracted from the corresponding AchE–inhibitor complex was taken as target for the docking simulation. The lowest RMSD values (calculated for the heavy atoms) between theoretically predicted and experimentally determined positions were for tacrine: 0.28 Å, for huperzine A: 0.51 Å, for edrophonium: 0.71 Å and for decamethonium: 1.15 Å [12].

The ability to accurately predict the binding conformation of tacrine, decamethonium, edrophonium and huperzine gave confidence that we could use our model to evaluate the binding conformation of the aminopyridazine compounds (Fig. 7.6). Since the aminopyridazine derivatives are comparable in size to decamethonium, and it is likely that they interact in a similar way with the binding site, we took the protein structure from the AchE–decamethonium complex for our further docking. Fig. 7.7 shows the predicted position of an aminopyridazine in comparison to the position of decamethonium observed in the corresponding crystal structure. The hydrophobic parts of the aminopyridazine inhibitors interact with the various aromatic residues at the binding pocket. The benzyl ring of the inhibitor displays classic π-π stacking with the aromatic ring of Trp84. It thus occupies the binding site for the quaternary ligands. The charged nitrogen of the piperidine makes a cation–π interaction with Phe330 and electrostatic interactions with Tyr121. No direct hydrogen bonds were observed between the polar groups of the inhibitor and the binding site. A similar binding orientation was observed for all other inhibitors (Fig. 7.8). The docking poses were subsequently extracted from the protein environment and were taken as input for a GRID/GOLPE analysis.

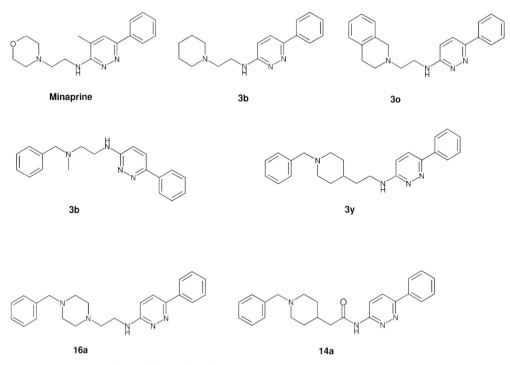

Figure 7.6. Examples of AChE inhibitors from the training set [12].

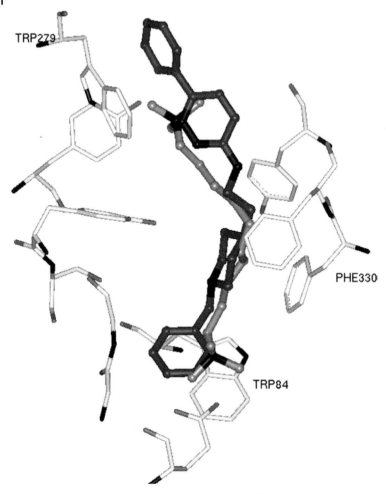

Figure 7.7. Comparison between the predicted position of the amino-pyridazine 3y (dark-grey) and the X-ray structure of decamethonium (grey).

The interaction fields between the aligned ligands and (i) a water and (ii) a methyl probe were calculated by applying the GRID program. A cut-off of +5 kcal mol^{-1} was applied in order to obtain a more symmetric distribution of energy values. The GRID calculation gave 17 160 variables for each compound. After the pretreatment the data set contained 5464 variables for the water and 4456 variables for the methyl probe. Subsequently, the SRD procedure was applied to carry out variable selection on the groups of variables chosen according to their positions in 3D space. The derived regions were subsequently used in a FFD variable selection procedure. The number of variables was reduced to 576 for the water and to 1097 for the methyl probe.

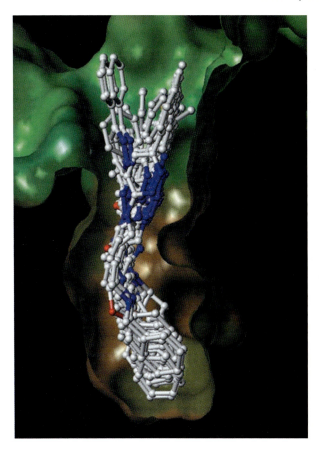

Figure 7.8. Receptor-based alignment of all investigated inhibitors as obtained by the docking study. The solvent accessible surface area is colored according the lipophilic potential (blue: polar, brown: lipophilic).

The LOO cross-validated q^2_{LOO} values for the initial models was 0.875 using the water probe and 0.850 using the methyl probe. The application of the SRD/FFD variable selection resulted in an improvement of the significance of both models. The analysis yielded a correlation coefficient with a cross-validated q^2_{LOO} of 0.937 for the water probe and 0.923 for the methyl probe. In addition we tested the reliability of the models by applying leave-20%-out and leave-50%-out cross-validation. Both models are also robust, indicated by high correlation coefficients of q^2 = 0.910 (water probe, SDEP = 0.409) and 0.895 (methyl probe, SDEP = 0.440) obtained by using the leave-50%-out cross-validation procedure. The statistical results gave confidence that the derived model could also be used for the prediction of novel compounds.

To get an impression which parts of the AChE inhibitors are correlated with variation in activity we analyzed the PLS coefficient plots (obtained using the water and the methyl probe) and compared them with the amino acid residues of the binding pocket. The plots indicate those lattice points where a particular property significantly contributes and thus explains the variation in biological activity data (Fig. 7.9). The plot obtained with the methyl probe indicated that close to the arylpyridazine part, a region with positive coefficients exists (region A in Fig. 7.9). The coefficients were superimposed with the original GRID field obtained for compound 4j with the methyl probe. The interaction energies in region A are positive, therefore the decrease in activity is due to overlapping of this region. Thus, it should be possible to get active inhibitors by reducing the ring size compared to compound 4j (which is shown in Fig. 7.8 together with the PLS coefficient maps). For that reason, several molecules containing hydrophobic groups were proposed (Table 7.4). A second interesting field was observed above the arylpyridazine moiety in the model obtained using the water probe. There exists a region where polar interaction increases activity (region B in Fig. 7.9). Together with an analysis of the entrance of the gorge (the interaction site for the arylpyridazine system) we got the idea to design compounds bearing polar groups. In the calculated AchE–aminopyridazine complexes we observed two polar amino acid residues (Asn280 and Asp285) located at the entrance of the gorge which could serve as an additional binding site for the substituted arylpyridazine system. To test this hypothesis, several inhibitors possessing polar groups with hydrogen bond donor and acceptor properties were synthesized and tested (Table 7.4). The designed inhibitors were docked in the binding pocket by applying the developed procedure and their biological activities were predicted using the GRID/GOLPE PLS models. Table 7.4 shows the predicted and experimentally determined inhibitor activities for these compounds. In general an excellent agreement between predicted and experimentally determined values was observed, indicated by the low $SDEP_{ext}$ values of 0.440 (water model) and 0.398 (methyl model). The reducing of the size of the aminopyridazine ring system resulted in highly potent inhibitors 4g–4i. The molecules of the second series of designed inhibitors containing polar groups were also accurately predicted. The gain in activity compared to the nonsubstituted compound 3y is moderate, indicating that the potential interaction with the two polar residues at the entrance does not play an important role. Since the two residues are located at the entrance of the binding pocket, it may be possible that these residues make stronger interaction with water molecules than with the protein sidechains.

In conclusion, we were able to design AChE inhibitors based on the docking and GRID/GOLPE study which seems to interact simultaneously with the cation–π subsite of the catalytic site and the peripheral site of the enzyme. Further support for our docking study came from the crystal structure of a novel AchE–inhibitor complex [78]. The crystal structure of AChE complexed with the marketed drug donepezil was solved in 1999. Like donepezil, our most potent inhibitors contain a benzylpiperidine moiety which shows a similar position and orientation when compared with the published crystal structure. The comparison of both AchE–inhibitor complexes revealed that both kind of inhibitors adopt a comparable conformation in the narrow binding pocket [12]. As we predicted for our aminopyridazine inhibitors, donepezil makes no direct hydrogen bond to any amino acid residue of the binding pocket. Only water-bridged hydrogen bonds have been detected for donepezil, as proposed for the described aminopyridazine compounds.

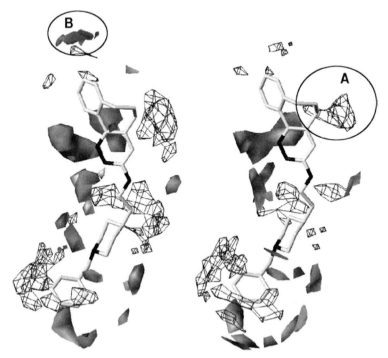

Figure 7.9. PLS coefficient maps obtained using the water probe (a) and the methyl probe (b). Opaque fields are contoured at –0.003, grid fields are contoured at +0.003 (compound 4j is shown for comparison).

7.4
Conclusion

In this chapter it was shown, that combination of GRID/GOLPE and receptor-based alignments can lead to highly predictive and meaningful 3D-QSAR models. Besides the good predictive ability, the received models are also able to indicate which interaction sites in the binding pocket might be responsible for the variance in biological activities. In this context, it must be considered that a PLS analysis indicates only where a variation in the interaction fields is correlated with a variation in the biological activities. If all molecules of a data set show a certain important interaction with the receptor, indicated by a similar interaction energy at a particular grid point for all compounds, this would not be reflected by the resulting PLS model. Thus the degree of correspondence depends strongly on the structural diversity of the studied ligands. If one considers these circumstances, important information can be obtained from a comparison of the contour maps and the binding site, which can then be integrated in the drug design process.

In the last decade, structure-based methods have become major tools in drug design, including lead finding and optimization. It has also been shown, that structure-based methods are able to predict fairly accurately the position of ligands in receptor binding sites. Apart from the accurate prediction of experimental data, modern docking methods are becoming more and more efficient. Meanwhile docking programs have been developed, which can perform the docking of highly flexible ligands in a few seconds/minutes on modern PCs. The major problem is still the prediction of the binding affinity, probably limited by the approximation used in today's scoring and force field methods [24, 26–28]. The application of 3D-QSAR methods – such as GRID/GOLPE – may facilitate the prediction of binding affinities if one has a series of compounds which bind in a similar way to a target protein.

Since a multivariate QSAR analysis considers only the information which applies to the considered data set, advantages are offered in comparison to the more rigorous methods. The rigorous methods have to consider all influences on ligand binding, and must calculate the corresponding amounts correctly. We guess, that a multivariate QSAR analysis is able to provide a kind of scoring function valid for a particular data set. Since the reported combined strategy is able to rapidly predict biological affinity, the method can be applied to large ligand series. As long as no methods are developed which are able to solve the affinity prediction problem, structure-based 3D-QSAR is an exciting strategy for drug design studies.

Acknowledgments

The author wishes to thank Jean-Marie Contreras and Camille Wermuth (Prestwick Chemical, Illkirch, France) for their support.

References

1 Hansch, C., Leo, A. *Exploring QSAR. Fundamentals and Applications in Chemistry and Biology*, American Chemical Society, Washington DC, **1995**.

2 Kubinyi, H. QSAR and 3D-QSAR in Drug Design, *Drug Discovery Today* **1997**, *2*, 457–467.

3 Cramer III, R. D., Patterson, D. E., Bunce, J. D. Comparative Molecular Field Analysis (CoMFA) 1. Effect of Shape on Binding of Steroids to Carrier Proteins, *J. Am. Chem. Soc.* **1988**, *110*, 5959–5967.

4 GRID/GOLPE, www.moldiscovery.com, Molecular Discovery, UK.

5 Kuntz, J.D. Structure-based Strategies for Drug Design and Discovery, *Science*, **1992**, *257*, 1078–1082.

6 Kim, K. Non-linear Dependencies in CoMFA. *J. Comput.-Aided Mol. Des.* **1993**, *7*, 71–82.

7 Folkers, G., Merz, A., Rognan, D. CoMFA: Scope and Limitations, in

3D-QSAR in Drug Design. Theory, Methods and Applications, Kubinyi, H., (ed.), ESCOM Science Publishers B.V., Leiden, **1993**.

8 Klebe, G., Abraham, U. On the Prediction of Binding Properties of Drug Molecules by Comparative Molecular Field Analysis, *J. Med. Chem.* **1993**, 36, 70–80.

9 Sippl, W. Development of Biologically Active Molecules by Combining 3D-QSAR and Structure-based Design Methods, *J. Comput.-Aided Mol. Des.* **2002**, 16, 825–830.

10 Höltje, H.-D., Sippl, W., Rognan, D., Folkers, G., *Molecular Modeling – Basic Principles and Applications*, Wiley-VCH, Weinheim, **2003**.

11 Pastor, M., Cruciani, G., Watson, K. A Strategy for the Incorporation of Water Molecules Present in a Ligand Binding Site into a Three-Dimensional Quantitative Structure-Activity Relationship Analysis, *J. Med. Chem.* **1997**, 40, 4089–4102.

12 Sippl, W., Contreras, J. M., Parrot, I., Rival, Y., Wermuth, C.G. Structure-based 3D-QSAR and Design of Novel Acetylcholinesterase Inhibitors, *J. Comput.-Aided Mol. Des.* **2001**, 15, 395–410.

13 Classen-Houben, D., Sippl, W., Höltje, H.-D. Molecular Modeling on Ligand/Receptor Complexes of Protein-Tyrosine-Phosphatase 1B, in *EuroQSAR2002 Designing Drugs and Crop Protectants: Processes, Problems and Solutions*, Ford, M., Livingstone, D., Dearden, J., Van de Waterbeemd, H., (eds.) Blackwell Publishing, Bournemouth, **2002**.

14 Cinone, N., Höltje, H.-D., Carotti, A. Development of a Unique 3D Interaction Model of Endogenous and Synthetic Peripheral Benzodiazepine Receptor Ligands, *J. Comput.-Aided Mol. Des.* **2000**, 14, 753–768.

15 Hammer, S., Spika, L., Sippl, W., Jessen, G., Kleuser, B., Höltje, H.-D., Schäfer-Korting, M. Glucocorticoid Receptor Interactions with Glucocorticoids: Evaluation by Molecular Modeling and Functional Analysis of Glucocorticoid Receptor Mutants, *Steroids* **2003**, 68, 329–339.

16 Audouze, K., Ostergaard Nielsen, E., Peters D. New Series of Morpholine and 1,4-Oxazepane Derivatives as Dopamine D4 Receptor Ligands: Synthesis and 3D-QSAR Model, *J. Med. Chem.* **2004**, 47, 3089–3104.

17 Terp, G. E., Cruciani, G., Christensen, I. T., Jorgensen, F. S. Structural Differences of Matrix Metalloproteinases with Potential Implications for Inhibitor Selectivity Examined by the GRID/CPCA Approach, *J. Med. Chem.* **2002**, 45, 2676-2684.

18 Tervo, A. J., Nyrönen, T. H., Rönkkö, T., Poso, A. A Structure-Activity Relationship Study of Catechol-O-Methyltransferase Inhibitors Combining Molecular Docking and 3D-QSAR Methods, *J. Comput.-Aided Mol. Des.* **2003**, 17, 797–810.

19 Kramer, B., Rarey, M., Lengauer, T. CASP2 Experiences with Docking Flexible Ligands Using FlexX, *Proteins*, **1997**, Suppl. 1, 221–225.

20 Morris, G. M., Goodsell, D. S., Huey, R., Olson, A. J. Distributed Automatic Docking of Flexible Ligands to Proteins, *J. Comput.-Aided Mol. Des.* **1994**, 8, 243–256.

21 Verdonk, M. L., Cole, J. C., Hartshorn, M., Murray, C., Taylor, R. D. Improved Protein-Ligand Docking using GOLD, *Proteins* **2003**, 52, 609–623.

22 Meng, E., Shoichet, B. K., Kuntz. I. D. Automated Docking with Grid-based Energy Evaluation, *J. Comput. Chem.* **1992**, 13, 505–524.

23 Kontoyianni, M., McClellan, L. M., Sokol, G. S. Evaluation of Docking Performance: Comparative Data on Docking Algorithms, *J. Med. Chem.* **2004**, 47, 558–565.

24 Tame, J. R. H. Scoring Functions: A View from the Bench, *J. Comput.-Aided Mol. Des.* **1999**, 13, 99–108.

25 Böhm, H. J. Prediction of Binding Constants of Protein Ligands: A Fast Method for the Prioritisation of Hits Obtained from De-novo Design or 3D Database Search Programs, *J. Comput.-Aided Mol. Des.* **1998**, 12, 309–323.

26 Wang, R., Lu, Y., Fang, X., Wang, S. An Extensive Test of 14 Scoring Functions Using the PDBbind Refined Set of 800 Protein-Ligand Complexes, *J. Chem. Inf. Comput. Sci.* **2004**, 44, 2114–2125.

27 Perola, E., Walters, W. P., Charifson, P.S. A Detailed Comparison of Current Docking and Scoring Methods on Systems of Pharmaceutical Relevance, *Proteins* **2004**, *56*, 235–249.

28 Gohlke, H., Klebe, G. Approaches to the Description and Prediction of the Binding Affinity of Small Molecule Ligands to Macromolecular Receptors, *Angew. Chem., Int. Ed.* **2002**, *41*, 2644–2676.

29 Masukawa, K. M., Kollman, P. A., Kuntz, I. D. Investigation of Neuraminidase-Substrate Recognition Using Molecular Dynamics and Free Energy Calculations, *J. Med. Chem.* **2003**, *46*, 5628–5637.

30 Huang, D., Caflisch, A. Efficient Evaluation of Binding Free Energy Using Continuum Electrostatics Solvation, *J. Med. Chem.* **2004**, *47*, 5791–5797

31 Waller, C. L., Oprea, T. I., Giolitti, A., Marshall, G.R. Three-Dimensional QSAR of Human Immunodeficiency Virus (I) Protease Inhibitors. 1. A CoMFA Study Employing Experimentally-Determined Alignment Rules, *J. Med. Chem.* **1993**, *36*, 4152–4160.

32 De Priest, S. A., et al. 3D-QSAR of Angiotensin-Converting Enzyme and Thermolysin Inhibitors: A Comparison of CoMFA Models Based on Deduced and Experimentally Determined Active-Site Geometries, *J. Am. Chem. Soc.* **1993**, *115*, 5372–5384.

33 Cho, S. J, Garsia, M. L., Bier, J., Tropsha, A. Structure-based Alignment and Comparative Molecular Field Analysis of Acetylcholinesterase Inhibitors, *J. Med. Chem.* **1996**, *39*, 5064–5071.

34 Vaz, R. J., McLEan, L. R., Pelton, J. T. Evaluation of Proposed Modes of Binding of (2S)-2-[4-[[(3S)-1-acetimidoyl-3-pyrrolidinyl]oxyl]phenyl]-3-(7-amidino-2-naphtyl)-propanoic Acid Hydrochloride and some Analogs to Factor Xa Using a Comparative Molecular Field Analysis, *J. Comput.-Aided Mol. Des.* **1998**, *12*, 99–110.

35 Sippl, W. Receptor-based 3D Quantitative Structure-Activity Relationships of Estrogen Receptor Ligands. *J. Comput.-Aided. Mol. Des.*, **2000**, *14*, 559–572.

36 Sippl, W. Binding Affinity Prediction of Novel Estrogen Receptor Ligands Using Receptor-based 3D-QSAR Methods, *Bioorg. Med. Chem.* **2002**, *10*, 3741–3755.

37 Ortiz, A. R., Pisabarro, M. T., Gago, F., Wade, R. C.. Prediction of Drug Binding Affinities by Comparative Binding Energy Analysis, *J. Med. Chem.* **1995**, *38*, 2681–2691.

38 Holloway, M. K. et al. A Priori Prediction of Activity for HIV-1 Protease Inhibitors Employing Energy Minimization in the Active Site, *J. Med. Chem.* **1995**, *38*, 305–317.

39 Lozano, J. J., Pastor, M., Cruciani, G., Gaedt, K., Centeno, N. B., Gago, F., Sanz, F. 3D-QSAR Methods on the Basis of Ligand-Receptor Complexes. Application of Combine and GRID/GOLPE Methodologies to a Series of CYP1A2 Inhibitors, *J. Comput.-Aided Mol. Des.* **2000**, *13*, 341–353.

40 Cramer III, R. D., Patterson, D. E., Bunce, J. D. Comparative Molecular Field Analysis CoMFA): 1. Effect of Shape on Binding of Steroids to Carrier Proteins, *J. Am. Chem. Soc.* **1988**, *110*, 5959–5967.

41 Goodford, P. J. A Computational Procedure for Determining Energetically Favorable Binding Sites on Biologically Important Macromolecules, *J. Med. Chem.* **1985**, *28*, 849–857.

42 Cruciani, G., Crivori, P., Carrupt, P.-A., Testa B. Molecular Fields in Quantitative Structure-Permeation Relationships, *J. Mol. Struct.* **2000**, *503*, 17–30.

43 Baroni, M., Constantino, G., Cruciani, G., Riganelli, D, Valigli, R., Clementi, S. Generating Optimal Linear PLS Estimations (GOLPE): An Advanced Chemometric Tool for Handling 3D-QSAR Problems, *Quant. Struct.-Act. Relat.* **1993**, *12*, 9–20.

44 Pastor, M., Cruciani, G., Clementi, S. Smart Region Definition: A New Way to Improve the Predictive Ability and Interpretability of Three-Dimensional Quantitative Structure-Activity Relationships, *J. Med. Chem.* **1997**, *40*, 1455–1464.

45 Cruciani, G., Watson, K. Comparative Molecular Field Analysis Using GRID Force Field and GOLPE Variable Selection Methods in a Study of Inhibitors of Glycogen Phosphorylase b, *J. Med. Chem.* **1994**, *37*, 2589–2601.

46 Oprea, T. I., Garcia, A. E. Three-Dimensional Quantitative Structure-Activity Relationships of Steroid Aromatase Inhibitors, *J. Comput.-Aided Mol. Des.* **1996**, *10*, 186–200.

47 Golbraikh, A., Trophsa, A. Beware of q2! *J. Mol. Graph. Model.* **2002**, *20*, 269–276.

48 Kubinyi, H., Hamprecht, F. A., Mietzner, T. Three-Dimensional Quantitative Similarity-Activity Relationships (3D QSiAR) from SEAL Similarity Matrices, *J. Med. Chem.* **1998**, *41*, 2553–2564.

49 Golbraikh, A., Shen, M., Xiao, Z., Xiao, Y., Lee, K.-H., Tropsha, A. Rational selection of training and test sets for the development of validated QSAR models, *J. Comput.-Aided Mol. Des.* **2003**, *17*, 241–253.

50 Norinder, U. Single and Domain Made Variable Selection in 3D QSAR Applications, *J. Chemomet.* **1996**, *10*, 95–105.

51 Evans, R. M. The Steroid and Thyroid Hormone Receptor Superfamily, *Science*, **1988**, *240*, 889–895.

52 Von Angerer, E. *The Estrogen Receptor as a Target for Rational Drug Design*, Landes, Austin, USA, **1995**.

53 Zeelen, F. J. *Medicinal Chemistry of Steroids*, Elsevier, Amsterdam, **1990**.

54 Anstead, G. M., Carlson, K. E., Katzenellenbogen, J. A. The Estradiol Pharmacophore: Ligand Structure-Estrogen Receptor Binding Affinity Relationships and a Model for the Receptor Binding Site, *Steroids*, **1996**, *62*, 268–303.

55 Höltje, H.-D., Dall, N. A Molecular Modeling Study on the Hormone Binding Site of the Estrogen Receptor, *Pharmazie* **1993**, *48*, 243–246.

56 Lewis, D. F. V., Parker, M. G. and King, R. J. B. Molecular Modeling of the Human Estrogen Receptor and Ligand Interaction Based on Site-Directed Mutagenesis and Amino Acid Sequence Homology, *J. Steroid. Biochem. Mol. Biol.* **1995**, *52*, 55–65.

57 Goldstein, R. A., Katzenellenbogen, J. A., Luthey-Schulten, Z. A., Seielstad, D. A., Wolynes, P. G. Three-Dimensional Model for the Hormone Binding Domains of Steroid Receptors, *Proc. Natl. Acad. Sci. USA* **1993**, *90*, 9949–9953.

58 Lemesle-Varloot, L., Ojasoo, T., Mornon, J. P., Raynaud, J. P. A Model for the Determination of the 3D-Spatial Distribution of the Functions of the Hormone- Binding Domain of Receptor that Bind 3-keto-4-ene Steroids, *J. Steroid. Biochem. Mol. Biol.* **1992**, *41*, 369–388.

59 Wurtz, J. M., Egner, U., Heinrich. N., Moras, D., Mueller-Fahrnow, A. Three-Dimensional Models of Estrogen Receptor Ligand Binding Domain Complexes, Based on Related Crystal Structures and Mutational and Structure-Activity Relationships Data, *J. Med. Chem.* **1998**, *41*, 1803–1814.

60 Sadler, B. R., Cho, S. J., Ishaq, K., Chae, K., Korach, K.S. Three-Dimensional Structure-Activity Relationship Study of Nonsteroidal Estrogen Receptor Ligands Using the Comparative Molecular Field Analysis/Cross-Validated r^2-Guided Region Selection Approach, *J. Med. Chem.* **1998**, *41*, 2261–2267.

61 Brzozowski, A. M. et al. Molecular Basis of Agonism and Antagonism in the Estrogen Receptor, *Nature* **1997**, *389*, 753–758.

62 Shiau, A. K., Barstad, D., Loria, P. M., Cheng, L., Kushner, P. J., Agard, D. A., Greene, G.L. The Structural Basis of Estrogen Receptor/Coactivator Recognition and the Antagonism of this Interaction by Tamoxifen, *Cell* **1998**, *95*, 927–937.

63 Goodsell, D. S., Morris G. M., Olson, A. J. Automated Docking of Flexible Ligands: Applications of AutoDock, *J. Mol. Recognit.* **1996**, *9*, 1–5.

64 Rao, M. J., Olson, A. J. Modeling of Factor Xa-Inhibitor Complexes: A Computational Flexible Docking Approach, *Proteins: Struct. Funct. Gen.* **1999**, *34*, 173–183.

65 Vedani, A., Huhta, D. W. A New Force Field for Modeling Metalloproteins. *J. Am. Chem. Soc.* **1990**, *112*, 269–280.

66 Vedani, A., Dunitz, J. D. Lone-Pair Directionality of H-Bond Potential Functions for Molecular Mechanics Calculations: The Inhibition of Human Carbonic Anhydrase II by Sulfonamides, *J. Am. Chem. Soc.* **1985**, *107*, 7653–7658.

67 Meyers, M. J., Sun, J., Carson, K. E., Katzenellenbogen, B. S., Katzenellenbogen, J. A. Estrogen Receptor Subtype Selective Ligands, *J. Med. Chem.* **1999**, *42*, 2456–2468.

68 Stauffer, S. R., Huang, Y., Coletta, C. J., Tedesco, R., Katzenellenbogen, J. A. Estrogen Pyrazoles: Defining the Pyrazole Core Structure and the Orientation of Substituents in the Ligand Binding Pocket of the Estrogen Receptor, *Bioorg. Med. Chem.* **2001**, *9*, 141–150.

69 Tedesco, R., Youngman, M. K., Wilson, S. R., Katzenellenbogen, J. A. Synthesis and Evaluation of Hexahydrochrysene and Tetrahydrobenzofluorene Ligands for the Estrogen Receptor, *Bioorg. Med. Chem. Lett.* **2001**, *11*, 1281–1284.

70 Jorgensen, A. S., Jacobsen, P., Christiansen, L. B., Bury, P. S., Kanstrup, A., Thorpe, S. M., Naerum, L., Wassermann, K. Synthesis and Estrogen Receptor Binding Affinities of Novel Pyrrolo[2,1,5-cd]indolizine Derivatives. *Bioorg. Med. Chem. Lett.* **2000**, *10*, 2383–2386.

71 Parnetti, L., Senin, U., Mecocci P. Cognitive Enhancement Therapy for Alzheimer's Disease. The Way Forward, *Drugs Future* **1997**, *53*, 752–768.

72 Brufani, M.; Filocamo, L., Lappa, S., Maggi, A. New Acetylcholinesterase Inhibitors, *Drugs Future* **1997**, *22*, 397–410.

73 Contreras, J. M., Rival, Y., Chayer, S., Bourguignon, J. J., Wermuth, C. G. Aminopyridazines as Acetylcholinesterase Inhibitors, *J. Med. Chem.* **1999**, *42*, 730–741.

74 Wermuth, C. G. et al. 3-Aminopyridazine Derivatives with Atypical Antidepressant Serotonergic, and Dopaminergic Activities, *J. Med. Chem.* **1989**, *32*, 528–537.

75 Sussman, J. L., Silman, I. Atomic Structure of Acetylcholinesterase from Torpedo Californica: A Prototypic Acetylcholine-Binding Protein, *Science* **1991**, *253*, 872–879.

76 Harel, M., Sussman, J.L. Quaternary Ligand Binding to Aromatic Residues in the Active Site Gorge of Acetylcholinesterase, *Proc. Natl. Acad. Sci. USA* **1993**, *90*, 9031–9035.

77 Raves, M. L., Harel, M., Pang, Y. P., Silman, I., Kozikowski, A. P., Sussman, J. L. Structure of Acetylcholinesterase Complexed with the Nootropic Alkaloid Huperzine A, *Nature Struct. Biol.* **1997**, *4*, 57–63.

78 Kryger, G., Silman, I., Sussman, J. L. Structure of Acetylcholinesterase Complexed with E2020 (Aricept): Implications for the Design of New Anti-Alzheimer Drugs, *Structure Fold. Des.* **1999**, *15*, 297–307.

III
Pharmacokinetics

8
Use of MIF-based VolSurf Descriptors in Physicochemical and Pharmacokinetic Studies

Raimund Mannhold, Giuliano Berellini, Emanuele Carosati, and Paolo Benedetti

8.1
ADME Properties and Their Prediction

Major causes for failure in drug development are unsuitable pharmacokinetic properties of drug candidates including absorption, distribution, metabolism, and excretion (ADME), which were traditionally measured at rather late stages of drug development. Nowadays, the testing of ADME properties is done much earlier; that is, before clinical evaluation of a compound is decided. At the same time, the rate at which biological screening data are obtained has dramatically increased, and high-throughput screening (HTS) facilities are now commonly used for hit-finding. In response to these developments, combinatorial chemistry has been adopted to feed the hit-finding machines.

Increased capacities for biological screening and chemical synthesis have in turn magnified the demands for large quantities of early information on ADME data. Various medium- and high-throughput *in vitro* ADME screens are therefore now in use. In addition, there is an increasing need for good tools for predicting these properties to serve two key aims: (i) at the design stage of new compounds and compound libraries so as to reduce the risk of late-stage attrition; and (ii) to optimize the screening and testing by looking at only the most promising compounds. For reviews see [1–3].

This is the framework where computational chemistry could play an important role in the prediction of these properties in order to obtain more efficient and faster drug discovery cycles. To obtain useful descriptors for ADME properties is not an easy task. A large number of descriptors have been developed [4], all of which have major limitations in terms of relevance, interpretability or speed of calculation.

Alternatively, calculated molecular properties from 3D molecular fields of interaction energies represent a valuable approach to correlate 3D molecular structures with physicochemical and pharmacodynamic properties. In contrast, their use in correlations with pharmacokinetic properties is still poorly explored and exploited. The rather new VolSurf approach [5–7] is able to compress the relevant information present in 3D maps into a few descriptors characterized by the simplicity of

Molecular Interaction Fields. Edited by G. Cruciani
Copyright © 2006 WILEY-VCH Verlag GmbH & Co. KGaA, Weinheim
ISBN: 3-527-31087-8

their use and interpretation. These descriptors can be quantitatively compared and used to build multivariate models correlating 3D molecular structures with biological responses.

8.2
VolSurf Descriptors

The interaction of molecules with biological membranes is mediated by surface properties such as shape, electrostatic forces, H-bonds and hydrophobicity. Therefore, the GRID [8] force field was chosen to characterize potential polar and hydrophobic interaction sites around target molecules by the water (OH2), the hydrophobic (DRY), and the carbonyl oxygen (O) probe. The information contained in the MIF is transformed into a quantitative scale by calculating the volume or the surface of the interaction contours. The VolSurf procedure is as follows: (i) in the first step, the 3D molecular field is generated from the interactions of the OH2, the DRY, and the O probe around a target molecule; (ii) the second step consists of the calculation of descriptors from the 3D maps obtained in the first step. The molecular descriptors obtained, called VolSurf descriptors, refer to molecular size and shape, to hydrophilic and hydrophobic regions and to the balance between them; (iii) finally, chemometric tools (PCA [9], PLS [10]) are used to create relationships of the VolSurf matrix with ADME properties. A scheme of the VolSurf programme steps is reported in Figure 8.1.

In the updated version, presented here, VolSurf descriptors are enlarged from 72 to 94. The new descriptors include elongation, diffusivity, log$P$, and the so-called best volumes. Definition of the original VolSurf descriptors is given in [5–7] and in Table 8.1; the novel descriptors are additionally defined in detail below.

Elongation descriptors give an idea of the maximum extension a molecule could reach if properly stretched. Starting from a statistical model developed for long chain compounds we modified it in order to take into account that part of the molecules may be fixed and hence we defined fixed and flexible contributions to the maximum extension a molecule can reach. "elongation" is the most probable extension of the molecule. "fixed elongation" is the portion of the extension given by the rigid part of a molecule. The new VolSurf descriptors are "elongation" and "ratio between elongation and fixed elongation".

Molecular diffusion is the migration of matter along a concentration gradient from a region of high concentration to a region of low concentration. Molecular diffusion controls, to a degree, the transport rate of chemicals at phase boundaries and in the slow-moving fluids found in porous media. To a lesser degree, it controls the dispersion of chemicals in turbulent fluids.

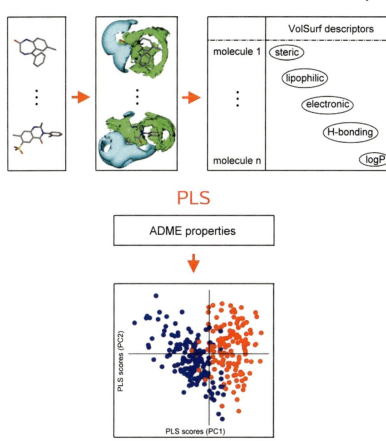

Figure 8.1. The sequence of steps in chemometric analyses using VolSurf descriptors is schematized.

Table 8.1 Detailed definition of VolSurf descriptors.

	Size and shape descriptors
Molecular volume	represents the water-excluded volume (in Å^3), i.e. the volume enclosed by the water-accessible surface computed at a repulsive value of +0.2 kcal mol^{-1}.
Molecular surface	represents the accessible surface (in Å^2) traced out by a water probe interacting at +0.2 kcal mol^{-1} when a water molecule rolls over the target molecule.
Rugosity	is a measure of a molecular wrinkled surface; it represents the ratio of volume/surface. The smaller the ratio, the larger the rugosity.
Molecular globularity	is defined as S/S_{equiv} with S_{equiv} = surface area of a sphere of volume V, where S and V are the molecular surface and volume described above, respectively. Globularity is 1.0 for perfect spherical molecules. It assumes values greater than 1.0 for real spheroidal molecules. Globularity is also related to molecular flexibility.
Elongation	represents the maximum extension a molecule could reach if properly stretched.
EEFR (elongation, elongation-fixed ratio)	represents the portion of the extension given by the rigid part of the molecule. In fact, within each molecule a fixed part is considered as the rigid core, and "EEFR" is the ratio between the elongation and the fixed elongation.
	Descriptors of hydrophilic regions
Hydrophilic descriptors	describe the molecular envelope which is accessible to and attractively interacts with water molecules. The volume of this envelope varies with the level of interaction energies. Hydrophilic descriptors computed from molecular fields of −0.2 to −1.0 kcal mol^{-1} account for polarizability and dispersion forces; descriptors from molecular fields of −1.0 to −6.0 kcal mol^{-1} account for polar and H-bond donor–acceptor regions.
Best volumes (OH2 probe)	are six new descriptors which refer to the best three hydrophilic interactions generated by a water molecule. The best volumes are measured at −1.0 and −3.0 kcal mol^{-1}. To understand the concept of best volumes, we refer to the definition of common group used by VolSurf. When an atom is mainly responsible for the attractive interaction energy of two or more contiguous nodes of the grid cage, these two points belong to the same group. In the 3D GRID map around the molecule, some groups may be identified at specific energy values (−1.0 and −3.0), and their volume can be calculated. BV descriptors refer to the first, second and third largest volumes among such groups. The contribution to the field produced by each atom of the molecule can be appreciated by using the specific options in 3D-plots.

Capacity factors	represent the ratio of the hydrophilic surface over the total molecular surface. In other words, it is the hydrophilic surface per surface unit. Capacity factors are calculated at eight different energy levels, the same levels used to compute the hydrophilic descriptors.

Descriptors of hydrophobic regions

Hydrophobic descriptors	GRID[a] uses a probe called DRY to generate 3D lipophilic fields. In analogy to hydrophilic regions, hydrophobic regions may be defined as the molecular envelope generating attractive hydrophobic interactions. VolSurf computes *hydrophobic descriptors* at eight different energy levels adapted to the usual energy range of hydrophobic interactions (i.e. from -0.2 to -1.6 kcal mol^{-1}).
Best volumes (DRY probe)	are six new descriptors which represent the best three hydrophobic interactions generated by the DRY probe. They are calculated as described for the water probe, but they are measured at -0.6 and -1.0 kcal mol^{-1}.

INTEraction enerGY (= INTEGY) moments

INTEGY moments	express the imbalance between the center of mass of a molecule and the barycenter of its hydrophilic or hydrophobic regions. When referring to hydrophilic regions, integy moments are vectors pointing from the center of mass to the center of the hydrophilic regions: high integy moments indicate a clear concentration of hydrated regions in only one part of the molecular surface, small moments indicate that the polar moieties are either close to the center of mass or they balance at opposite ends of the molecule, so that their resulting barycenter is close to the center of the molecule. When referring to hydrophobic regions, integy moments measure the unbalance between the center of mass of a molecule and the barycenter of the hydrophobic regions.

Mixed descriptors

Local interaction energy minima	represent the energy of interaction (in kcal mol^{-1}) of the best three local energy minima between the water probe and the target molecule. Alternatively, the minima can refer to the three deepest local minima in the molecular electrostatic potential. They are produced for both probes OH2 and DRY.
Energy minima distances	represent the distances between the best three local energy minima when a water probe interacts with a target molecule. They are produced for both probes OH2 and DRY.
Hydrophilic-lipophilic balance	is the ratio between hydrophilic regions measured at -4.0 kcal mol^{-1} and hydrophobic regions measured at -0.8 kcal mol^{-1}. The balance describes which effect dominates in the molecule, or if they are roughly equally balanced.

Amphiphilic moment	is defined as a vector pointing from the center of the hydrophobic domain to the center of the hydrophilic domain. The vector length is proportional to the strength of the amphiphilic moment, and it may determine the ability of a compound to permeate a membrane.
Critical packing parameter	defines a ratio between the hydrophilic and lipophilic part of a molecule. In contrast to the hydrophilic–lipophilic balance, critical packing refers just to molecular shape. It is defined as: volume(lipophilic part)/[(surface(hydrophilic part) · (length of lipophilic part)] Lipophilic and hydrophilic calculations are performed at –0.6 and –3.0 kcal mol^{-1}, respectively. Critical packing is a good parameter to predict molecular packing such as in micelle formation, and may be relevant in solubility studies in which the melting point plays an important role.
Hydrogen bonding parameter	describes the H-bonding capacity of a molecular target, as obtained with a polar probe. The water probe presents an optimal ability to donate and accept hydrogen bonds to and from the target. If a different polar probe is used, the interaction may be less favorable, and in any case the H-bonding parameters will differ depending on the nature of the polar probe used.
Polarizability	is not computed from 3D-molecular fields. Polarizability is an estimate of the average molecular polarizability, calculated according to Miller[b]. This method is based on the structure of the compounds (and not any molecular field) and is therefore independent of the number and type of probes used. The correlation between the experimental molecular polarizability and the polarizability calculated with VolSurf for more than 300 chemicals is very good ($r = 0.99$).
Diffusivity	computed using a modified Stokes–Einstein equation, controls the dispersion of chemical in water fluid.
Molecular weight	is simply computed by summing the atomic weights.
Log P	is computed via a linear equation derived by fitting VolSurf descriptors to experimental data on n-octanol/water partition coefficients.

a Reference [8].
b K.J. Miller, *J. Am. Chem. Soc.* **1990**, *112*, 8533–8542.

A method of estimating molecular diffusivity in water is represented by the Stokes–Einstein equation which describes the theoretical relationship between a chemical's diffusivity in water and its molecular size:

$$D_W = RT / 6\pi\eta r \tag{1}$$

where D_W is the chemical diffusivity in water, R is the Boltzmann constant, T the temperature, η the solution viscosity and r is the solute hydrodynamic radius. It shows that it is inversely proportional to solvent viscosity and molecular hydrodynamic radius.

A similar equation is used in VolSurf to estimate chemical diffusivity. However in VolSurf the hydrodynamic radius is precisely estimated from the molecular shape (globularity descriptor).

The so-called "best volumes" are new descriptors identifying the three largest volumes of hydrophilic (probe OH2) and hydrophobic (probe DRY) regions derived by individual chemical groups. Figure 8.2 represents two molecules with the same hydrophobic total volume and different "best volumes" (identified with different colors). The best volume descriptors represent the largest (hydrophibic or hydrophilic) regions around a molecule. Instead of collapsing them into an unique descriptor giving their overall sum, the single region volumes are estimated and stored.

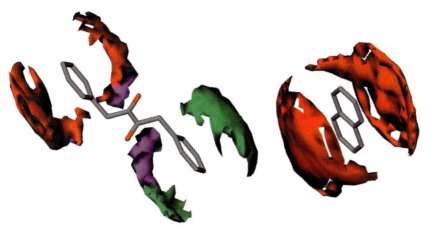

Figure 8.2. Two molecules with an identical total hydrophobic volume (60–Å^3 at –0.6 kcal mol^{-1}), but different "best volumes" identified by different colours. The molecule on the left exhibits three "best volumes" of 25 Å^3 (red), 25 Å^3 (green) and 10 Å^3 (purple), while naphthalene represents a unique group which has only the first "best volume", equal to the total hydrophobic volume, and zero for the second and third ones.

8.3
Application Examples

In the following we describe successful applications of the updated VolSurf software in modeling physicochemical and pharmacokinetic drug properties, comprising aqueous solubility, octanol/water partition coefficients, volume of distribution, and metabolic stability.

8.3.1
Aqueous Solubility

Aqueous solubility [11,12] is an important characteristic of drug candidates; it can influence not only drug delivery, but also metabolism and pharmacodynamic properties such as receptor affinity. In synthetic chemistry, low solubility can be problematic for homogeneous reactions, and in preclinical studies low solubility may produce erroneous experimental determinations or precipitation.

Solubility is extremely difficult to calculate. Dozens of methods exist, but none is reliable enough to be used in the entire chemical diversity space populated by infinite drug candidates. Experimental solubility errors are relatively high and frequent. Moreover, solubility can change dramatically with the purity of the compounds, stability, and time. Solubility of liquid substances differs from that of solid phase compounds. Solubility is thermodynamically affected by crystal packing, influencing the process of crystal lattice disruption and hence polymorphism, amorphous solid compounds lead to imprecise experimental measures. Finally, publically available databases of solubility values contain a lot of errors.

Does the above described complexity allow one to reliably predict thermodynamic solubility for drug candidates? While we think that this is impossible at present, we believe it is realistic to make compound rankings for solubility behavior, provided that the model used is trained appropriately. VolSurf offers a solubility model developed with controlled literature data and in-house solubility data. Although the solubility error in the prediction phase can be evaluated in 0.7 log units (not suitable to rank the solubility of very similar compounds), the model can still be valid to filter compounds with calculated solubility below a certain threshold.

A training set of 1028 diverse chemical structures was used to build a quantitative model for thermodynamic solubility. The structures were extracted from the literature and the dataset was further completed using solubility data produced in-house. Solubility values are given as $-\log[\text{mol L}^{-1}]$ at 25 °C. A three-component PLS model was used to correlate the chemical structures and solubility values. Statistics give an $r^2 = 0.74$, a $q^2 = 0.73$, and a SDEC (standard deviation of the error of calculation) [13] value = 0.89. Figure 8.3 shows the plot of calculated versus experimental log$S$ values for the 1028 training set molecules used for library model building (grey dots). The object pattern nicely proves that a differentiation between very low, low, medium, high, and very high solubility compounds is possible; more quantitative predictions, however, will be difficult to achieve.

The average error in the external prediction is about ± 0.7 log unit. While this range is not suitable for the prediction of solubility values of external compounds, it is still sufficient to rank compounds in different categories and to use for the filtering of compounds in virtual databases. Overall, it seemed unlikely that this model could be improved upon, and any attempt made to do so resulted in dangerous overfitting. Many factors can play a role in solubility, and most of these are virtually imposssible to control.

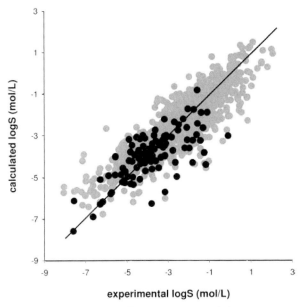

Figure 8.3. Plot of calculated versus experimental –logS values for the 1028 training set molecules used for building the VolSurf solubility library model (grey dots). Projections of the predictions for the 105 compounds of the test set are shown as black dots.

On the basis of the above considerations we used a test set of 105 compounds for external validation of the VolSurf model. Test set compounds are listed together with their experimental and calculated aqueous solubility values in Table 8.2. Projection of the test set predictions into the VolSurf training set model is documented in Fig. 8.3; the SDEP (standard deviation of the error of prediction) value amounts to 0.99. The black dots nicely prove that the majority of the 105 test set structures were well predicted.

Table 8.2 Predicted solubilities for test set compounds ($n = 105$).

Compound	exp. logS	pred. logS	Compound	exp. logS	pred. logS
11-hydroxyprogesterone	−3.82[a]	−3.82	ethinylestradiol	−3.95[b]	−3.99
betamethasone	−3.77[a]	−3.41	famotidine	−2.49[b]	−2.48
betamethasone-17-valer.	−4.71[a]	−4.99	felodipine	−5.89[b]	−4.90
cortisone	−3.27[a]	−3.23	flurbiprofen	−4.36[b]	−3.16
cortisone acetate	−4.21[a]	−3.88	fluvastatin	−3.83[b]	−4.23
dexamethasone	−3.59[a]	−3.27	furosemide	−4.75[b]	−3.47
hydrocortisone	−3.11[a]	−3.40	glyburide	−4.82[b]	−5.31
hydrocortisone acetate	−4.34[a]	−3.94	griseofulvin	−4.83[b]	−3.79
prednisolone	−3.18[a]	−3.04	halcion	−4.10[b]	−4.68
prednisolone acetate	−4.37[a]	−3.90	hydrochlorothiazide	−2.70[b]	−2.35
progesterone	−4.42[a]	−4.09	hydrocortisone	−3.09[b]	−3.09
testosterone	−4.08[a]	−3.65	ibuprofen	−3.62[b]	−2.74
triamcinolone	−3.68[a]	−3.00	ibutilide	−1.81[b]	−4.29
triamcinolone aceto	−4.31[a]	−4.07	imipramine	−4.52[b]	−4.52
triamcinolone diac	−4.13[a]	−4.31	indomethacin	−5.20[b]	−4.55
decadron	−4.90[a]	−4.20	ketoconazol	−3.80[b]	−6.26
corticosterone	−3.24[a]	−3.57	ketoprofen	−3.43[b]	−3.43
deoxycorticosterone acet.	−3.45[a]	−4.54	labetalol	−3.41[b]	−3.42
deoxycorticosterone	−4.63[a]	−3.92	linezolid	−2.07[b]	−3.22
cholesterol	−5.29[a]	−4.86	melengestrol acetate	−5.57[b]	−4.87
cholic acid	−3.16[a]	−3.48	methotrexate	−4.10[b]	−4.06
deydrocholic acid	−3.35[a]	−4.04	metolazone	−4.33[b]	−4.10
glycholic acid	−3.15[a]	−3.96	metoprolol	−1.43[b]	−3.25
ovabagenin	−3.10[a]	−2.99	miconazole	−5.79[b]	−5.79
estrone	−3.95[a]	−3.67	minoxidil	−2.04[b]	−1.72
fludrocortisone	−3.43[a]	−3.46	nadolol	−1.57[b]	−2.42
fluocortolone	−3.30[a]	−3.52	naproxen	−4.22[b]	−2.99
triamcinolone hexa	−5.12[a]	−5.25	omeprazole	−3.42[b]	−3.41
acebutolol	−2.20[b]	−3.60	paclitaxel	−6.63[b]	−6.89

Table 8.2 Continued.

Compound	exp. logS	pred. logS	Compound	exp. logS	pred. logS
aciclovir	−2.27 [b]	−2.20	phenazopyridine	−4.53 [b]	−3.08
almokalant	−1.17 [b]	−4.39	phenytoin	−4.12 [b]	−3.13
alprenolol	−2.83 [b]	−3.08	pindolol	−3.88 [b]	−3.73
alprostadil	−3.67 [b]	−3.66	prazosin	−5.08 [b]	−4.71
amiloride	−3.18 [b]	−2.22	primaquine	−2.52 [b]	−2.75
amitryptilin	−5.14 [b]	−5.14	probenecid	−4.90 [b]	−3.19
amoxicillin	−2.09 [b]	−2.62	promethazine	−4.34 [b]	−4.34
aspirin	−1.75 [b]	−1.74	propoxyphene	−4.96 [b]	−4.96
atenolol	−1.30 [b]	−2.63	propranolol	−3.92 [b]	−3.92
atropine	−1.80 [b]	−3.24	quinine	−2.77 [b]	−3.73
benzoic acid	−1.58 [b]	−0.82	tamoxifen	−7.55 [b]	−6.12
benzydamine	−3.78 [b]	−4.60	terfenadine	−6.17 [b]	−6.17
buprenorphine	−4.37 [b]	−5.08	testosterone	−4.20 [b]	−3.52
chlorpheniramine	−2.65 [b]	−4.27	theophylline	−1.39 [b]	−1.89
chlorpromazine	−5.22 [b]	−5.22	tipranavir	−6.30 [b]	−6.29
cimetidine	−1.62 [b]	−2.97	tirilazad	−7.59 [b]	−7.58
ciprofloxacin	−3.79 [b]	−3.58	tolterodine	−2.58 [b]	−4.96
clonidine	−0.10 [b]	−3.02	trimipramine	−6.29 [b]	−5.09
delavirdine	−5.74 [b]	−5.72	trovafloxacin	−4.43 [b]	−3.72
desipramine	−3.76 [b]	−3.98	verapamil	−4.69 [b]	−5.34
diclofenac	−5.56 [b]	−4.00	warfarin	−4.74 [b]	−4.23
enalapril	−1.25 [b]	−3.83	xanax	−3.60 [b]	−4.22
eperezolid	−1.97 [b]	−3.61	zidovudine	−1.10 [b]	−1.88
erythromycin	−3.15 [b]	−5.71			

Experimental and predicted solubilities are given in mol L^{-1}.
a The Merck Index, 12th edition.
b C. A. S. Bergström et al., *J. Chem. Inf. Comput. Sci.* **2004**, *44*, 1477–1488.

8.3.2
Octanol/Water Partition Coefficients

The importance of lipophilicity as a descriptive parameter in biostudies [14–16] is nowadays acknowledged by its frequent use in an increasing number of research fields. Lipophilicity is an important factor affecting the distribution and fate of drug molecules. Increased lipophilicity has been shown to correlate with poorer aqueous solubility, increased storage in tissues, more rapid metabolism and elimination, increased rate of skin penetration, increased plasma protein binding, and faster rate of onset of action, to mention a few. An emerging field of application is in combinatorial chemistry. In the design of compound libraries, lipophilicity data can be used as estimates for oral absorption as an important contribution to bioavailability. Consequently, log$P$ is included as a parameter in the well-known "rule of five" work of Lipinski et al. [17], dedicated to define the drug-likeness of compounds.

Log$P$ has been introduced as an additional descriptor in the new release of VolSurf. A training set of 7871 diverse chemical structures was used to build a linear equation to calculate the log$P$ values by fitting the structures with the other VolSurf descriptors. Using a five-component PLS regression, statistics give an $r^2 = 0.82$, a $q^2 = 0.82$, and a SDEC [13] value = 0.74. The structures and data stem from Hansch et al. [18].

A test set of 330 compounds, obtained from different literature sources, was used for external validation of the VolSurf equation. Test set compounds are reported in Fig. 8.4 and listed together with their experimental and calculated log$P$ values in Table 8.3; the SDEP value amounts to 0.81; this closely resembles

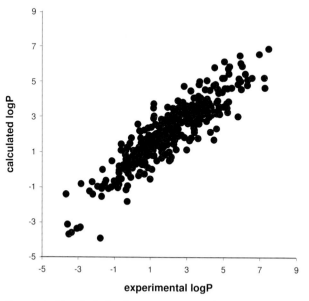

Figure 8.4. Plot of calculated versus experimental log$P$ values for the 330 test set molecules.

the experimental variability and underlines the quality of predictions. Finally, it should be realized that the use of VolSurf for calculating log$P$ circumvents any fragmentation procedure, which often produces ambiguities. Taken together, VolSurf-based log$P$ calculation is fast, reliable and easy to compute.

Table 8.3 Prediction of experimental log$P$ for literature test set compounds (n = 330).

Compound	log$P$exp[a]	log$P$calc	Compound	log$P$exp[a]	log$P$calc
17-α-hydroxyprogesterone	3.15	2.81	mebendazole	3.06	2.65
17-methyltestosterone	3.74	3.16	medazepam	4.68	4.63
abecarnil	5.09	4.66	medroxyprogesterone acetate	4.21	3.85
acadesine	−2.90	−3.32	medroxyprogesterone	3.67	2.95
acenocoumarol	2.53	3.52	mefruside	1.47	1.77
acetazolamide	−0.77	−1.02	megestrol acetate	3.78	4.04
albendazole	3.66	2.64	melatonin	0.88	2.40
alfuzosin	1.32	2.07	meloxicam	2.87	1.85
allopurinol	0.45	−0.15	mephenesin	0.86	0.81
alprazolam	2.78	3.87	mephenytoin	1.74	1.54
altretamine	1.58	2.33	meprobamate	0.28	0.75
aminoglutethimide	0.67	1.37	mercaptopurine	0.72	0.41
amrinone	−0.59	1.46	mesterolone	4.07	2.79
anisindione	2.70	2.70	mestranol	4.06	3.62
aprobarbital	1.10	0.03	metenolon	3.74	2.89
ascorbic acid	−2.17	−1.40	methaqualone	2.48	2.91
atovaquone	5.86	5.22	methimazole	−0.34	1.11
azathioprine	0.25	0.87	methocarbamol	−0.05	0.70
azaurine	−1.74	−3.91	methohexital	1.81	0.85
azintamide	1.40	2.08	methylphenobarbital	1.85	0.91
beclamide	1.29	2.61	methylprednisolone	1.64	1.01
beclometasone	1.83	1.89	methyproscillaridin	2.34	1.87
bemetizide	1.94	1.75	methylthiouracil	1.21	0.81
bendroflumethiazid	1.69	2.67	metolazone	2.02	1.89
benorylate	2.22	2.71	metronidazole	−0.70	−0.17
benzarone	4.64	3.87	miconazole	5.96	6.09
benzbromarone	5.95	5.07	midazolam	3.70	5.05

Table 8.3 Continued.

Compound	logPexp[a]	logPcalc	Compound	logPexp[a]	logPcalc
benziodarone	6.33	5.16	mifepristone	4.10	5.11
benznidazole	0.48	1.82	milrinone	−0.04	1.82
betamethasone	1.49	1.50	misoprostol	3.07	2.81
bromazepam	2.02	3.10	mitotane	5.91	6.54
brotizolam	2.70	3.67	mofebutazone	4.49	1.72
budesonide	2.29	1.46	nabilone	6.57	5.26
bunazosin	2.97	2.90	nabumetone	2.83	3.56
busulfan	−0.59	1.12	niacinamide	−0.21	0.16
butizide	1.58	1.28	niclosamide	4.32	3.67
caffeine	−0.06	0.19	nicorandil	0.29	0.56
calcifediol	7.27	4.70	nifedipine	2.35	3.31
calcipotriol	4.21	3.02	nifurtimox	0.08	1.10
calcitriol	5.18	3.81	nilvadipine	2.26	3.20
calusterone	4.26	3.38	nimodipine	3.09	3.59
camazepam	4.77	3.61	nisoldipine	3.81	4.05
carbamazepine	1.98	2.57	nitrazepam	2.63	2.80
carbimazole	3.00	1.15	nitrendipine	2.43	3.33
carisoprodol	1.67	1.87	nitrofurantoin	−0.47	0.28
carmustine	1.32	1.39	nitroglycerin	1.02	1.36
cefetametpivoxil	2.33	1.68	nivaldipine	2.26	3.12
cefpodoxim	0.78	1.16	nordazepam	3.33	3.68
cefuroxime	0.33	0.73	norethindrone acetate	3.34	3.59
chloralhydrate	0.72	0.27	norethindrone	2.39	2.45
chloramphenicol	0.69	0.67	novobiocin	3.84	3.29
chlormadinone acetate	3.64	4.22	odansetron	2.64	3.64
chlormezanone	1.55	1.57	omeprazole	2.53	2.34
chlorothiazide	−0.31	−0.04	ornidazole	−0.33	0.59
chlorthalidone	0.32	1.16	oxandrolone	3.90	3.19
chlorzoxazone	1.87	1.94	oxazepam	3.45	2.61
cicletanine	3.09	3.47	oxcarbazepine	1.21	1.48
clioquinol	3.70	3.90	oxiconazole	6.03	5.93
clobazam	2.44	2.86	oxyphenbutazone	2.50	3.09

Table 8.3 Continued.

Compound	logPexp[a]	logPcalc	Compound	logPexp[a]	logPcalc
clofazimine	7.47	6.93	pantoprazole	1.87	2.28
clofibrate	3.68	3.75	papaverine	3.22	4.53
clomethiazole	1.22	1.97	paracetamol	0.49	1.53
clonazepam	2.70	3.19	paraldehyde	1.00	−0.52
clopamide	2.35	1.75	paramethasone	1.04	1.25
cloprednol	1.07	1.70	pentaerythritol tetranitrate	1.69	2.20
clotiazepam	3.36	3.67	pentobarbital	2.11	0.59
clotrimazole	5.05	6.19	pentoxyfylline	0.10	0.81
colchicine	0.32	2.62	phenacemide	0.87	0.12
corticosterone	1.16	1.49	phenacetin	1.77	2.15
cortisol	0.54	1.02	phenazone	0.41	2.23
cortisone acetate	0.47	2.19	phenobarbital	1.37	0.51
cortisone	0.14	1.05	phenolphthalein	3.67	4.88
coumarin	1.41	2.10	phenprocoumon	4.75	4.40
cyclandelate	4.64	3.22	phensuximide	0.96	1.12
cyclophosphamide	0.80	1.64	phenylbutazone	3.17	3.20
cyproterone acetate	3.39	4.11	phenylethylmalonamide	2.25	3.52
cyproterone	2.85	3.25	phenytoin	2.09	2.21
cytarabine	−3.05	−3.37	pimobendan	5.00	3.61
danazol	3.54	3.66	piracetam	−1.49	−0.78
dantrolen	1.63	2.31	piroxicam	2.70	1.99
dapsone	1.07	1.72	pivmecillinam	5.80	3.06
deflazacort	2.27	2.12	polythiazide	2.29	2.22
desogestrel	5.29	3.90	prazepam	4.26	4.56
dexamethasone	1.49	1.53	praziquantel	3.43	2.62
diazepam	3.29	3.72	prazosin	2.45	3.13
diazoxide	1.20	1.64	prednimustin	5.28	5.78
dichlorphenamide	0.22	0.33	prednisolone	1.12	0.83
didanosine	−1.90	−0.96	prednisone	0.72	1.11
diethylstilbestrol	4.96	5.24	prednylidene	1.16	1.12
diloxanide	1.62	2.90	primidone	1.74	1.31
dipyridamole	2.04	1.02	progabide	2.51	3.96

Table 8.3 Continued.

Compound	logPexp[a]	logPcalc	Compound	logPexp[a]	logPcalc
disulfiram	3.88	3.84	progesterone	3.78	3.68
doxazosin	4.06	3.39	propofol	4.33	3.57
enprofylline	0.36	−0.34	propylthiouracil	2.27	1.49
estradiol	3.78	3.28	propyphenazone	5.23	3.29
estradiol-valerate	6.32	4.84	proquazone	3.65	3.82
estriol	2.55	1.93	proscillaridin	1.68	1.74
estrone	3.38	3.60	protionamide	2.26	2.33
ethacridine	3.69	2.89	protirelin	−2.83	−0.81
ethanol	−0.24	−0.92	proxyphylline	−0.58	−0.51
ethinylestradiol	3.47	3.43	psicofuranine	−3.58	−3.13
ethionamide	1.73	2.08	pyrantel	4.63	2.36
ethosuximide	−0.33	0.34	pyrazinamide	−0.71	−0.50
etofenamate	4.36	3.15	pyridoxine	−0.70	−0.04
etofibrate	3.75	4.13	pyrimethamine	3.38	2.41
etofylline	−0.89	−0.77	quazepam	4.52	5.85
etoposide	−1.10	0.50	ramosetron	1.98	3.12
etretinate	6.97	6.62	retinol	6.20	5.48
famciclovir	−0.36	0.85	ribavirin	−3.36	−3.60
fampridine	0.50	0.63	riboflavin	−0.25	−1.81
felbamate	−0.29	0.70	rifaximin	5.46	4.69
felodipine	4.00	4.59	riluzole	3.41	2.15
fenofibrate	5.23	4.65	rolipram	1.22	3.08
fenoximone	2.26	1.82	rutoside	−3.66	−1.40
finasteride	3.01	3.66	secnidazole	−0.39	0.04
floctafenine	3.85	3.44	secobarbital	2.16	0.74
fluconazole	−0.11	1.50	silymarin	1.18	2.95
flucytosine	−1.43	−0.61	simvastatin	5.21	3.54
fludrocortisone	0.38	1.28	spironolactone	3.19	3.51
flumazenil	1.06	2.45	stavudine	−1.14	−1.00
flunisolide	1.15	1.51	stiripentol	2.40	2.70
flunitrazepam	2.11	2.95	strophanthin	−2.35	−1.23
fluocortolone	1.66	1.86	sulfinpyrazone	1.44	2.54

Table 8.3 Continued.

Compound	logPexp[a]	logPcalc	Compound	logPexp[a]	logPcalc
fluorouracil	–0.97	–0.64	sulfisoxazole	0.28	1.13
flupirtine	3.23	2.85	sultiam	0.03	0.56
flutamide	3.19	3.05	tacrine	3.45	3.02
fluticasone	1.45	2.51	temazepam	3.71	3.21
fructose	–3.50	–3.69	tenidap	3.41	2.28
ftorafur	–0.21	–0.21	tenoxicam	2.42	1.16
gestodene	2.44	2.88	terazosin	2.58	2.00
gestrinone	2.04	2.99	terizidone	0.55	0.30
griseofulvin	1.76	2.73	testolactone	3.79	3.13
halazepam	4.58	4.89	testosterone	3.22	2.69
hexobarbital	1.63	0.67	testosterone propionate	4.69	3.95
hydrochlorothiazide	–0.40	0.17	tetrahydrocannabinol	7.24	5.27
hydroflumethiazide	–0.25	1.29	tetrazepam	3.69	3.63
hydroxyprogesterone caproate	5.80	4.90	tetroxoprim	0.55	1.56
hymecromone	2.11	2.32	theophylline	–0.06	–0.43
ifosfamide	0.92	1.97	thiamazole	0.02	0.10
indapamide	4.29	2.16	thiamphenicol	–0.70	0.21
iodoquinol	4.11	4.07	tiabendazol	2.35	1.70
ipriflavone-yambolap	4.17	3.59	tinidazole	0.03	0.79
isocarboxazid	0.97	1.60	tolnaftate	5.34	5.91
isoniazid	–0.71	–0.84	topiramate	0.68	–0.24
isosorbide-2-mononitrate	–1.67	–0.97	trapidil	1.94	1.44
isosorbide-5-mononitrate	–1.67	–1.07	treosulfan	–2.20	–0.70
isosorbide-dinitrate	–1.62	0.01	triamcinolone	–0.24	0.44
isoxicam	2.40	1.89	triamterene	2.06	0.52
isradipine	3.14	3.34	triazolam	2.85	4.79
kebuzone	0.68	2.27	tribenoside	4.94	4.45
ketazolam	3.71	3.48	trichlormethiazide	0.85	1.11
ketonazole	4.48	4.83	trimazosin	4.51	3.30
lacidipine	5.31	5.50	trimethadione	0.08	0.33
lamivudine	–1.67	–1.53	trimethoprim	0.80	1.78
lamotrigine	3.24	1.82	trimetrexate	1.81	2.87

Table 8.3 Continued.

Compound	logPexp[a]	logPcalc	Compound	logPexp[a]	logPcalc
lansoprazole	3.07	3.05	trofosfamide	1.18	3.53
letrozole	1.20	3.77	troglitazone	5.99	4.61
levamisole	3.61	1.80	velnacrine	1.82	1.53
levonorgestrel	2.92	2.79	vesnarinone	1.95	3.59
loratadine	5.05	5.26	warfarin	2.79	3.73
lorazepam	3.52	3.31	xipamide	2.19	2.48
lormetazepam	3.77	3.56	xylose	−2.86	−3.29
lornoxicam	3.15	1.68	zalcitabine	−1.29	−1.08
lovastatin	4.81	3.19	zidovudine	−0.33	−1.14
lynoestrenol	4.73	3.41	zolpidem	2.82	3.78

8.3.3
Volume of Distribution (VD)

The half-life of a drug is a major contributor to the dosing regimen, and it is a function of the clearance and apparent volume of distribution (VD), each of which can be predicted and combined to predict the half-life. Drugs with short half-lives are more likely to be required to be administered more frequently than those with long half-lives. Much attention has been focused on the prediction of human half-life. Good success is attained if the two major components of half-life, clearance and VD, are predicted separately and combined to generate a half-life prediction.

Volume of distribution represents a complex combination of multiple chemical and biochemical phenomena. It is a measure of the relative partitioning of a drug between plasma (the central compartment) and the tissues. Thus, the VD term considers all of the tissues as a single homogeneous compartment.

The VD of a drug accounts for the total dose administration based on the observed plasma concentration. VD is indicative of the extent of distribution of a drug. The larger the VD, the greater is the extent of the distribution. The plasma volume of an average adult is approximately 3 L. Thus, an apparent VD larger than the plasma compartment (i.e. > 3 L) indicates that the drug is also present in tissue or fluid outside the plasma compartment. Although the VD cannot be used to determine the actual site of distribution of a drug in the body, it is of extreme importance in estimating the loading dose necessary to rapidly achieve a desired plasma concentration.

In this section, we describe a quantitative VD model on 118 chemically diverse drugs comprising neutral and basic compounds. VD data were collected from literature by Lombardo et al. [19]. In the vast majority of cases, these data represent VD_{ss} values, i.e. volume of distribution at steady state.

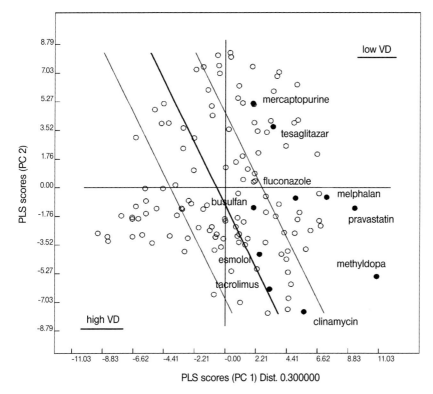

Figure 8.5. The 2D PLS scores plot of the VD VolSurf library model (open grey points) as well as the projection of the predictions for the ten test set compounds (filled points) are shown. The middle bar represents the best discrimination between training set compounds (with high and low VD).

After converting the VD data (L kg^{-1}) into −log[VD] values, PLS discriminant analysis was used to build the statistical model. Two significant latent variables emerged from the PLS discriminant analysis; statistics give an $r^2 = 0.61$, a $q^2 = 0.53$, and a SDEC [13] value = 0.33. Figure 8.5 shows the 2D PLS scores of the VolSurf VD library model (empty dots).

It is extremely difficult to find compounds with experimental VD equivalent to those collected by Lombardo. We could detect only 10 compounds, which were in turn used as test set for external validation of the VolSurf library model. Test set compounds are listed together with their experimental and calculated VD values in Table 8.4; the projection of their predictions is plotted in Figure 8.5 (filled dots); the SDEP value amounts to 0.53.

Table 8.4. Prediction of the volume of distribution (VD) for the test set ($n = 10$).

Compound	$VD_{experimental}$ (L kg^{-1})	$-logVD_{experimental}$	$VD_{predicted}$ (L kg^{-1})	$-logVD_{predicted}$
busulfan	0.84 [a]	0.08	2.15	−0.33
clindamycin	0.79 [b]	0.10	2.13	−0.33
esmolol	1.19 [c]	−0.08	2.68	−0.43
fluconazole	0.63 [d]	0.20	1.06	−0.02
melphalan	0.50 [e]	0.30	0.64	0.19
mercaptopurine	5.62 [f]	−0.75	1.02	−0.01
methyldopa	0.46 [g]	0.34	0.54	0.26
pravastatin	0.46 [h]	0.34	0.46	0.34
tacrolimus	1.26 [i]	−0.10	3.02	−0.48
tesaglitazar	0.13 [j]	0.89	0.85	0.07

a Cremers, S. et al. *Br. J. Clin. Pharmacol.* **2002**, *53*, 386–389.
b Gatti, G. et al. *Antimicr. Ag. Chemother.* **1993**, *37*, 1137–1143.
c Sum, C.Y. et al. *Clin. Pharmacol. Ther.* **1983**, *34*, 427–434.
d Cutler, R.E. et al. *Clin. Pharmacol. Ther.* **1978**, *24*, 333–342.
e Physician Desk Reference (Alkeran for injection, Celgene).
f Van Os, E.C. et al. *Gut* **1996**, *39*, 63–68.
g Skerjanec, A. et al. *J. Clin. Pharmacol.* **1995**, *35*, 275–280.
h Singhvi, S. M. et al. *Br. J. Clin. Pharmacol.* **1990**, *29*, 239–243.
i Mancinelli, L.M. *Clin. Pharmacol. Ther.* **2001**, *69*, 24–31.
j Ericsson et al. *Drug. Met. Disp.* **2004**, *32*, 923–929.

8.3.4
Metabolic Stability

Metabolic trasformations tend to reduce the bioavailability of compouds and, in turn, their pharmacological profile. The family of human P450 cytochromes comprises many different enzymes including 3A4, 3A5, 2C9, 2C19, 2D6, etc., among which the 3A4 and 2C9 subtypes are involved in the great majority of the metabolic transformations of drugs.

Metabolic stability in human CYP3A4 cDNA-expressed microsomal preparation offers a suitable approach to predict the metabolic stability of external compounds. From a dataset ($n = 1507$) from Pharmacia Corporation, each compound was incubated at a fixed concentration for 60 min with a fixed concentration of protein at 37 °C. The reaction was stopped by adding acetonitrile to the solution and, after centrifugation to remove the protein, the supernatant was analyzed using LC/MS and MS. Compounds with a final concentration > 90% of the corresponding control sample were defined as stable, whereas compounds with final concentrations below 20% of the control were defined as unstable. The solubility of stable compounds was used as a primary filter. All the compounds with a solubility lower than 10 µmol L^{-1} were removed from the analysis. Insoluble compounds are always metabolically stable.

The above described dataset was used to build a quantitative model for metabolic stability. Two significant latent variables were extracted; statistics give an $r^2 = 0.44$, a $q^2 = 0.43$, and a SDEC (13) value = 0.74. Figure 8.6 shows the plot of calculated versus experimental metabolic stability for the 1507 training set molecules used for the VolSurf Metabolic Stability model building (open grey dots).

The model can be used to evaluate the metabolic stability from the 3D structure of drug candidates prior to experimental measurements. Thus, we have used a test set of 1346 compounds from Johnson & Johnson [20].

Figure 8.6 shows the projection of 1346 compounds from Johnson & Johnson on the VolSurf metabolic stability model. The projected compounds are color-coded according to their percentage experimental metabolic stability (%MS): the red color defines compounds with %MS > 95, while the blue color defines compounds with %MS < 40. The figure shows that the great majority of projected compounds with %MS > 95 are predicted to be of medium or high stability for the metabolic activity of the CYP3A4, while compounds with low %MS < 40 are predicted to be unstable. The presence of outliers may be explained by the fact that the experimental %MS of the test set compounds is obtained from the activity of all CYP family enzymes (2C9, 3A4, 2D6 etc.), while the model uses only 3A4 mediated information.

8.4
Conclusion

The complex and often uncertain outcome of drug discovery and development processes requires the simultaneous optimization of several properties. It has now long been recognized that favorable potency and selectivity characteristics are not the sole hallmarks of a successful drug discovery program, nor is the safety profile considered to be the only hurdle to be overcome, although it is of paramount importance. The ability to prospectively predict the pharmacokinetics of new chemical entities in humans is a powerful means by which one can select for further development only those compounds with the potential to be successful therapeutic agents.

In the present chapter, we have used an updated version of the VolSurf procedure to exemplify its validity for predicting physicochemical (solubility and log P) and pharmacokinetic properties (volume of distribution and metabolic stability) of drug molecules.

However, the VolSurf procedure is an evolving tool. New MIF-derived descriptors are planned and will soon be developed. Moreover, the descriptor space and the chemometrical (the model) space will be much more integrated. After draw-

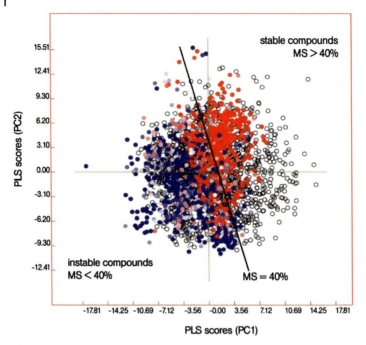

Figure 8.6. Projection of 1346 compounds from Johnson & Johnson on the 2D PLS scores plot of the VolSurf metabolic stability model (open grey points). The black line discriminates between unstable and stable compounds in the metabolic stability model: projected compounds on the left of the black line are predicted as unstable (%MS < 40), while projected compounds on the right are predicted as stable (%MS > 40%). The projected compounds are color-coded according to their percentage experimental metabolic stability (%MS): red color for compounds with %MS > 95; blue color for compounds with %MS < 40.

ing a 2D structure, the user will add or remove groups and see instantly its projection on the ADME library model plot. So it will be simple to add a series of chemical groups to a drug structure to follow its movement on the model property-plots in real time. The gap between chemists, 3D molecular representation and ADME propeties has never been so small.

References

1. H. van de Waterbeemd, H. Lennernäs, P. Artursson (eds.), *Drug Bioavailability. Estimation of Solubility, Permeability and Absorption, Methods and Principles in Medicinal Chemistry*, vol. 18, Wiley-VCH, Weinheim, **2003**.
2. F. Lombardo, E. Gifford, M.Y. Shalaeva, In Silico ADME Prediction: Data, Models, Facts and Myths, *Mini-Rev. Med. Chem.* **2003**, *3*, 861–875.
3. H. van de Waterbeemd, E. Gifford, ADMET *in silico* Modelling: Towards Prediction Paradise? *Nature Rev. Drug Discov.* **2003**, *2*, 192–204.
4. R. Todeschini, V. Consonni, *Handbook of Molecular Descriptors; Methods and Principles in Medicinal Chemistry*, vol. 11, VCH, Weinheim, **2000**.
5. G. Cruciani, P. Crivori, P.-A. Carrupt, B. Testa, Molecular fields in quantitative structure-permeation relationships: the VolSurf approach, *J. Mol. Struct.: Theo. Chem.* **2000**, *503*, 17–30.
6. P. Crivori, G. Cruciani, P.-A. Carrupt, B. Testa, Predicting blood-brain barrier permeation from three-dimensional molecular structure, *J. Med. Chem.* **2000**, *43*, 2204–2216.
7. G. Cruciani, M. Meniconi, E. Carosati, I. Zamora, R. Mannhold. VOLSURF: A Tool for Drug ADME-properties Prediction, in *Drug Bioavailability. Estimation of Solubility, Permeability and Absorption, Methods and Principles in Medicinal Chemistry*, vol. 18, (eds. H. van de Waterbeemd, H. Lennernäs, P. Artursson), Wiley-VCH, Weinheim, pp. 406–419, **2003**.
8. P. Goodford. A computational procedure for determining energetically favourable binding sites on biologically important macromolecules, *J. Med. Chem.* **1985**, *28*, 849–856; E. Carosati, S. Sciabola, G. Cruciani, Hydrogen Bonding Interactions of Covalently-Bonded Fluorine Atoms: From Crystallographic Data to a New Angular Function in the GRID Force Field, *J. Med. Chem.* **2004**, *47*, 5114–5125.
9. S. Wold, M. Sjöström, in *Chemometrics: Theory and Application*, B.R. Kowalski (ed.) pp. 243–282, ACS Symposium Series, Washington DC, **1977**.
10. S. Wold, C. Albano, W.J. Dunn III, U. Edlund, K. Esbensen, P. Geladi, S. Hellberg, E. Johansson, W. Lindberg, M. Sjöström, in *Chemometrics*, B.R. Kowalski (ed.), pp. 17–94. Reidel, Dordrecht, **1984**.
11. S.H. Yalkowsky, S.C. Valvani. Solubility and partitioning I: Solubility of nonelectrolytes in water. *J. Pharm. Sci.* **1980**, *69*, 912–922.
12. N. Jain, S.H. Yalkowsky, Estimation of the aqueous solubility I: application to organic nonelectrolytes, *J. Pharm. Sci.* **2001**, *90*, 234–252.
13. G. Cruciani, M. Baroni, S. Clementi, G. Costantino, D. Riganelli, B. Skagerberg, Predictive Ability of Regression Models. Part I: Standard Deviation of Prediction Errors (SDEP), *J. Chemomet.* **1992**, *6*, 335–346.
14. J. Sangster, *Octanol-Water Partition Coefficients: Fundamentals and Physical Chemistry*, Wiley Series in Solution Chemistry, vol. 2, Wiley, New York, **1997**.
15. V. Pliska, B. Testa, H. van de Waterbeemd (eds.), *Lipophilicity in Drug Action and Toxicology. Methods and Principles in Medicinal Chemistry*, vol. 4, VCH, Weinheim, **1996**.
16. P.-A. Carrupt, B. Testa, P. Gaillard, Computational Approaches to Lipophilicity: Methods and Applications, in *Reviews in Computational Chemistry*, vol. 11, K. B. Lipkowitz, D. B. Boyd (eds.), pp. 241–315, Wiley-VCH, Weinheim, **1997**.
17. C.A. Lipinski, F. Lombardo, B.W. Dominy, P.J. Feeney, Experimental and computational approaches to estimate solubility and permeability in drug discovery and development settings, *Adv. Drug Delivery Rev.* **1997**, *23*, 3–25.
18. C. Hansch, A. Leo, D. Hoekman, *Exploring QSAR. Hydrophobic, Electronic, and Steric Constants*, ACS Professional Reference Book, ACS, Washington DC, **1995**.

19 F. Lombardo, R.S. Obach, M.Y. Shalaeva and F. Gao, Prediction of Human Volume of Distribution Vallues for Neutral and Basic Drugs. 2. Extended Data Set and Leave-Class-Out Statistics, *J. Med. Chem.* **2004**, *47*, 1242–1250.

20 Experimental data from K. Ethyrajuly, C. Makie, B. De Boeck, Johnson & Johnson, Beerse, Belgium.

9
Molecular Interaction Fields in ADME and Safety

Giovanni Cianchetta, Yi Li, Robert Singleton, Meng Zhang, Marianne Wildgoose, David Rampe, Jiesheng Kang, and Roy J. Vaz

9.1
Introduction

The GRID [1] force field has been widely used in structure based drug design, in crystallography and 3D QSAR. Moreover, there are several programs utilizing the MIFs produced by GRID that are now being used increasingly. The first, META-SITE, is used to predict the site of metabolism by cytochrome P450 enzymes and is described very well in Chapter 12. The second program is VolSurf which is largely used to describe and predict physical properties and measurements such as permeability etc., as described in Chapter 8. The third program is FLAP which utilizes 3 or 4 point pharmacophores derived from GRID fields which are then used in either database searching or library design. This program has been developed to work on proteins and ligands at the same time and is well described in Chapter 4.

Finally, there is a fourth program which encapsulates GRIND descriptors named ALMOND. ALMOND theory is described in Chapter 6. What we will describe below refers to our involvement in utilizing the GRIND descriptors to model and understand various ADME and safety properties.

Accordingly, we will briefly outline the use of MIFs in three research areas. In absorption, the use of MIFs to understand and model permeability through PAMPA, Caco-2, MDCK cells as well as BBMEC cells has been demonstrated. However, this approach is limited to a passive mechanism.

Outliers in these models, as well in biological assays, are due to several reasons which encompass issues related to transporters in the cells which could be efflux or uptake transporters. We have tried to address the problem using MIFs to understand P-glycoprotein (PGP) efflux, especially as it pertains to substrates.

Secondly, we will describe the use of GRIND descriptors to obtain a general model for inhibition of the K channel in the heart, known as the Human Ether-a-go-go Gene (HERG).

Finally MIFs will be used to explain cytochrome P450 (CYP) 3A4 inhibition. The use of METASITE in understanding and predicting the site of metabolism within a particular molecule has already been described. However, METASITE has not yet been used to predict inhibition for various CYPs. We will describe a model which has been obtained from experimental data (inhibition of the formation of 6β-testosterone formation from testosterone itself).

9.2
GRID and MIFs

Molecular interaction fields are produced by Peter Goodford's GRID software (see Chapter 1). GRID alignment independent descriptors (GRIND) [2] were chosen due to their ability to represent pharmacodynamic properties in such a way that they are no longer dependent upon their positions in the 3D space. The GRIND calculation starts by computing several molecular interaction fields (MIFs) using the GRID program. The GRIND approach aims to extract the information enclosed in the MIFs and compress it into new types of variables whose values are independent of the spatial position of the molecule studied. Most relevant regions are extracted from the MIFs using an optimization algorithm that uses the intensity of the field at a node and the mutual node–node distances between the chosen nodes as a scoring function. A discrete number of categories, each one representing a small rank of distances, are considered.

The innovative autocorrelation algorithm used in ALMOND allows the representation of the descriptors in the original 3D space as a line linking two specific MIF nodes.

GRIND variables are then grouped into blocks representing interactions between couples of nodes generated by the same probe (autocorrelograms) or combination of probes (cross-correlograms). Such variables constitute a matrix of descriptors that can be analyzed using multivariate techniques, such as principal component analysis (PCA) and partial least squares (PLS) regression analysis.

All the calculations were done by means of the program ALMOND 3.2.0. In this version a new kind of descriptor has been added that is able to describe the shape of the molecule using the same GRIND formalism. Shape descriptors are represented in a correlogram-like form where the autocorrelograms describe the distance between certain regions defining the spatial extent of the molecule and the cross-correlograms describe the distance between these regions and other regions representing relevant interactions of the compounds.

9.3
Role of Pgp Efflux in the Absorption

9.3.1
Materials and Methods

9.3.1.1 Dataset

A dataset comprised of 129 molecules, 100 Sanofi-Aventis proprietary compounds and 29 publicly available compounds, was studied in order to obtain a 3D quantitative structure–property relationship (3D-QSPR) model able to identify the structural features that a molecule should possess in order to be recognized as a substrate of Pgp. All the chosen compounds have an efflux ratio in a Caco-2 assay greater than one, which normally implies that the molecules are Pgp substrates.

The degree of inhibition of Pgp activity, calculated from the fluorescence increase of the Calcein-AM assay was used as activity data [3]. The experimental inhibition data are normalized to the value obtained for Cyclosporine A (which is a competitive inhibitor of Pgp [4])so that the value for untreated cells is 0% and 100% is the value for Cyclosporine A. The inhibition value of each substrate was then transformed in the logarithm in order to reduce the residuals for the larger values. The activity range spans from 2.32 to 0.37, covering 1.95 log units.

All the molecules were divided into a training set (109 compounds) and a test set (20 molecules). Activity values of the test set molecules range between 1.86 and 0.59 log units. The test set was chosen in such a way as to fully cover the activity range; the dataset was divided into four classes according to the activity value (2.32–1.63, 1.62–1.23, 1.22–0.84, and 0.83–0.37). Then five molecules for each class were randomly chosen, ensuring that the chosen molecules were similar to the training set.

9.3.1.2 Computational Methods

Molecular modeling and subsequent geometry optimization were performed using the molecular modeling software package SYBYL version 6.9.244 [5]. 3D structures were obtained from smiles notation by means of the Unity [6] program included in the SYBYL package. CONCORD [7] was used to generate a single conformation that was used for the model development and to analyze the test set compounds. Then energy minimization was performed with the standard TRIPOS force field using the Powell method with initial Simplex optimization. Gradient termination was set to 0.05 $kcal\,(mol*A)^{-1}$. When needed, conformational analysis was performed with software MOE 2004.03 release [8], limiting the number of conformers to 50 and used as described. Except for compounds with quaternary nitrogen atoms, structures were considered to be uncharged. The high structural diversity of the molecules that form the data set made it very difficult to find rules for superimposition of the structures; hence, an alignment independent method was required to analyze the dataset.

9.3.1.3 ALMOND Descriptors

940 ALMOND descriptors have been obtained using 4 GRID probes: DRY (which represents hydrophobic interactions), O (sp2 carbonyl oxygen, representing an H bond acceptor), N1 (neutral flat NH like in amide, an H bond donor) and the TIP probe (molecular shape descriptor). The grid spacing was set to 0.5 Å and the smoothing window to 0.8. The number of filtered nodes was set to 100 with 50% of relative weights. 10 groups of variables were produced by ALMOND: 4 autocorrelograms and 6 cross-correlograms.

9.3.2 Results

The structural variance of the dataset was analyzed with principal component analysis (PCA) [9] performed on the complete set of ALMOND descriptors calculated for the compounds which comprised the training and test sets. The first two components explained 35% of the structural variance of the dataset. Figure 9.1 shows that no structural outliers are present in the dataset and that the training and test sets share similar chemical space.

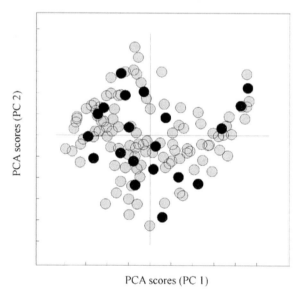

Figure 9.1. PCA score plot derived from the analysis of the GRIND descriptors calculated for the entire training set of 129 compounds. The gray objects represent the compounds that form the training set and the black dots represent the molecules of the test set.

In order to analyze in detail the pharmacophoric aspect of the interaction between the compounds that formed the training set and the protein, the PLS multivariate data analysis correlating the activity with the complete set of variables

(940) was carried out using the algorithm implemented in the ALMOND program. In the first instance, five of the Sanofi-Aventis compounds were strong outliers (data not shown); conformational analysis was performed on these molecules using the software MOE. The final chosen conformation was that which gave the best correlation value in the model. The PLS analysis resulted in a three latent variables model with $r^2=0.82$. The cross validation of the model using LOO yielded q^2 values of 0.72.

A variable selection was applied to reduce the variable number using FFD factorial selection implemented in the ALMOND program using all the default values suggested by the program. The resulting number of active variables decreased from 653 to 576. A new PLS multivariate data analysis was performed yielding a three latent variables model with $r^2=0.83$. The cross validation of the model using LOO yielded q^2 values of 0.75.

The quality and robustness of the obtained model were tested predicting the activity of the test set previously defined. Figure 9.2 shows the plot of the experimental versus calculated biological activities. The entire set could be modeled without any significant outlier behavior, in spite of the fact that structurally different classes of compounds are present in the dataset. This fact confirms a common mechanism of action and consequently the common structural features required.

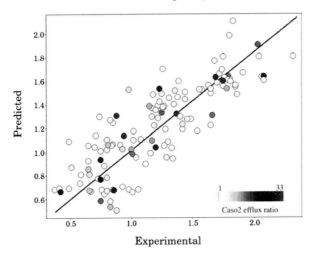

Figure 9.2. The overall quality of the obtained model is demonstrated in the plot of the experimental activities (experimental Y) versus the data calculated with our model (calculated Y). The plot shows that the entire training set could be modeled without any significant outlier behavior. The color of the objects in the chart is based upon the Caco-2 efflux ratio value for the molecule.

9.3.3
Pharmacophoric Model Interpretation

The PLS pseudo-coefficients profile of the third component of the PLS model, highlights the descriptors that have a greater importance in the chemometric model. The most important 3D-pharmacophoric descriptors in the PLS model suggest a common pharmacophore for all the substrates. The activity increases strongly in molecules with a high value of the descriptors: 33-23, 11-33, 13-8, 14-41, 44-43. The descriptors are explained in detail in Table 9.1. The most important descriptors in the PLS model can be arranged to obtain an approximate pharmacophore valid for molecules actively transported by Pgp. The pharmacophore consists of two H-bond acceptor groups, two hydrophobic areas and the size of the molecule that plays a major role in the interaction (Fig. 9.3).

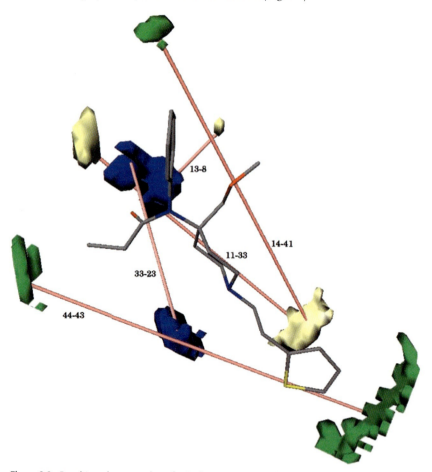

Figure 9.3. Resulting pharmacophore for P-glycoprotein actively transported molecules. The depicted molecule is the analgesic (narcotic) sufentanyl. The colored areas around the molecules are the GRID fields produced by the molecule: yellow for DRY probe, green for TIP probe and blue for N1 probe.

Table 9.1 Structural interpretation of the ALMOND descriptors that are highly correlated to the variance of the experimental data.

33-23	Is related to the presence of two H-bond acceptor atoms placed 11.5 Å apart.
11-33	Are the distances between two hydrophobic regions: they have to be 16.5 Å apart.
13-8	Describes a distance of 4 Å between the hydrophobic region previously described and an H-bond acceptor group.
14-41	Describes a distance of 20.5 Å between a hydrophobic region of the molecule and one of the edges of the same molecule.
44-43	Is mainly related to the size, being the distance of 21.5 Å that is required between two edges of the molecule.

9.3.4 Discussion

Several papers proposing multiple recognition sites for Pgp have been presented in the past. In this work, the pharmacophoric analysis of the dataset shows that the requirements to interact with Pgp are the same for all 129 compounds. Since in our database we have not included known R-site binders and anthracyclines, we cannot say definitively that the pharmacophore found represents one of the binding sites that have been described in the literature. Two of the molecules present in the database, verapamil and dipyridamole, are known to bind in the H-site described by Shapiro and Ling [10]. We also cannot definitely state that the pharmacophore defines any functional site within the transporter. Further work, to try to define the location of the corresponding amino acids in a protein homology model, is in progress.

This work supports the two-step process proposed by Seelig [11]. The strong correlation, highlighted by PLS statistical analysis, between pharmacophoric descriptors and inhibition values suggests that substrate interaction with the protein plays a key role in the efflux process, yielding a model in which diffusion across the membrane (first step) is less important than substrate–protein interaction (second step).

In our hypothesis, Pgp substrates, being prevalently lipophilic, can easily cross the membrane and tend to accumulate in the bilayer. Here they will interact with the protein by means of pharmacophoric recognition. Interaction will trigger a sequence of transformations in the protein (conformational changes, ATP hydrolysis etc.) that have a great impact on the rate of the efflux process. The high concentration that substrates reach in the bilayer can help to explain the broad specificity showed by Pgp. A recent review [12] suggests that binding of substrates to the TMDs initiates the transport cycle by facilitating ATP-dependent closed dimer formation, the first step in the ATP switch model for transport by ABC transporters.

Our hypothesis would be in agreement with the work of Dye et al. [13] since the mouth of the protein may be seen as the ON-site that first interacts with the substrate.

The model would be in accordance even with the suggestion of Homolya et al. [14], the pioneers in the use of Calcein-AM as a Pgp substrate, namely, that Calcein-AM and other fluorescent methyl esters are expelled directly from the cell membrane, before reaching the cytoplasmic phase. Preemptive pumping of Calcein-AM is also in accordance with the theoretical analysis of Stein [15] who showed that the initial rate of substrate accumulation is reduced by pumping only if such pumping is preemptive.

9.4
HERG Inhibition

9.4.1
Materials and Methods

9.4.1.1 Dataset

Our dataset consisted of 882 compounds with experimentally measured IC50 values for HERG channel inhibition. A standard whole-cell patch clamp electrophysiology method was used to record the currents of HERG channels stably expressed in Chinese hamster ovary cells [16]. The number of compounds with zero, one, or two ionizable basic nitrogen atoms was 338, 499, and 45, respectively. Amide, aromatic and aniline nitrogen atoms were considered as nonbasic. Activity data were reported as $-\log_{10}$ of IC50 (pIC50), and ranged from -2.5 to 2.5.

9.4.1.2 Computational Methods

To develop 3D pharmacophore models, the geometries of all molecules were initially generated by Concord and subsequently energy minimized using Tripos' software package SYBYL with the standard TRIPOS force field. MIFs were then calculated using the program GRID to determine energetically favorable interactions between the molecule and a probe group. MIFs obtained from GRID calculations (DRY, carbonyl oxygen, NH amide and TIP probes) were then transformed into alignment independent descriptors (GRIND).

Correlations between the HERG channel inhibition and GRIND descriptors were analyzed using multivariate techniques such as principal component analysis (PCA) and PLS regression analysis.

9.4.2
Results

In order to assess the importance of the presence of a charged nitrogen atom, the dataset was divided into two subsets. The first subset contained 338 compounds without any basic nitrogen atoms while the second set consisted of 544 molecules with one or two basic nitrogen atoms.

9.4.2.1 Nonbasic Nitrogen Subset

For the nonbasic nitrogen subset, the 338 molecules were split into a training set (322 compounds) and a test set (16 molecules). To select the test set, the dataset was divided into three groups according to the activity value of pIC50 (2>0.5; 0.5>–1; –1>–2.5). The test set compounds were randomly chosen from each group to cover the activity range uniformly. Principal component analysis of the complete dataset was performed to analyze the structural variance of both the training and test sets. The PCA scores plot (Fig. 9.4) showed that there was no structural outlier present in the dataset and that the training set and test set shared the same chemical space.

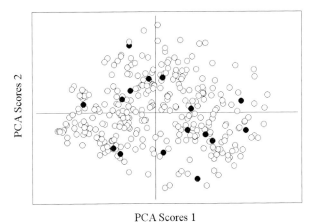

Figure 9.4. 2D plot of PCA showing the descriptor space of training set (open circle) and test set (filled dot) for the HERG data set (neutral molecules).

The PLS multivariate data analysis of the training set was carried out on the descriptors matrix to correlate the complete set of variables with the activity data. From a total of 710 variables, 559 active variables remained after filtering descriptors with no variability by the ALMOND program. The PLS analysis resulted in four latent variables (LVs) with $r^2 = 0.76$. The cross validation of the model using the leave-one-out (LOO) method yielded q^2 values of 0.72. As shown in Table 9.2, the GRIND descriptors 11-36, 44-49, 12-28, 13-42, 14-46, 24-46 and 34-45 were found to correlate with the inhibition activity in terms of high coefficients.

Table 9.2 Salient GRIND descriptors in PLS models.

GRIND descriptor	GRID MIFs	Subset I		Subset II	
		Step	Å	Step	Å
11	dry–dry	36	18	36	18
44	tip–tip	49	24.5	58	29
12	dry–H-bond donor	28	14	41	20.5
13	dry–H-bond acceptor	42	21	40	20
14	dry–tip	46	23	47	23.5
24	H-bond donor–tip	46	23	52	26
34	H-bond acceptor–tip	45	22.5	49	24.5

9.4.2.2 Ionizable Nitrogen Subset

The predictive quality and robustness of the model were examined using 16 molecules of the test set defined previously. Figure 9.5 plots the predicted HERG activities of the test set molecules versus experimental measured values, showing good agreement between the two.

For the ionizable basic nitrogen subset, all 544 molecules were protonated with a formal charge on the basic nitrogen. They were divided into a training set (518 compounds) and a test set (26 molecules). To select the test set, the dataset was divided into three groups according to the activity value of pIC50 (2.5>1; 1>–0.5;

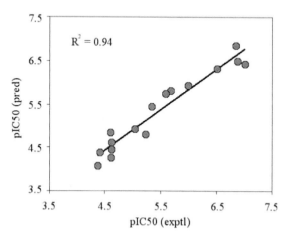

Figure 9.5. Calculated vs. experimental pIC50 for 16 molecules of the test set using model for molecules with nonionizable basic nitrogen.

–0.5>–2.1). The test set compounds were randomly selected from each group to represent the activity span uniformly. The PCA of the datasets was performed to assess structural variance of both training and test sets. The PCA scores plot (Fig. 9.6) showed the absence of any structural outlier in the dataset and that the training set and test set shared the same chemical space.

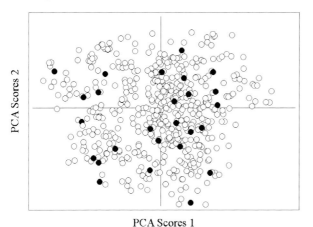

Figure 9.6. 2D plot of PCA showing the descriptor space of training set (open circle) and test set (filled dot) for the HERG data set (charged molecules).

PLS analysis was performed on the reduced set to identify a correlation between the complete set of variables and the activity data. The ALMOND program kept 624 active variables out of a total of 750 variables after filtering out descriptors with no variability. The 45 compounds with two basic nitrogen atoms were protonated on just one of the two basic centers. For these compounds, two models were built using one of the two isomers at a time. The isomer that produced the model with the best r^2 was chosen. The PLS analysis resulted in a model with three LVs and $r^2 = 0.77$. The cross validation of the model by the LOO method yielded q^2 values of 0.74. As shown in Table 9.2, the GRIND descriptors that had high coefficients in the PLS model were 11-36, 44-58, 12-41, 13-40, 14-47, 24-52, and 34-49.

The predictive quality and robustness of the model were assessed using the test set of 26 molecules selected previously. Figure 9.7 plots the predicted HERG activities of the test set molecules versus experimental measured values.

Confirming the results of a previous CoMSiA model [16], the charged nitrogen was found to be a relevant feature correlating the variance of the structural data to the activity data. PLS data analysis on the same training set but with all the basic nitrogen atoms in the neutral (nonprotonated) form resulted in a PLS model of statistically lower quality than that derived from the basic nitrogen in a protonated form. This illustrated the importance of the positive charge to correlate the variance of the structural data to the activity value for this basic nitrogen subset, in spite of the fact that the same kinds of descriptors were found to correlate with the activity data.

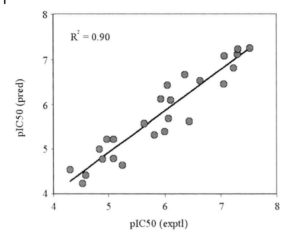

Figure 9.7. Calculated vs. experimental pIC50 for 26 molecules of the test set using the model for the charged basic nitrogen.

9.4.2.3 Interpretation of Pharmacophoric Models

According to the two PLS models, most of the descriptors that correlated with the variance of the activity were very similar. For example, the distance between two hydrophobic MIFs is 18 Å, and the distance between the MIFs produced by a hydrophobic area and a H-bond acceptor group is ~21 Å in both the models.

Between the two models, there are differences in terms of the descriptor distance between the edges of the molecule (tip–tip or GRIND descriptor 44) and the space between a hydrophobic MIF and a field generated by a H-bond donor group (dry-H-bond donor or GRIND descriptor 12). According to the PLS model built on the first subset (molecules devoid of any basic nitrogen atom), the optimal spacing between the fields generated by two edges of the molecule was ~25 Å, while the ideal size suggested by the second model was ~29 Å.

The distance between the MIFs produced by a hydrophobic and a H-bond donor group showed the greatest difference between the two models. This also revealed an important pharmacophoric feature that defines the position of the groups that generate the dry–H-bond donor descriptors. For a molecule devoid of any ionizable basic nitrogen group, the distance was ~14 Å, while the optimal distance between the same two MIFs was ~21 Å for molecules with a basic nitrogen atom. Both pharmacophoric models assigned statistical importance to MIFs produced by H-bond donor groups situated on the edge of the molecule. Figure 9.4 shows two inhibitors and the relative GRIND descriptors 12. The distance between the two MIFs for the molecule with no positive charge (mol31672) is different from the other (sertindole). For both compounds, high statistical relevance was always assigned to the field generated by the H-bond donor group close to the edge of the structure.

Our studies also suggested that H-bond acceptors play an important role for compounds that bind the HERG channel (GRIND descriptors 13 and 34 in both the pharmacophoric models, as shown in Fig. 9.8). The statistical relevance of the MIFs generated by the hydrophobic probe confirmed the assumptions that were made in a previous CoMSiA model [17] regarding the presence of a hydrophobic feature.

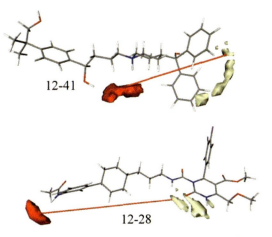

Figure 9.8. The GRIND descriptors from field generated by a H-bond donor group and hydrophobic area in mol11 (top) and mol316172 (bottom).

In the absence of a high quality X-ray structure of the HERG channel or experiments with compounds not containing a basic nitrogen blocking the HERG channel containing mutations, one can, at best, speculate how the two classes of compounds interact with the protein. The PLS multivariate analysis was able to identify two pharmacophores that showed some degree of similarity.

9.5
CYP 3A4 Inhibition

9.5.1
Materials and Methods

9.5.1.1 **Dataset**

The starting dataset used to develop the 3D quantitative structure property relationship (3D-QSPR) model consisted of 370 commercially available compounds. Activity data and 2D structures were retrieved from the Cerep database [18]. Inhibition of CYP 3A4 was reported as inhibition of the formation of 6β-hydroxy-testosterone [19]. Ketoconazole was used as reference compound so that all values are expressed as percentages. The log of the normalized CYP3A4 inhibition percentage was used as the activity value. The activity range spanned from 0.3 to 2.0, covering 1.7 log units.

9.5.1.2 Computational Methods

Molecular structures and geometry optimizations were obtained with the modeling software package SYBYL 6.9.2. Initial 3D structures were obtained using the Concord program included in the SYBYL package running on an SGI workstation. All the compounds were considered in their neutral form. The energy minimization was performed with the standard TRIPOS force field using the Powell method with simplex initial optimization, gradient termination of 0.05 kcal (-mol*A)$^{-1}$ and the number of iterations set to 1000.

All QSAR analysis was carried out using the program ALMOND 3.2.0 and grid alignment independent descriptors (GRIND).

9.5.1.3 Ligand GRIND Descriptors

A total of 640 GRIND descriptors was obtained for the set of ligands using 4 GRID probes: DRY (representing hydrophobic interactions), O (sp2 carbonyl oxygen, representing H bond acceptor), N1 (neutral flat NH like in amide, for H-bond donor) and the TIP probe (molecular shape descriptor). The grid spacing was set to 0.5 Å and the smoothing window to 0.8. The number of filtered nodes was set to 100 with 50% of relative weights. ALMOND produced ten groups of variables, four autocorrelograms and six cross-correlograms.

9.5.1.4 Protein GRIND Descriptors

The four CYP 3A4 crystal structures (pdb code 1W0E, 1W0F, 1W0G, 1TQN) [20, 21] were aligned to the 1W0E file using the algorithm included in the SYBYL package. C-alpha was used as the atom type for the fitting process, which yielded a weighted root mean square (WRMS) of 0.45, 0.53 and 0.77 for structures 1w0f, 1w0g and 1tqn, respectively.

GRIND descriptors were calculated inside the binding sites of the proteins to find descriptors that could match those highlighted by the PLS model. GRID computations on the structures were carried out using DRY, N1 and O probes. The order of the last two probes was inverted in order to facilitate the interpretation of cross- and auto-correlograms. The grid spacing was set to 0.5 Å and the directive ALMD was set to 1 to reproduce the ALMOND settings. The GRID fields were imported into ALMOND program and the GRIND descriptors were calculated with the smoothing window set to 0.8, the number of filtered nodes set to 100 and the relative weights set to 50%. Since ALMOND calculates TIP MIFs only for the convex part of the molecule, no TIP evaluation was performed for the binding sites of the proteins.

9.5.1.5 Overlap of Structures

Recently a crystal structure of CYP 2B4 has been published (pdb code 1P05) [22]. A peculiarity of this structure is that the enzyme shows a big cavity that is open to the outer space. The common parts of the crystal structures of CYP 2B4 and CYP3A4 were superimposed in order to assess whether the residues highlighted by the ALMOND model belonged to flexible or rigid domains. Figure 9.9 shows that the five residues belong to domains that show different degrees of flexibility. PHE304 is found on a helix that is almost rigid (I helix). The domain that contains the C and B' helices bears PHE108 and ILE120 and is moderately flexible. PHE213 and PHE215 are present on the "lid" domain that is composed of helices F and G, the motion of which controls substrate entry. This domain is extremely flexible which could help to explain the broad substrate specificity shown by CYP 3A4.

9.5.2 Results

Molecules in the dataset were divided into a training set (331 compounds) and a test set (39 molecules). The test set was chosen to cover the activity data span of the two subsets of data uniformly. The dataset was divided into three groups according to the activity value. (2.02–1.45, 1.43–0.95, 0.9 0.3) Then 13 molecules from each group were randomly chosen. The principal component analysis of the complete dataset was performed to analyze the structural variance of both the training and test sets. The PCA scores plot (Fig. 9.10) showed that there was no structural outlier present in the dataset and that the training set and test set shared the same chemical space.

PLS multivariate data analysis of the training set was carried out on the descriptors matrix to correlate the complete set of variables (640) with the activity data. The PLS analysis resulted in a three latent variables (LVs) model whose statistical parameters are summarized in Table 9.3. The predictive quality and robustness of the model obtained were assessed using the test set previously defined. Figure 9.11 is a plot of the predicted inhibitory activities of the test set molecules versus the experimental measured values, showing good agreement between the two.

Table 9.3 Statistical data of the models. Three different cross-validated correlation coefficients were calculated using the leave one out (LOO), leave two out (LTO) and the five random group (5RG) methods.

LV	r^2	q^2 LOO	q^2 LTO	q^2 5RG
1	0.7078	0.6896	0.6895	0.6863
2	0.7263	0.7042	0.7042	0.7035
3	0.7621	0.7150	0.7150	0.7106
4	0.7885	0.7097	0.7096	0.7003
5	0.8063	0.6949	0.6948	0.6846

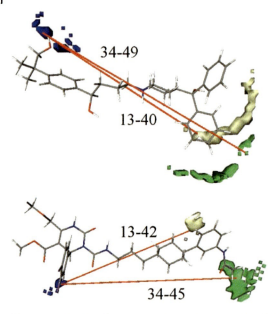

Figure 9.9. The GRIND descriptors for mol11 (top) and mol316172 (bottom) related to the presence of field generated by a H-bond acceptor group (blue), hydrophobic area (yellow), and TIP field (green).

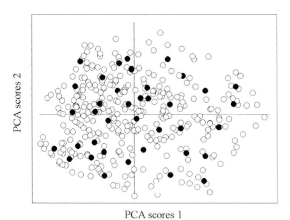

Figure 9.10. 2D plot of the PCA scores showing the descriptor space of training set (open circle) and test set (filled dot).

Table 9.4 summarizes the loadings of the first LV of the PLS model. According to the loadings of the PLS model, strong CYP3A4 inhibitors are characterized by intense hydrophobic interactions, by the presence of an H-bond acceptor and by an optimal distance between the edges of the molecule. GRIND descriptor 11-23 represents the optimal span of ~11.5 Å that separates two MIFs generated by

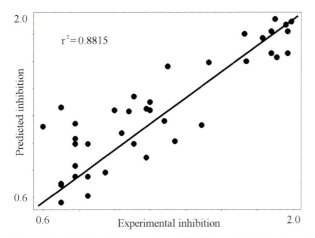

Figure 9.11. Calculated vs. experimental inhibitory activity for the molecules of the test set.

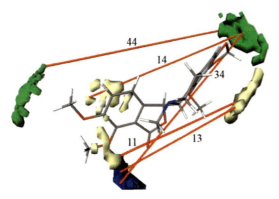

Figure 9.12. Projection of the GRIND descriptors identified by the PLS model around the 3D structure of the antitussive molecule Noscapine. The red lines between the MIFs represent the GRIND descriptors.

hydrophobic areas of the same molecule. Descriptor 14-33 represents the distance of ~16.5 Å between the fields produced by a hydrophobic area and by one of the edges of the molecule.

The presence of a MIF generated by a H-bond acceptor group separated by ~12.5 Å from the field produced by a hydrophobic area (GRIND descriptor 13-25) and ~19.5 Å from that generated by an edge (GRIND descriptor 34-39) is important to achieve CYP3A4 inhibition. The PLS model highlights that there is an optimal spacing of ~20.5 Å (GRIND descriptor 44-41) between the convex parts of the molecule. The GRIND approach allows one to project the descriptors in the space to obtain a spatial arrangement of the energetically favorable interaction points around the molecule studied. If we project simultaneously the descriptors that have a high impact on the model, we can obtain a representation of the structural features that are responsible for the interaction between the molecule and CYP3A4, Fig. 9.12.

Table 9.4 Salient GRIND descriptors in PLS models.

GRIND descriptor	GRID MIFs	Step	Å
11	dry–dry	23	11.5
44	tip–tip	41	20.5
13	dry–H-bond acceptor	25	12.5
14	dry–tip	33	16.5
34	H-bond acceptor–tip	39	19.5

9.5.2.1 Distances in the Protein Pocket

Since high resolution structures of CYP3A4 have been determined by X-ray crystallography, it has allowed us to confirm the pharmacophoric distances highlighted by the PLS model. Although two enzyme structures (1W0F and 1W0G) had ligands bound in different positions, no significant conformational variation was found in the four protein structures. Inspection of the GRIND descriptors profiles for the four crystal structures revealed that the residues responsible for the spatial arrangement of the MIFs match the descriptors highlighted by the PLS analysis of the ligand dataset.

A distance of 11.5 Å between two hydrophobic areas (GRIND descriptor 11-23) was formed by the MIFs generated by PHE-215, PHE-213, PHE-304 and PHE-108 and by a field that lies above two of the pyrrole rings of the heme on the same side of the PHE cluster, as shown in Fig. 9.13. GRIND descriptor 13-25 is defined as a distance of 12.5 Å between a hydrophobic area and a region that shows favorable interaction fields for an H-bond acceptor/electron donor group. Note that the order of N1 and O probes was inverted compared to the order used for the ligands. The distance in the pocket between a MIF that is located approximately 2.5 Å above the Fe atom of the heme and the PHE cluster previously described could be found.

9.5.3
Discussion

The predicted activity values for the test set compounds (Table 9.3) show the overall quality of the pharmacophoric model. Evidently, the pharmacophoric model was able to correlate the variance of the experimental data with a set of distances between interaction points present in the compounds. Descriptors 11-23 and 14-33 are related to hydrophobic interactions, consistent with the previous notion that CYP3A4 inhibition has been described in terms of descriptors related to the hydrophobicity of the compounds. Riley et al. [23], using N-demethylation of erythromycin as probe, demonstrated the importance of hydrophobic interactions in CYP3A4 inhibition. Working on metabolic stability, Crivori et al. [24] found that the descriptors of the DRY autocorrelograms (11 descriptors from 23 to 44) represented the optimal distances that separated the hydrophobic regions of compounds interacting with CYP3A4.

Analysis of the GRIND descriptors generated for the protein structures, allowed us to identify the hydrophobic cavity as formed by the cluster of PHE-215, PHE-213, PHE-304 and PHE-108. Domanski et al. [25] have demonstrated the pivotal role of PHE-304 in CYP3A4 kinetics along with LEU-211 and ASP-214. It is unclear how these residues are involved in the binding processes. While the structure of CYP3A4 cocrystallized with progesterone shows that ASP-214 can interact with the ligand outside the active site pocket, all four crystal structures showed that PHE-304 was located inside the pocket with little or no variation in the position. The crystal structure of CYP3A4 with progesterone provides little help in explaining the role of LEU-211.

Homotropic and heterotropic modulation of P450 3A4 substrate oxidation reactions have been reported [26–37] Several hypotheses have been proposed regarding multiple binding sites [38] or multiple ligands binding to the same site. Many authors have proposed that cooperativity may be due to the binding of multiple molecules to the enzyme, either within the active site [39–42] or at separate, distant locations on the enzyme [43–45]. Baasa et al. [46] have shown that at least three testosterone molecules can bind to each CYP3A4, and that the observed spin shift is caused almost exclusively by the second and third binding events with the very low signal or spin shift caused by the first binding event.

Kenworthy et al. [45] proposed a model with three subsites in order to explain the binding of testosterone (TS) and diazepam (DZ). One site binds diazepam, another binds testosterone, and the third is capable of binding either diazepam or testosterone. They found that testosterone caused extensive activation of diazepam metabolism, whereas diazepam caused inhibition of testosterone metabolism. Diazepam is present in the inhibitor database used to obtain the pharmacophoric model and its inhibitory activity is well explained by the model (y-residual = 0.27 log units). The model with multiple binding sites proposed by Kenworthy helped to explain the results of the obtained pharmacophoric model if one considers that competition will occur in the catalytic site that can bind either DZ or TS and that competition in the third site will be of low importance considering the low affinity of testosterone for this site.

With the solving of the two crystal structures of 2B4 [47, 48], an overlap of 3A4 with the 2B4 closed and open forms was performed. This was done simply to highlight the flexibility of the enzyme as well as the potential size of a molecule or multiple molecules that could fit into the enzyme. The overlap of CYP3A4 and CYP2B4 structures together with the pertinent (corresponding to the pharmacophoric elements) MIFs are shown in Fig. 9.5.

Many authors [49–51] have demonstrated that inhibition of catalysis by CYP3A4 is substrate dependent and that interactions observed with one CYP3A4 probe may not be representative of those observed with other CYP3A4 substrates. The proposed model could help to explain these data considering that different probes could bind to different sites and therefore show different competition patterns.

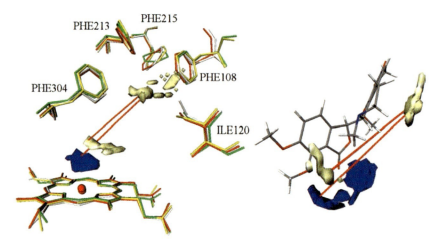

Figure 9.13. Analysis of the GRIND descriptors produced by the crystal structures of the proteins allowed us to find the distances between MIFs that could match GRIND descriptors 11-23 and 13-25. On the left are depicted the residues responsible for the MIFs (white = 1TQN, yellow = 1W0E, red 1 =W0F, green = 1W0G) on the right the same kind of descriptors generated by Noscapine.

9.6
Conclusions

The GRIND approach has proved useful in three ADME-TOX studies. A 3D-QSAR was generated using inhibition data for 339 inhibitors of CYP3A4 using testosterone metabolism (7-hydroxy testosterone formation) as a substrate A single pharmacophore emerged which consisted of two hydrophobic elements as well as an H-bond acceptor element. The elements corresponded well with the MIFs generated from the published crystal structure of CYP3A4.

For Pgp it was shown that, for a diverse set of substrates, a pharmacophore could be identified. This pharmacophore played a large role in explaining the variance in PGP inhibition data from a Calcein AM fluorescent assay, suggesting that interacting with the protein is important to inhibition to the efflux transporter.

Finally, two general 3D-QSAR models superposition independent for datasets with and without ionizable nitrogen were obtained for HERG inhibitors.

References

1. Goodford, P. J. A computational procedure for determining energetically favorable binding sites on biological important macromolecules, *J. Med. Chem.* **1985**, *28*, 849–857.
2. Fontaine, F.; Pastor, M.; Sanz, F. Incorporating Molecular Shape into the Alignment-free GRid-INdependent Descriptors, *J. Med. Chem.* **2004**, *47*, 2805–2815.
3. Reference to our work
4. Saeki T.; Ueda K.; Tanigawara Y.; Hori R.; Komano, T. Human Pglycoprotein transports cyclosporin A and FK506, *J. Biol. Chem.* **1993**, *268*, 6077–6080.
5. Sybyl Molecular Modeling System (version 6.9.2) Tripos Inc.; St. Louis; MO.
6. Martin, Y. C. 3D Database Searching in Drug Design, *J. Med. Chem.* **1992**, *35*, 2145–2154.
7. Pearlman, R. S. 3D Molecular Structure: Generation and Use in 3D Searching, in *3D QSAR in Drug Design-Theory Methods and Applications*, Kubyni, H. Ed., Escom Science Publishers, Leiden, 1993, pp. 41–79.
8. MOETM, www.chemcomp.com/Corporate_Information/MOE.html
9. Wold, S.; Esbensen, K.; Geladi, P. Principal component analysis, *Chemom. Intell. Lab. Syst.* **1987**, *2*, 37–52.
10. Shapiro, A. B.; Ling, V. The Mechanism of ATP-Dependent Multidrug Transport by P-Glycoprotein, *Acta. Physiol. Scand.* **1993**, *163*, Suppl 643, 227–234.
11. Seelig, A.; Landwojtowicz, E. Structure-activity Relationship of P-glycoprotein substrates and modifiers, *Eur. J. Pharm. Sci.* **2000**, *12*, 31–40.
12. Higgins, C. F.; Linton, K. J. The ATP switch model for ABC transporters, *Nat. Struct. Mol. Biol.* **2004**, *11*, 918–926.
13. Dey, S.; Ramachandra, M.; Pastan, I.; Gottesman, M. M.; Ambudkar, S. V. Evidence for two nonidentical drug-interaction sites in the human P-glycoprotein, *Proc. Natl. Acad. Sci. USA* **1997**, *94*, 10594–10599.
14. Homolya, L.; Hollo, Z.; Germann, U. A.; Pastan, I.; Gottesman, M. M.; Sarkadi, B. Fluorescent cellular indicators are extruded by the multidrug resistance protein, *J. Biol. Chem.* **1993**, *268*, 21493–21496.
15. Stein, W. D. Kinetics of the multidrug transporter (P-glycoprotein) and its reversal, *Physiol. Rev.* **1997**, *77*, 545–590.
16. Kang, J.; Wang, L.; Cai, F.; Rampe, D. *Eur. J. Pharmacol.* **2000**, *392*, 137–140.
17. Pearlstein, R. A.; Vaz, R. J.; Kang, J.; Chen, X. L.; Preobrazhenskaya, M.; Shchekotikhin, A. E.; Korolev, A. M.; Lysenkova, L. N.; Miroshnikova, O. V.; Hendrix, J.; Rampe, D. *Bioorg. Med. Chem. Lett.* **2003**, *13*, 1829–1835.
18. Cerep, BioPrint® database (http://www.cerep.fr).
19. Lin, Y.; Lu, P.; Tang, C.; Mei, Q.; Sandig, G.; Rodrigues, A. D.; Rushmore, T. H.; Shou, M. *Drug Metab. Dispos.* **2001**, *29*, 368–374.
20. Williams, P. A.; Cosme, J.; Vinkovic, D. M.; Ward, A.; Angove, H. C.; Day, P. J.; Vonrhein, C.; Tickle, I. J.; Jhoti, H. *Science* **2004**, *305*, 683–686.
21. Yano, J. K.; Wester, M. R.; Schoch, G. A.; Griffin, K. J.; Stout, C. D.; Johnson, E. F. *J. Biol. Chem.* **2004**, *279*, 38091–38094.
22. Scott, E. E.; He, Y. A.; Wester, M. R.; White, M. A.; Chin, C. C.; Halpert, J. R.; Johnson, E. F.; Stout, C. D. *Proc. Natl. Acad. Sci. USA* **2003**, *100*, 13196–13201.
23. Riley, R. J.; Parker, A. J.; Trigg, S.; Manners, C. N. *Pharm. Res.* **2001**, *18*, 652–655.
24. Crivori, P.; Zamora, I.; Speed, B.; Orrenius, C.; Poggesi, I. *J. Comput.-Aided Mol. Des.* **2004**, *18*, 155–166.

25 Domanski, T. L.; He, Y. A.; Harlow, G. R.; Halpert, J. R. *J. Pharmacol. Exp. Ther.* **2000**, *293*, 585–591.

26 Harlow, G. R.; Halpert, J. R. *Proc. Natl. Acad. Sci. U S A* **1998**, *95*, 6636–6641.

27 Ueng, Y. F.; Kuwabara, T.; Chun, Y. J.; Guengerich, F. P. *Biochemistry* **1997**, *36*, 370–381.

28 Buening, M. K.; Fortner, J. G.; Kappas, A.; Corney, A. H. *Biochem. Biophys. Res. Commun.* **1978**, *82*, 348–355.

29 Huang, M. T.; Johnson, E. F.; Muller-Eberhard, U.; Koop, D. R.; Coon, M. J.; Conney, A. H. *J. Biol. Chem.* **1981**, *256*, 10897–10901.

30 Thakker, D. R.; Levin, W.; Buening, M.; Yagi, H.; Lehr, R. E.; Wood, A. W.; Conney, A. H.; Jerina, D. M. *Cancer Res.* **1981**, *41*, 1389–1396.

31 Kapitulnik, J.; Poppers, P. J.; Buening, M. K.; Fortner, J. G.; Conney, A. H. *Clin. Pharmacol. Ther.* **1977**, *22*, 475–484.

32 Schwab, G. E.; Raucy, J. L.; Johnson, E. F. *Mol. Pharmacol.* **1988**, *33*, 493–499.

33 Lee, C. A.; Lillibridge, J. H.; Nelson, S. D.; Slattery, J. T. *J. Pharmacol. Exp. Ther.* **1996**, *277*, 287–291.

34 Wang, R. W.; Newton, D. J.; Scheri, T. D.; Lu, A. Y. *Drug Metab. Dispos.* **1997**, *25*, 502–507.

35 Domanski, T. L.; Liu, J.; Harlow, G. R.; Halpert, J. R. *Arch. Biochem. Biophys.* **1998**, *350*, 223–232.

36 Ludwig, E.; Schmid, J.; Beschke, K.; Ebner, T. *J. Pharmacol. Exp. Ther.* **1999**, *290*, 1–8.

37 Lee, C. A.; Manyike, P. T.; Thummel, K. E.; Nelson, S. D.; Slattery, J. T. *Drug Metab. Dispos.* **1997**, *25*, 1150–1156.

38 Hosea, N. A.; Miller, G. P.; Guengerich, F. P. *Biochemistry* **2000**, *39*, 5929–5939.

39 Domanski, T. L.; He, Y. A.; Harlow, G. R.; Halpert, J. R. *J. Pharmacol. Exp. Ther.* **2000**, *293*, 585–591.

40 Korzekwa, K. R.; Krishnamachary, N.; Shou, M.; Ogai, A.; Parise, R. A.; Rettie, A. E.; Gonzalez, F. J.; Tracy, T. S. *Biochemistry* **1998**, *37*, 4137–4147.

41 Shou, M.; Grogan, J.; Mancewicz, J. A.; Krausz, K. W.; Gonzalez, F. J.; Gelboin, H. V.; Korzekwa, K. R. *Biochemistry* **1994**, *33*, 6450–6455.

42 Shou, M.; Mei, Q.; Ettore, M. W., Jr.; Dai, R.; Baillie, T. A.; Rushmore, T. H. *Biochem. J.* **1999**, *340* (3), 845–853.

43 Ueng, Y. F.; Kuwabara, T.; Chun, Y. J.; Guengerich, F. P. *Biochemistry* **1997**, *36*, 370–381.

44 Schwab, G. E.; Raucy, J. L.; Johnson, E. F. *Mol. Pharmacol.* **1988**, *33*, 493–499.

45 Kenworthy, K. E.; Clarke, S. E.; Andrews, J.; Houston, J. B. *Drug Metab. Dispos.* **2001**, *29*, 1644–1651.

46 Baasa, B. J.; Denisov, I. G.; Sligar, S. G. *Arch. Biochem. Biophys.* **2004**, *430*, 218–228.

47 Scott, E. E.; He, Y. A.; Wester, M. R.; White, M. A.; Chin, C. C.; Halpert, J. R.; Johnson, E. F.; Stout, C. D. *Proc. Natl. Acad. Sci. USA* **2003**, *100*, 13196–13201.

48 Scott, E. E.; White, M. A.; He, Y. A.; Johnson, E. F.; Stout, C. D.; Halpert, J. R. *J. Biol. Chem.* **2004**, *279*, 27294–27301.

49 Kenworthy, K. E.; Bloomer, J. C.; Clarke, S. E.; Houston, J. B. *Br. J. Clin. Pharmacol.* **1999**, *48*, 716–727.

50 Wang, R. W.; Newton, D. J.; Liu, N.; Atkins, W. M.; Lu, A. Y. *Drug Metab. Dispos.* **2000**, *28*, 360–366.

51 Stresser, D. M.; Blanchard, A. P.; Turner, S. D.; Erve, J. C.; Dandeneau, A. A.; Miller, V. P.; Crespi, C. L. *Drug Metab. Dispos.* **2000**, *28*, 1440–1448.

10
Progress in ADME Prediction Using GRID-Molecular Interaction Fields

Ismael Zamora, Marianne Ridderström, Anna-Lena Ungell, Tommy Andersson, and Lovisa Afzelius

10.1
Introduction: ADME Field in the Drug Discovery Process

There are several aspects regarding the absorption, distribution, metabolism and excretion (ADME) properties of new chemical entities (NCE) that are relevant during the drug discovery process. The ADME properties of NCEs will define the plasma profile and potential drug–drug interactions when used in patients. The preclinical ADME studies contribute to the drug discovery process by considering the optimal plasma concentration profile with regards to the pharmacokinetics and pharmacodynamics, together with properties such as the frequency of dosing (often once daily) and drug–drug interaction risks (Fig. 10.1).

The interpretation of the pharmacokinetic profile (concentration vs. time) based on the structure is a complex task since several biological and physicochemical processes are taking place in parallel in the human body. The *in vivo* data is inter-

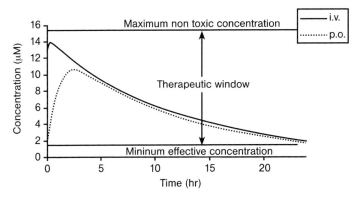

Figure 10.1. Simulation of a pharmacokinetic profile (concentration vs. time) for a drug administered intravenously (i.v.) or orally (p.o.) at (7.14 mg kg^{-1}) and with a half life of 12 h and a volume of distribution of 5 L. The therapeutic window shows the minimum concentration in plasma that produces the desired effect and the maximum concentration that does not produce toxic effects.

Molecular Interaction Fields. Edited by G. Cruciani
Copyright © 2006 WILEY-VCH Verlag GmbH & Co. KGaA, Weinheim
ISBN: 3-527-31087-8

preted based on a number of calculated parameters such as clearance, half life, bioavailability, C_{max}, T_{max}, volume of distribution and area under the curve (AUC). There have been various attempts to predict some of these composite parameters from the structure [1, 2] but in most cases the multifactorial nature of the *in vivo* data makes it impossible to build a global model.

In order to address the complexity of the system and to increase the throughput, several *in vitro* assays have been developed to study the individual processes. Absorption can be studied using a cell-based system like the transport across Caco-2 [3] or MDCK [4] cell monolayers, or even a simpler physicochemical system like parallel artificial membrane permeation assay (PAMPA) [5], where a lipid layer is built on top of a filter to perform the transport experiment. The distribution of a compound into the body is indirectly analyzed using *in vitro* systems to measure unspecific binding to plasma proteins like albumin or a1 acid glycoprotein [6, 7] and the solubility in water [8] or from *in vivo* derived parameters like volume of distribution [7, 9]. The drug metabolism is generally studied using recombinant cytochromes [10], liver microsomes [10, 11] or isolated hepatocyte [12]. Excretion processes are usually related to the previous ones and are not routinely screened by *in vitro* systems.

In parallel to the experimental ADME field, computational models are developed for specific *in vitro* assays or even, in some cases, for specific pathways represented in each system (Fig. 10.2). The computational models are based on the experimental data, and since there are still not consistent data available for some of the properties shown in Fig. 10.2, they have not yet been addressed.

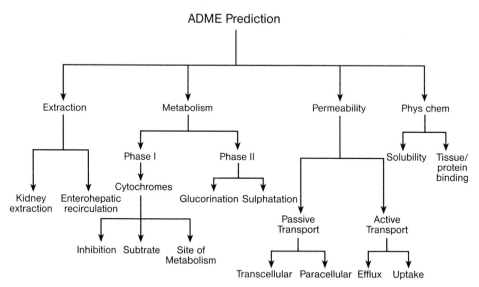

Figure 10.2. Scheme of different experimental properties for which models can be derived.

One aspect that makes the ADME modeling different from a pharmacological target-based study is that, in general, the proteins involved in the ADME field have a broad substrate specificity. The low specificity of these enzymes is probably related to the evolutionary function of these systems as a defense against various toxic environments. Moreover, these enzymes often have lower affinity for the substrates than for most of the pharmacological targets. This unspecificity makes it difficult to develop global models that cover the entire chemical space. Therefore in many cases it is more successful to build local models based on a closely related series of compounds where similar factors influence their biological behavior.

The GRID molecular interaction fields (MIFs) have been used extensively within the ADME area to compute molecular descriptions for compounds or proteins (Fig. 10.3) [13]. For example, the interaction of the hydrophobic probe in GRID (the DRY probe) can be used to compute the hydrophobic surface exposed by the compound to the environment, which is related to the lipophilicity and therefore to the passive transport or to the solubility. The program VolSurf [14] derives several molecular descriptors from the MIFs using the water (H2O), hydrophobic (DRY) and hydrogen bond acceptor (O) probes, with the aim being to describe the general features of the molecules (Fig. 10.4). These descriptors were designed to be used in the ADME field, and conceptually are quite original. Recently, Todeschini et al. [15] published a book reporting thousands of molecular descriptors. The VolSurf descriptors, however, were not included, which demonstrates the originality of the approach. The description of the VolSurf parameters is shown elsewhere in this book. Nevertheless, in order to show their originality some descriptors are presented here (Fig. 10.5):

- Hydrophilic–lipophilic balances (HL1 HL2) are the ratio between the hydrophilic regions measured at –3 and –4 kcal mol^{-1} and the hydrophobic regions measured at –0.6 and –0.8 kcal mol^{-1}. These descriptors represent the balance between both interaction types.
- Amphiphilic moment (*A*) is the vector pointing from the center of the hydrophobic domain to the center of the hydrophilic domain. This parameter measures the distribution of the polar and nonpolar groups in the molecule.
- Critical packing parameter (CP) is the ratio between the hydrophobic and lipophilic parts of a molecule. In contrast to HL balance, CP refers just to the molecular shape.

222 | *10 Progress in ADME Prediction Using GRID-Molecular Interaction Fields*

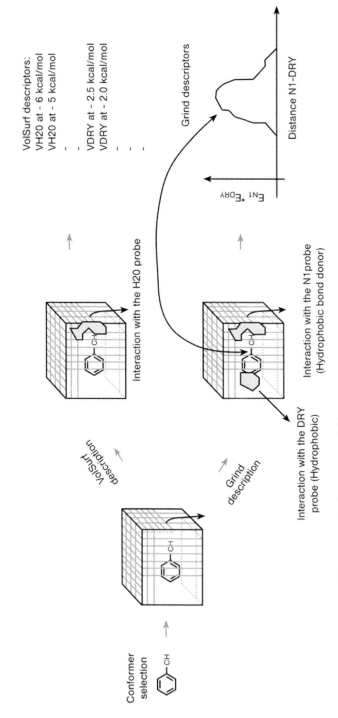

Figure 10.3. Transformation of the GRID molecular interaction fields into (a) global properties description and (b) interaction distance based representation.

Another descriptor that has been used in the literature to measure the interaction of a compound with a polar environment is the polar surface area (PSA). The PSA is described as the surface area of all nitrogen and oxygen and the hydrogen atoms attached to them. However, this descriptor does not take account of the fact that different nitrogen and/or oxygen atoms can interact differently with the hydrophilic environment. For example nitrobenzene and benzoic acid have very similar PSA (40.128 and 45.825 Å^2 respectively) [16], although the interaction pattern of both compounds with water is quite different (Fig. 10.6). VolSurf descriptors are able to capture this difference in the hydrophilic regions at –3, –4, –5 and –6 kcal mol^{-1}, where nitrobenzene has a lower volume of interaction than benzoic acid.

In addition to the VolSurf treatment of the GRID fields, the information from the MIF can also be transformed to obtain a pharmacophoric type of representation, which is useful in the modeling of metabolic stability, cytochrome inhibition or even the direct study of the ADME related proteins (Fig. 10.3). The Almond software [17] transforms the MIF into a distance-based representation of the molecule interaction. These parameters describe the geometry of the interaction and QSAR models can be derived where the interaction with a protein is essential. Detailed information on these descriptors is presented elsewhere in this book.

10.2
Absorption

The permeability is one of the key factors that determines the percentage of the compound absorbed through the gastrointestinal tract or other physical barriers, including the blood brain barrier (BBB). The permeability is a kinetic parameter that measures the amount of the compound transported per unit time considering the accessible surface area and the initial concentration. Among the different experimental techniques that can be used to measure the permeability, the one based on the transport of the compounds across Caco-2 cells has become the industrial standard. This measurement is affected by a number of experimental conditions like the cell culture day, the cell passage, the speed at which the cells are shaken during the experiment, the pH in the apical or basolateral sides of the cells and the temperature [3]. Therefore, the use of experimental data from different published sources as a basis for computational models should be avoided for quantitative structure–activity relationships (QSAR), especially when standard conditions are not reported.

The compounds can cross the membranes by passive processes, which depend only on the concentration gradient on both sides of the barrier, or by active ones, which are mediated by the interaction of the compound with a protein. The passive processes of the epithelial cells in the gastrointestinal tract include passive transport through the cell (trans-cellular pathway) or in the space between the cells (para-cellular pathway) [18].

There are an increasing number of proteins of interest for drug discovery that are described as mediating the transport of compounds across the cell monolayer

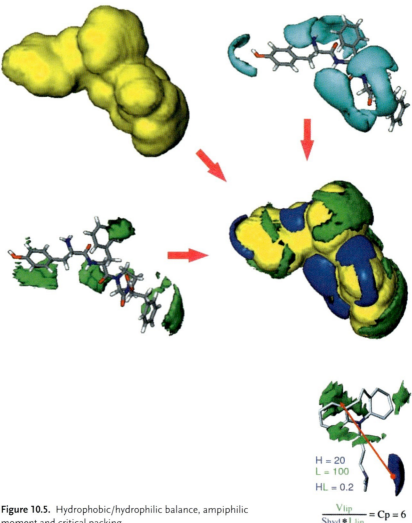

Figure 10.5. Hydrophobic/hydrophilic balance, ampiphilic moment and critical packing

$H = 20$
$L = 100$
$HL = 0.2$

$$\frac{V_{lip}}{S_{hyd} * L_{lip}} = C_p = 6$$

or the blood brain barrier. These proteins can be divided into those that facilitate the net flux from the gastrointestinal tract to the blood, called uptake systems like the PepT1 (dipeptide carrier system), and those which go against a net absorptive flux like P-glycoprotein. Some of these systems are relevant not only for the absorption of the compound, but also for the excretion of the compound in the kidney or the liver and also for the distribution of compounds between the tissues. The Caco-2 cell-based system considers both passive pathways and some of the active mediated transporters.

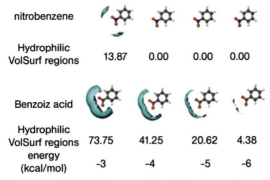

Figure 10.6. Hydrophilic regions for nitrobenzene and benzoic acid at −3, −4, −5 and −6 kcal mol^{-1}.

10.2.1
Passive Transport, Trans-cellular Pathway

The key physicochemical processes that determine the passive trans-cellular transport are the water desolvation of the compound before entering the membrane and the solvation in a lipophilic environment. In order to account for these processes the interaction of the compound with the water and with a lipophilic environment has to be computed. These interactions have been shown to be well described by VolSurf descriptors based on the GRID program [13] using the H2O and the DRY probes. The volumes and surfaces at different energy levels are able to model the passive absorption across Caco-2 cell monolayer [19, 20] (Fig. 10.7(a)), the blood–brain barrier [21] (Fig. 10.7(b)) or permeability across the physicochemical system like PAMPA [22] using the multivariate technique discriminative-PLS. Since data from Caco-2 experiments consider not only the passive pathway, but also a combination of the different absorption routes that may affect each, a discriminative model was developed (Table 10.1). Moreover, the collection of the experimental data from different sources makes it difficult to combine them in a quantitative manner. In the case of the blood–brain barrier a discriminative model was derived because of the data availability. The interpretation of both models based on the PLS coefficients (Fig. 10.7(c) and (d)) shows that a good interaction of the compounds with the water probe has a negative impact on the permeability, while the interaction with the DRY probe has a positive contribution.

226 | 10 Progress in ADME Prediction Using GRID-Molecular Interaction Fields

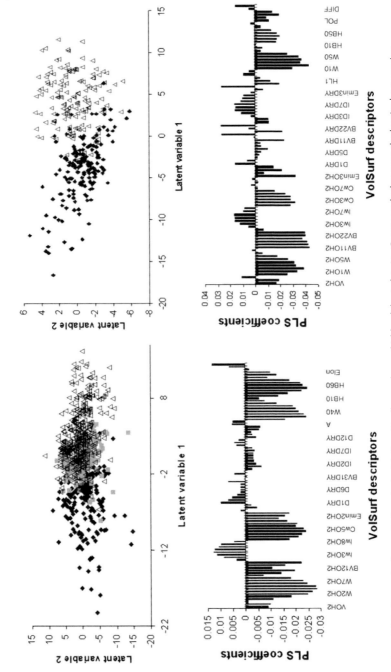

Figure 10.7. Score plot from a PLS analysis. (a) Caco-2 cell monolayer; (b) blood–brain barrier. Compounds that pass the barrier by a passive mechanism are shown in gray, and those that do not are shown in black. PLS coefficients: (c) Caco-2 cell monolayer; (d) blood–brain barrier.

Table 10.1 Percentage of the compounds predicted by the Caco-2 (713 compounds, 2 latent variables) and BBB (313 compounds, 2 latent variables) discriminative models. (High: permeable compound, medium: intermediate permeable compounds, low: nonpermeable compounds.)

Caco-2 cell-based VolSurf model				BBB discriminative-PLS VolSurf model			
	High	Medium	Low		High	Medium	Low
High	78.24	19.09	2.65	High	80.42	16.08	3.49
Medium	43.33	52	4.66	Medium	33.33	44.44	22.22
Low	4.9	28.57	66.51	Low	1.44	9.93	88.19
Total		69.50		Total		69.50	

10.2.2
Active Transport

The main factor that governs the transport of a compound by an active carrier system is the interaction of this compound with a carrier protein. In this case the description of the molecular structure should be similar to that used in ligand-based design to describe the interaction of the compound with any other protein using pharmacophoric representation or 3D-QSAR.

The GRID-based molecular representation has been used to develop models for different carrier systems. For example: the GRID-GOLPE [2] approach or comparative molecular field analysis (CoMFA) [23] has been used in modeling the PepT1 transport system (Fig. 10.8). The experimental data used as a basis for these modeling efforts was the inhibition of the transport of a known substrate for PepT1 by 43 close related cepfalosporines [24]. However, this CoMFA-like study requires the superimposition of the different compounds and the definition of alignment rules. Therefore, the use of this modeling technique to develop global models will be very difficult.

Another GRID-based approach to the modeling of active transport is obtained from the GRIND descriptors computed by the Almond software [17]. In this case, the GRID interaction fields are transformed to a distance-based structure representation called correlograms. The new descriptor set is alignment independent and therefore can be used in a larger chemical space than the classical 3D QSAR approach. This technique has also been applied in the P-glycoprotein case [25].

Sometimes it is useful to apply a combination of the descriptors based on the global properties of the molecule and those based on a pharmacophoric representation. Conceptually, the global properties would better describe the initial passive membrane permeation required to reach the site of action. Then, the specific protein interactions could be explained by the pharmacophoric descriptors. This has been demonstrated successfully in the P-glycoprotein case [26], where two processes are important for the transport; passive transport to the cell and active transport from inside the cell. Cianchetta et al. [25] obtained an initial model for the inhibition of the P-glycoprotein transport using the VolSurf descriptor set, with good fitting and cross-validation parameters. However, the model was not able to predict the most active compounds. Conversely, when these authors combined VolSurf descriptors with the GRIND descriptors they obtained a model with better statistical parameters and a very good prediction rate over the entire activity range.

10.3
Distribution

The distribution of a compound in the human body can also be partially related to the absorption properties. There are specific transport systems that are expressed in certain tissues that can influence the distribution of the compound. For example, rosuvastatin, a new member of the statin family is transported by the OATP-C carrier system, which is selectively expressed in the liver, making this compound selectively distributed into this organ [27]. In general it is not possible to derive computational models for these selective transport systems since there is not yet enough experimental information and data to support the model building and validation. Nevertheless, there are three properties that are commonly used to describe the distribution of a compound in the human body: the solubility, the unspecific binding of the compound to plasma proteins and the volume of distribution.

10.3.1
Solubility

Solubility is a key property in the distribution of the compound from the gastrointestinal tract to the blood. There have been several modeling efforts to predict the solubility, based on different type of descriptors. The intrinsic solubility (thermodynamic solubility of the neutral species) for a set of 1028 compounds has been modeled using the VolSurf descriptors based on GRID-MIFs (Fig. 10.9(a)) and PLS multivariate analysis [20]. The interpretation of the model can be based on the PLS coefficients: the ratio of the surface that has an attractive interaction with the water probe contributes positively to the solubility, while the hydrophobic interactions and log P have a negative contribution.

Although several models to predict the solubility have been published, none take into account the crystal packing of the compounds. Neglecting the crystal packing could be relevant for some compounds, causing the solubility prediction to fail.

During the drug discovery process the solubility in a mixture of water with another cosolvent like DMSO is commonly measured. The compounds are usually dissolved first in DMSO and then diluted in water and/or buffer solution. Therefore, a classification model has been successfully derived to predict the solubility in this solvent mixture using the VolSurf type of descriptors (Fig. 10.9(b)) for a set of 150 compounds with two latent variables (Table 10.2).

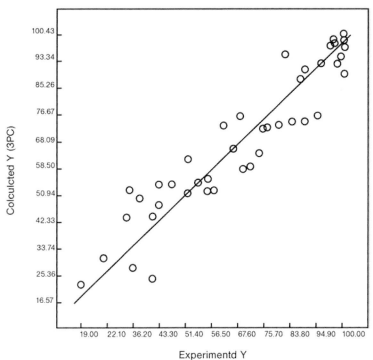

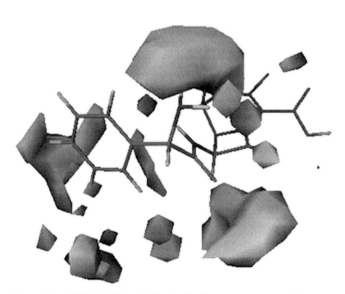

Figure 10.8. GRID-GOLPE model for the PepT1 transport system (2 latent variables, $r^2 = 0.88$, $q^2 = 0.66$). (a) Prediction vs. experimental. (b) Activity contribution to the PepT1 for a GRID-GOLPE.

Table 10.2 Percentage of the compounds predicted by the discriminative models for solubility in a mixture of 2% DMSO in water. (High: soluble compound, high–medium: intermediate to high solubility compounds, medium–low: intermediate to low solubility compounds, low: considered as insoluble compounds.)

	High	High–medium	Medium–low	Low
High	70	30	0	0
High–medium	25	32.5	33.5	0
Medium–low	0	29.5	65.91	4.54
Low	0	5.26	55.36	39.47
Total		59.33		

10.3.2
Unspecific Protein Binding

Binding to proteins in plasma could be an important factor that influences the distribution of the compound. Generally, this process correlates to the compound: lipophilic compounds bind to lipophilic domains of proteins like albumin. Since this property depends on the global characteristic of the structure, VolSurf descriptors describe properly the unspecific binding of the compounds to plasma. A quantitative PLS model was obtained based on a set of 408 compounds with two latent variables with good prediction capacity. The interpretation of the model indicates that a better interaction of the compound with the hydrophobic environment leads to better unspecific binding to albumin. Nevertheless, not all the compounds bind unspecifically to human serum albumin. There is at least one binding site in albumin, called the warfarin binding site, where the interaction driving force is not only linked to hydrophobicity, but also to ligand–protein complementarity. No model has been developed for this type of binding to albumin, since there is little experimental information about specific binding to plasma proteins.

10.3.3
Volume of Distribution

The volume of distribution (VD) for a drug is the apparent volume that accounts for the total dose administration based on the observed plasma concentration. The plasma volume of the average adult is approximately 3 L. Therefore, an apparent volume of distribution larger than the plasma compartment (i.e. greater than 3 L) indicates that the drug is also present in tissue or fluid outside the plasma compartment. The volume of distribution represents a complex combination of multiple chemical and biochemical phenomena. Nevertheless, in general terms and considering only the passive distribution processes, the distribution of a compound in the body depends on the partition of the compound between the differ-

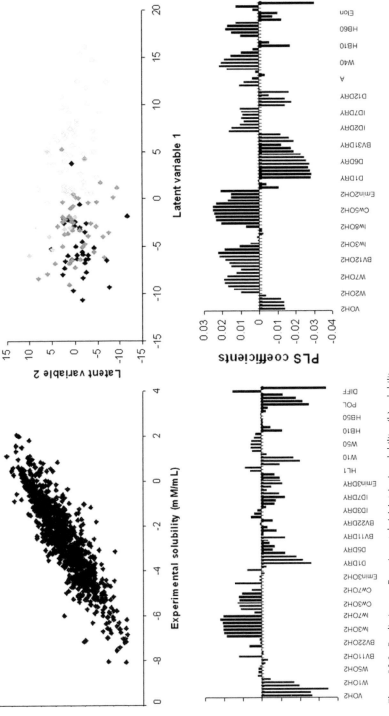

Figure 10.9. Prediction vs. Experimental. (a) Intrinsic solubility; (b) solubility in a 2% DMSO solution. PLS coefficients (c) intrinsic solubility and (d) solubility.

ent tissues and the plasma. The passive partition processes can be analyzed by the lipophilicity–hydrophilicity balance [7]. In this case the application of the VolSurf descriptors to build a computational model could be appropriate [20].

10.4
Metabolism

In the drug discovery process the study of metabolism of new chemical entities is carried out to clarify several aspects such as: the rate and site of metabolism, enzymes and tissues selectivity and enzyme inhibition and induction responsible for drug–drug interactions. The complexity of the area represents a great challenge for the pharmaceutical industry. These issues have been addressed by different approaches ranging from *in vitro* experiments (using different experimental systems, expressed enzymes, tissue fractions, and hepatocytes) to more labor intensive *in vivo* experiments in animals. Recently, many publications have shown computational methods trying to address metabolic issues. GRID based models have contributed in the field of cytochrome inhibition [28–31], site of metabolism prediction [32], selectivity analysis [33, 34], selective site of metabolism prediction and metabolic stability [35]. However, there are still relevant aspects that have not been analyzed using any computational technique, like the substrate specificity for different cytochromes and the correlation between the metabolism in recombinant cytochromes and human liver microsomes. This is mainly due to the fact that there is insufficient consistent experimental data on which to base the studies and produce the models.

There are different families of enzymes involved in the xenobiotic metabolism, but the most relevant one in drug discovery is the cytochrome P450 (CYP) family. The major CYP enzymes contributing to drug metabolism are:

- CYP3A4, responsible for metabolism of approximately 50% of drugs cleared by hepatic metabolism.
- CYP2D6, usually described as the second most relevant CYP enzyme responsible for about 20% of drugs on the market. It is particularly relevant for drug classes like serotonin re-uptake inhibitors (i.e. fluoxetine, fluvoxamine, paroxetine or sertraline).
- CYP2C subfamily, responsible for the metabolism of 20% of the drugs on the market. It covers a wide variety of therapeutical classes.
- Finally there is a set of enzymes that are responsible for the metabolism of certain chemical classes like CYP1A2 or CYP2E1.

The percentage of drugs designated to be metabolized by each CYP is obtained from historical data [36]. In the future other CYPs may increase in relevance for new chemical entities.

The interaction between a substrate and a cytochrome can be described in the same way as the interaction of a compound with any pharmacological target. The crystal structures of several enzymes are published in the protein data bank (pdb):

CYP2C9 (pdb codes: 1OG2, 1OG5 and 1R09), CYP 2C8 (pdb code:1PQ2) and CYP3A4 (pdb code: 1TQN,1 WOE, 1WOF and 1WOG). Moreover several homology models are also available for CYP1A2 [37], CYP2C9 [33,34] CYP2C19 [33], CYP2D6 [37] and CYP3A4 [37]. The availability of the homology models and crystal structures of the cytochromes leads to the possibility of using structure-based design strategies like docking or GRID-based protein selectivity analysis, in addition to the ligand-based modeling.

10.4.1
Cytochrome P450 Inhibition

The inhibition of several cytochrome enzymes has been studied using different MIF-based techniques. One of the experimental factors that has to be considered in the modeling of this property is the type of inhibition. Therefore, it is relevant to study a homologous series of compounds which, hypothetically, has the same inhibition type, or it has to be experimentally checked.

There are several published studies based on CoMFA [38–42] or GRID/GOLPE methodology [28, 43, 44] in the CYP inhibition area. In this case the alignment is the most difficult step in the modeling process. Usually, in order to obtain an objective alignment rule the studies are focused on one particular series of compounds that can be easily aligned. In other cases structural-based design techniques have been applied to select the conformation and the alignment used later in the GRID/GOLPE analysis [28, 43].

The same heterogeneous series of CYP2C9 competitive inhibitors used in the GRID/GOLPE analysis described by Afzelius et al. [28] has been utilized using an alignment independent GRID analysis [30]. Figure 10.10 reports the predicted versus experimental value for both modeling types. The interpretation of both models pointed to the same regions in the molecule being relevant for the drug–cytochrome interaction.

10.4.2
Site of Metabolism Prediction

The elucidation of the site in a molecule at which the metabolism occurs, is one of the most time and sample consuming experimental tasks in the ADME field. In some cases, the experimental methodology does not help one to determine the precise location of metabolism. Nevertheless, the experimental information could be very relevant for exploration of the pharmacological activity or the toxic effect of the formed metabolites. Moreover, knowledge of the site of metabolism could help in the chemical protection of the molecule, making the compound less liable to metabolic reactions.

There are two main factors that determine the site of metabolism: (i) the chemical reactivity and (ii) the preferred orientation of the compound inside the cytochrome cavity. A new technique called MetaSite [32] has been developed in order to consider at the same time, the substrate–cytochrome interaction and the chemical reactivity of the compounds towards oxidation. The recognition part compares the interaction profile of the enzyme based on GRID-MIFs and different conformations of the potential ligands. The reactivity part comes from precomputed reactivity values of fragments that are recognized in the structure under consideration. The prediction rate for the site of metabolism for five cytochromes (CYP1A2, CYP2C9, CYP2C19, CYP2D6 and CYP3A4) has been validated using more than 900 metabolic reactions.

10.4.3
Metabolic Stability

Together with the site of metabolism, the rate of metabolism is a fundamental factor since this will contribute to the rate of elimination of the compound from the body. The rate of metabolism can be measured using different *in vitro* systems (cytochrome recombinant, human liver microsome, human hepatocites, etc). Normally, the data are produced using different compounds and the rate values are extracted from complex systems like liver microsomes, or human hepatocites.

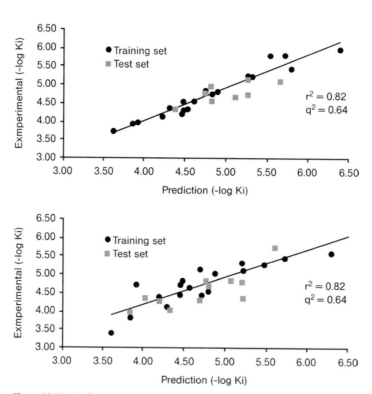

Figure 10.10. Prediction vs. experimental values using (a) GRID/GOLPE modeling, $r^2 = 0.94$; $q^2 = 0.73$ (training set: blue dots; validation set: red dots) and (b) GRIND independent descriptor, $r^2 = 0.73$; $q^2 = 0.54$.

Often the lack of homogenity of the data source and experimental methods makes derivation of predictive computational models too complex.

Predictive models can be better produced when recombinant cytochrome data are available, an experimental technique which may increase the probability of obtaining consistent and predictive models, since one protein is involved in the metabolic reaction. The most accurate data to describe the rate and affinity of the ligand towards an enzyme are the kinetic parameters Vmax and Km. Nevertheless, the calculation of these parameters is time consuming. A less precise parameter is the determination of the compound percentage remaining in a cytochrome incubation after a certain period of time. These metabolic data are less accurate and can only be used to classify the compounds in a metabolic system as stable or unstable. This type of data was the basis for a predictive model of metabolic stability towards CYP3A4 [35].

10.4.4
Selectivity Analysis

Another interesting aspect that can be computationally analyzed starting from the protein structures is the differences between the most relevant cytochromes [33]. This analysis can clarify the specific interaction pattern of each cytochrome which can be used to predict the selective sites of metabolism when a subfamily of cytochromes are compared. It may also help to analyze the differences between the different homology and crystal structures [34].

CYP1A2, CYP2C9, CYP2C19, CYP2D6 and CYP3A4 were analyzed using this technique (unpublished data). Five different probes, which correspond to five different types of interactions were used: DRY (hydrophobic interaction), N1 (hydrogen bond donor), O (hydrogen bond acceptor), N3+ (positive electrostatic interaction) and O- (negative electrostatic interaction). The grid step was 1 Å and the flexible option (MOVE = 1) was applied [45]. All other GRID directives were used at default values. The use of the flexible option enables the computation of the energies not only at the actual atomic positions but also at the potential positions where the amino acid sidechains could interact with a chemical group (probe).

The molecular interaction fields were subjected to multivariate data analysis using consensus principal component analysis (cPCA).

The relevant points for the different proteins and probes were extracted from the 3D representation of the loading values. Five loading fields were extracted per protein, one for each probe, describing the selective interactions that made one protein compared to the others. The loading field for CYP2D6 and the N1 probe is shown in Fig. 10.11.

Using a similar approach to that already used in MetaSite the loading fields can be transformed into a correlogram fingerprint.

This technique was applied to a study of the selective site of metabolism for the CYP 2C subfamily (CYP2C9, CYP2C8, CYP2C19 and CYP2C18). Tables 10.3–10.6 summarize the results obtained for the selective site of metabolism. For all cases

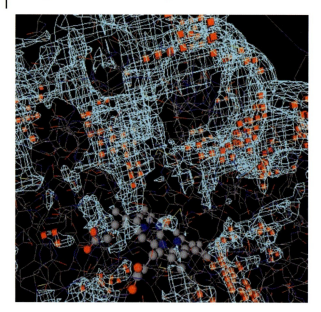

Figure 10.11. Loading fields for the cytochrome CYP2D6 and N1. Red Cubes: selective loadings.

analyzed for CYP2C9 the methodology predicted the correct site of metabolism within the first three options highlighted by the method.

Out of the ten substrates tested for CYP2C19, seven were well predicted by the methodology within the first three options reported by the method and three were mis-predicted by this selective site of metabolism method.

Four substrates out of six are well predicted by this methodology in the case of CYP2C8 substrates.

The overall prediction rate for the selective site of metabolism within the CYP2C subfamily based on the selective interaction profiling using the loading plots from a cPCA analysis based on flexible GRID interaction fields was 72.4%.

Selectivity analysis has also been used to compare the crystal structures and the homology models for CYP2C5 and CYP2C9. The conclusions from this study showed that the homology model of CYP2C5 built from CYP2C9 is more similar to the CYP2C9 crystal structure than to the CYP2C5 crystal and, similarly, the homology model for CYP2C9 based on the CYP2C5 crystal structure is more similar to the CYP2C5 than to the CYP2C9 crystal structure. The only homology model that is more similar to the crystal structure than the template was based on several CYP structures and this methodology is therefore recommended by the authors [34].

Table 10.3 CYP2C9 substrates. The experimentally determined site of metabolism.

Compound	Structure	Reaction	Ki (µM)	Km (µM)	Vmax (mmol min^{-1} nmol^{-1})	Reference
S-Warfarin		hydroxylation	13.6	4/9.9		46, 47
Progesterone		hydroxylation	5.5			
Diclofenac		hydroxylation		15/8.3	40/15	48, 49
S-Ibuprofen		hydroxylation				50
Naproxen		O-desmethylation				51
Piroxicam		hydroxylation				52

Table 10.3 Continued.

Compound	Structure	Reaction	Ki (μM)	Km (μM)	Vmax (mmol min⁻¹ nmol⁻¹)	Reference
Torasemide		hydroxylation				53
Phenytoin		hydroxylation	6.0			54
Fluvastatin		hydroxylation	2.2/3.3			55
Gemfibrozil		hydroxylation	5.8			56
S-Miconazole		hydroxylation	6.0			
R-Fluoxetine		N-desmethylation		13.6	17.0	57

10.4 Metabolism | 239

Table 10.4 CYP2C19 substrates. The experimentally determined site of metabolism.

Compound	Structure	Reaction	Ki (µM)	Km (µM)	Vmax (mmol min^{-1} nmol^{-1})	Reference
R-Lansoprazole		hydroxylation				58, 59
Moclobemide		hydroxylation				60
Imipramine		N-desmethylation				61
R-Omeprazole		O-desmethylation				66
Proguanil		hydroxylation				62
Bufuralol		hydroxylation	36	37		63

Table 10.4 Continued.

Compound	Structure	Reaction	Ki (µM)	Km (µM)	Vmax (mmol min^{-1} nmol^{-1})	Reference
Ticlopedine		hydroxylation	1.2			64
S-Mephenytoin		N-desmethylation		54	2.1	65
R-Warfarin		hydroxylation		55		66
R-Mephobarbital		hydroxylation		34		67

Table 10.5 CYP2C8 substrates. The experimentally determined site of metabolism.

Compound	Structure	Reaction	Km (µM)	Vmax (nmol min^{-1} nmol^{-1})	Reference
Retinoic acid	(42-44) 25/26 23/24	hydroxylation	6.1 33.6 22.8	10.8 74 100	68
R-Troglitazone	(39-45) 46-48 (54) (51)	quinone type	2.7 ± 2.5 3.6 ± 2.9 3.1 ± 1.8 120 ± 5	4.2 ± 1.0 0.6 ± 0.1 2.8 ± 0.3 47 ± 1	69
S-Rosiglitazone	15 32 34 33 12-14	hydroxylation N-desmethylation	4 ± 17 30 ± 12 10 ± 1.3 18 ± 3.7	(nmol/h/mg) 174 ± 49 7.7 ± 2.0 146 ± 8.4 54 ± 6.0	70
S & R-Zopiclone	(42-44) (36-39) 29	N-desmethylation (N-oxide)			71
Diazepan		N-desmethylation			72
Amiodarone	(54/55,59/60)	N-deethylation	54.2 ± 26.1 8.6 ± 2.5 41.9 ± 12.2 21.0 ± 3.1 17.5 ± 5.6	18.9 ± 3.5 2.3 ± 0.2 5.9 ± 0.6 2.2 ± 0.1 1.3 ± 0.1	73

Table 10.6. CYP2C18 substrates. The experimentally determined site of metabolism.

Compound	Structure	Reaction	K_m (µM)	k_{cat} (min^{-1})	Kcat/Km (min^{-1} µM^{-1})	Reference
TA derivative	(structure shown with positions 25, 26)	hydroxylation	9 ± 1	125 ± 25	13	74
			75 ± 25	1.7 ± 0.3	0.02	
			180 ± 70	0.4 ± 0.1	0.002	

10.5 Conclusions

There are several ways in which GRID molecular interaction fields can be used in the ADME area:

1. They can be used for series of related compounds to obtain 3D-QSAR models, that usually describe the special regions around the molecule as contributing positively or negatively to the experimental activity.
2. They can be used directly in protein families in order to perform selectivity analysis based on consensus principal component analysis methodology. This technique helps to identify regions in the protein space that are selective for one enzyme. The selective region may describe the selective pattern of the target proteins. Moreover, this information can be used to compare different models of the same enzyme.
3. The GRID fields can be transformed into a set of molecular descriptors describing the interaction of the compounds with different environments such as biological membranes. These descriptors are particularly useful in the modeling of experimental ADME properties like passive trans-cellular permeability, solubility, unspecific binding to plasmatic proteins or volume of distribution. They can also contribute to the description of the interaction of compounds with proteins with broad substrate specificity like CYP3A4 or P-glycoprotein.

The GRID MIFs can also be transformed to alignment independent descriptor (GRIND) aimed to give a distance/interaction profile which has proven to be useful in the description of CYP inhibition, metabolic stability and in the active transporter and recognition area.

However, there are certain properties for which no modeling attempts have been made due to the lack of the appropriate experimental data. Cytochrome substrate specificity or phase II metabolism are two examples.

Although MIF seems to be very appropriate for application in the ADME area, the interpretation of the MIF-related descriptors must be improved. Then they will be more used by medicinal chemists in the design of new analogs with better pharmacokinetic profiles.

Nevertheless, the prediction of the pharmacokinetic profile from the chemical structure reported in Fig. 10.1 is still a challenging task that remains partially unsolved. GRID-MIF descriptors were successfully applied to a lot of related sub-problems helping to address practical work during the drug discovery process.

References

1 Yoshida F.; Topliss J. G. QSAR Model for Drug Human Oral Bioavailability, *J. Med. Chem.* **2000**, *43 (24)*, 4723–4723.

2 Zamora, I.; Oprea, T. I.; Ungell A.-L. Prediction of Oral Drug Permeability, in *Rational Approaches to Drug Design*, Höldje, H. D., Sippl, W. (eds.), Prous Science, S .A. **2001**, pp.271–280.

3 Neuhoff, S.; Ungell, A.-L; Zamora,I.; Artursson, P. pH-Dependent Bidirectional Transport of Weakly Basic Drugs Across Caco-2 Monolayers: Implications for Drug–Drug Interactions, *Pharm. Res.* **2003**, *20 (8)*, 1141–1148.

4 Haller, T; Völk, H.; Deetjen, P.; Dietl, P. The lysosomal Ca^{2+} pool in MDCK cells can be released by $Ins(1,4,5)P_3$-dependent hormones or thapsigargin but does not activate store-operated Ca^{2+} entry, *Biochem. J.* **1996**, *319*, 909–912.

5 Bermejo, M.; Avdeef, A.; Ruiz, A.; Nalda, R.; Ruell, J. A.; Tsinman, O.; González, I.; Fernández, C.; Sánchez, G.; Garrigues, T. M.; Merino, V. PAMPA – a Drug Absorption in vitro Model. 7.a Comparing Rat in situ, Caco-2, and PAMPA Permeability of Fluoroquinolones, *Eur. J. Pharm. Sci.* **2004**, *21 (4)*, 429–441.

6 Nakai, D.; Kumamoto, K.; Sakikawa, C.; Kosaka, T.; Tokui, T. Evaluation of the protein binding ratio of drugs by a micro-scale ultracentrifugation method, *J. Pharm. Sci.* **2004**, *93*, 847–854.

7 Lombardo, F.; Obach, R. S.; Shalaeva, M. Y.; Gao, F. Prediction of Volume of Distribution Values in Humans for Neutral and Basic Drugs Using Physicochemical Measurements and Plasma Protein Binding Data, *J. Med. Chem.* **2002**, *45 (13)*, 2867–2876.

8 Huuskonen, J.; Salo, M.; Taskinen, J. Neural network modeling for estimation of the aqueous solubility of structurally related drugs, *J Pharm Sci.* **1997**, *86 (4)*, 450–454.

9 Lombardo, F.; Obach, R. S.; Shalaeva, M. Y.; Gao, F. Prediction of Human Volume of Distribution Values for Neutral and Basic Drugs. 2. Extended Data Set and Leave-Class-Out Statistics *J. Med. Chem.* **2004**, *47 (5)*, 1242–1250.

10 Gan, J.; Skipper, P. L.; Tannenbaum, S. R. Oxidation of 2,6-Dimethylaniline by Recombinant Human Cytochrome P450s and Human Liver Microsomes, *Chem. Res. Toxicol.* **2001**, *14 (6)*, 672–677.

11 MacDougall, J. M.; Fandrick, K.; Zhang, X.; Serafin, S. V.; Cashman, J. R. Inhibition of Human Liver Microsomal (S)-Nicotine Oxidation by (–)-Menthol and Analogues, *Chem. Res. Toxicol.* **2003**, *16 (8)*, 988–993.

12 Langouet, S.; Welti, D. H.; Kerriguy, N.; Fay, L. B.; Huynh-Ba, T.; Markovic, J.; Guengerich, F. P.; Guillouzo, A.; Turesky, R. J. Metabolism of 2-Amino-3,8-dimethylimidazo[4,5-*f*]- quinoxaline in Human Hepatocytes: 2-Amino-3-methylimidazo[4,5-*f*]quinoxaline-8-carboxylic Acid Is a Major Detoxication Pathway Catalyzed by Cytochrome P450 1A2, *Chem. Res. Toxicol.* **2001**, *14 (2)*, 211–221.

13 Goodford, P.J., A Computational Procedure for Determining Energetically Favorable Binding Sites on Biologically Important Macromolecules, *J. Med. Chem.* **1985**, *28*, 849–857.

14 Cruciani, G.; Pastor, M.; Clementi, S. Handling information from 3D grid maps for QSAR studies, in *Molecular*

Modeling and Prediction of Bioactivity, K. Gundertofte, K; Jørgensen, F. E. (eds.) Kluwer Academic/ Plenum Publishers, New York **2000**, pp. 73–82.

15 Todeschini, R.; Consonni, V. *Handbook of Molecular Descriptors, Methods and Principles in Medicinal* Chemistry, Mannhold, R.; Kubinyi, H.; Timmerman, H. (eds.) John Wiley & Sons, Weinheim, **2001**.

16 Ertl, P.; Rohde, B.P. Fast calculation of molecular surface area as sum of fragments based contributions and its appilication to the prediction of drug transport properties, *J.Med.Chem.*, **2000**, *43*, 3714–3717.

17 Pastor, M.; Cruciani, G.; McLay, I.; Pickett, S.; Clementi, S. GRid-INdependent descriptors (GRIND): a novel class of alignment-independent three-dimensional molecular descriptors, *J. Med. Chem.* **2000**, *43 (17)*, 3233–3243.

18 Lacombe, O.; Woodley, J.; Solleux, C.; Delbos, J. M.; Boursier-Neyret, C.; Houin, G. Localisation of drug permeability along the rat small intestine, using markers of the paracellular, transcellular and some transporter routes, *Eur. J. Pharm. Sci.* **2004**, *23 (4–5)*, 385–391.

19 Cruciani, G.; Pastor, M.; Guba, W. VolSurf: a new tool for the pharmacokinetic optimization of lead compounds, *Eur. J. Pharm. Sci.* **2000**, *11* Suppl 2, S29–39.

20 Cruciani, G.; Meniconi, M.; Carosati, E.; Zamora, I.; Mannhold R. VOLSURF: A Tool for Drug ADME-Properties Prediction, in *Drug Bioavailability*, Waterbeemd, H.; Lennernäs, H.; Artursson, P. (eds.) Wiley-VCH, Weinheim **2004**, pp. 406–419.

21 Crivori, P.; Cruciani, G.; Carrupt, P.-A.; Testa, B. Predicting Blood-Brain Barrier Permeation from Three-Dimensional Molecular Structure, *J. Med. Chem.* **2000**, *43 (11)*, 2204–2216.

22 Ano, R.; Kimura, Y.; Shima, M.; Matsuno, R.; Tamio Ueno, T.; Akamatsu, M. *Bioorg. Med. Chem.* **2004**, *12 (1)*, 257–264.

23 Gebauer, S.; Knutter, I.; Hartrodt, B.; Brandsch, M.; Neubert, K.; Thondorf, I. Three-dimensional quantitative structure-activity relationship analyses of peptide substrates of the mammalian H+/peptide cotransporter PEPT1, *J. Med. Chem.* **2003**, *46 (26)*, 5725–5734.

24 Snyder, H. J.; Tabbas, L. B.; Berry, D. M.; Duckworth, D. C.; Spry, D. O., Dantzing, A. Structure-activity relationship of carbacephalosporines and cephalosporines: Antibacterial activity and interaction with intestinal proton-dependent dipeptide transport carrier of caco-2 cells, *Antimicrobial Agents Chemother.* **1997**, *41 (8)*, 1649–1657.

25 Cianchetta, G.; Singlenton R. W.; Zhang, M.; Wildgoose, M.; Giesing, D.; Fravolini, A.; Cruciani, G.; Vaz, R. J. *J. Med. Chem.* in press.

26 Sharom, F. Probing of conformational changes, catalytic cycle and ABC transporter function in ABC proteins, in *From Bacteria to Man*, Holland, I. B.; Cole, S. P. C.; Kuchler, K.; Higgins, C. F. (eds.), Academic Press, New York **2003**, pp.107–133.

27 Nezasa, K.-i.; Higaki, K.; Matsumura, T.; Inazawa, K.; Hasegawa, H.; Nakano, M.; Koike, M. Liver-Specific Distribution of rosuvastatin in rats: Comparison with Pravastatin and Simvastatin, *Drug Metab. Disp.* **2002**, *30 (11)*, 1158–1163.

28 Afzelius, L.; Zamora, I.; Ridderström, M.; Andersson, T. B.; Karlén, A.; Masimirembwa, C. M. Competitive CYP2C9 inhibitors: enzyme inhibition studies, protein homology modeling, and three-dimensional quantitative structure-activity relationship analysis, *Mol. Pharmacol.* **2001**, *59*, 909–919.

29 Ekin,S.; Bravi, G. Three- and four dimensional-quantitative structure activity relationship (3D/4D-QSAR) analyses of CYP2C9 inhibitors, *Drug Metab. Disp.* **2000**, *28*, 994–1002.

30 Afzelius, L., Andersson, T. B.; Karlén, A.; Masimirembwa, C. M.; Zamora, I. Discriminant and quantitative PLS analysis of competitive CYP2C9 inhibitors versus non-inhibitors using alignment independent GRIND descriptors, *J. Comput.-Aided Mol. Des.*, **2002**, *16*, 443–458.

31 Afzelius, L.; Andersson, T. B.; Karlén, A.; Masimirembwa, C. M.; Zamora, I. Conformer- and alignment-independent model for predicting structurally diverse

competitive CYP2C9 inhibitors, *J. Med. Chem.* **2004**, *47 (4)*, 907–914.
32 Zamora, I.; Afzelius, L.; Cruciani, G. Predicting drug metabolism: a site of metabolism prediction tool applied to the cytochrome P450 2C9, *J. Med. Chem.* **2003**, *46*, 2313–2324.
33 Ridderström, M.; Zamora, I.; Fjäström, O.; Andersson, T.B. Analysis of Selective Regions in the Active Sites of Human CYP 2C8, 2C9, 2C18 and 2C19 Homology models Using GRID/CPCA, *J. Med. Chem.* **2001**, *44*, 4072–4081.
34 Afzelius, L.; Raubacher, F.; Karlen, A.; Jorgensen, F.S.; Andersson, T.B.; Masimirembwa, C.M.; Zamora, I. Structural analysis of CYP2C9 and CYP2C5 and an evaluation of commonly used molecular modeling techniques, *Drug Metab. Disp.* **2004**, *32 (11)*, 1218–29.
35 Crivori, P.; Zamora, I.; Speed, B.; Orrenius, C.; Poggesi, I. Model based on GRID-derived descriptors for estimating CYP3A4 enzyme stability of potential drug candidates, *J. Comput.-Aided Mol. Des.* **2004**, *18 (3)*, 155–166.
36. Rendic, S.; Di Carlo, F. J. Human cytochrome P450 enzymes: a status report summarizing their reactions, substrates, inducers, and inhibitors, *Drug Metab. Rev.* **1997**, *29*, 413–580.
37. DeRienzo F.; Fanelli F.; Menziani M. C.; De Benedetti P. G. *J.Comput.-Aided Mol. Des.* **2000**, *14*, 93–116.
38. Locuson, C.W. 2nd; Suzuki, H.; Rettie, A.E.; Jones J.P. Charge and Substituent Effects on Affinity and Metabolism of Benzbromarone-Based CYP2C19 Inhibitors, *J. Med. Chem.* **2004**, *47 (27)*, 6768–6776.
39. Suzuki, H.; Kneller, M. B.; Rock, D. A.; Jones, J. P.; Trager, W. F.; Rettie, A. E. Active-site characteristics of CYP2C19 and CYP2C9 probed with hydantoin and barbiturate inhibitors, *Arch Biochem Biophys.* **2004**, *429 (1)*, 1–15.
40 Haji-Momenian, S.; Rieger, J. M.; Macdonald, T. L.; Brown, M. L. Comparative molecular field analysis and QSAR on substrates binding to cytochrome p450 2D6, *Bioorg Med Chem.* **2003**, *11 (24)*, 5545–5554.
41 Asikainen, A.; Tarhanen, J.; Poso, A.; Pasanen, M.; Alhava, E.; Juvonen, R. O. Predictive value of comparative molecular field analysis modelling of naphthalene inhibition of human CYP2A6 and mouse CYP2A5 enzymes, *Toxicol In Vitro* **2003**, *17 (4)*, 449–455.
42 He, M.; Korzekwa, K. R.; Jones, J. P.; Rettie ,A. E.; Trager, W. F. Structural forms of phenprocoumon and warfarin that are metabolized at the active site of CYP2C9, *Arch. Biochem. Biophys.* **1999**, *372 (1)*, 16– 28.
43 Lozano, J. J.; Pastor, M.; Cruciani, G.; Gaedt, K.; Centeno, N. B.; Gago, F.; Sanz, F. 3D-QSAR methods on the basis of ligand-receptor complexes. Application of COMBINE and GRID/ GOLPE methodologies to a series of CYP1A2 ligands, *J. Comput.- Aided Mol. Des.* **2000**, *14 (4)*, 341–353.
44 Poso, A.; Gynther, J.; Juvonen, R. A comparative molecular field analysis of cytochrome P450 2A5 and 2A6 inhibitors, *J. Comput.-Aided Mol. Des.* **2001**, *15 (3)*, 195–202.
45 Goodford, P. J. Atom movement during drug-receptor interactions. *Rational Mol. Des. Drug Res.* **1998**, *42*, 215–230.
46 Rettie, A. E.; Korsekwa, K. R.; Kunze, K. L.; Lawrence, R. F.; Eddy, A. C.; Aoyama, T.; Gelboin, H. V.; Gonzalez, F. J.; Trager W. F. Hydroxylation of warfarin by human cDNA-expressed cytochrome P-450: A role for P-4502C9 in the etiology of (S)-warfarin-drug interactions, *Chem. Res. Toxicol.* **1992**, *5*, 54–59.
47 Thijssen, H. H.; Flinois, J. P.; Beaune, P. H. Cytochrome P4502C9 is the principal catalyst of racemic acenocoumarol hydroxylation reactions in human liver microsomes, *Drug Metab. Disp.* **2000**, *28*, 1284–1290.
48 Transon, C.; Leemann, T.; Vogt, N. Dayer, P. In vivo inhibition profile of cytochrome P450$_{tb}$ (CYP2C9) by (±)-fluvastatin. *Clin. Pharmacol. Ther.* **1995**, *58*, 412–417.
49 Poli-Scaife, S.; Attias, R.; Dansette, P. M.; Mansuny, D. The substrate binding site of human liver cytochrome P450 2C9, *Biochemistry,* **1997**, *36*, 12672– 12682.
50 Hamman, M. A.; Thompson, G. A.; Hall, S. D. Regioselective and stereoselective metabolism of ibuprofen by

human cytochrome P450 2C, *Biochem. Pharm.* **1997**, *54*, 33–41.
51 Miners, J. O.; Coulter, S.; Tukey, R. H.; Veronese, M. E.; Birkett, D. J. Cytochromes P450, 1A2, and 2C9 are responsible for the human hepatic O-demethylation of R- and S-naproxen, *Biochem. Pharm.* **1996**, *51*, 1003–1008.
52 Tracy, T. S.; Marra, C.; Wrighton, S. A.; Gonzalez, F. J.; Korzekwa, K. R. Studies of flurbiprofen 4'-hydroxylation. Additional evidence suggesting the sole involvement of cytochrome P450 2C9 *Biochem Pharmacol.* **1996**, *52*, 1305–1309.
53 Miners, J. O.; Rees, D. L. P.; Valente, L.; Veronese, M. E.; Birkett, D. J. Human hepatic Cytochrome P450 catalyses the rate-limiting pathway of torsemide metabolism, *J. Pharmacol. Exp. Ther.* **1995**, *272*, 1076–1081.
54 Veronese, M. E.; Mackenzie, P. I.; Doecke, C. J.; McManus, M. E.; Miners, J. O.; Birkett, D. J. Tolbutamide and phenytoin hydroxylations by cDNA-expressed human liver cytochrome P4502C9, *Biochem. Biophys. Res. Commun.* **1991**, *175*, 1112–1118.
55 Fischer, V.; Johanson, L.; Heitz, F.; Tullman, R.; Graham, E.; Baldeck, J. P.; Robinson, W. T. The 3-hydroxy-3-methylglutaryl coenzyme A reductase inhibitor fluvastatin: effect on human cytochrome P-450 and implications for metabolic drug interactions, *Drug Metab. Disp.* **1999**, *27*, 410–416.
56 Wen, X.; Wang, J. S.; Backman, J. T.; Kivistö, K. T.; Neuvonen, P. J. Gemfibrozil is a potent inhibitor of human cytochrome P450 2C9, *Drug. Metab. Disp.* **2001**, *29*, 1359–1361.
57 Margolis, J. M.; O'Donnell, J. P.; Mankowski, D. C.; Ekins, S.; Obach, R. S. (R)-, (S)-, and racemic fluoxetine N-demethylation by human cytochrome P450 enzymes, *Drug Metab. Disp.* **2000**, *28*, 1187–1191.
58 Pearce, R. E.; Rodrigues, A. D.; Goldstein, J. A.; Parkinson. A. Identification of the human P450 enzymes involved in lansoprazole metabolism, *J. Pharmacol. Exper. Therap.* **1996**, *277*, 805–816.
59 Äbelö, A.; Andersson, T. B.; Antonsson, M.; Knuts-Naudot, A.; Skånberg, I.; Weidolf, L. Stereoselective metabolism of omeprazole by human cytochrome P450 enzymes, *Drug Metab. Disp.* **2000**, *28*, 966–972.
60 Gram, L. F.; Guentert, T. W.; Grange, S.; Vistisen, K.; Brøsen, K. Moclobemide: a substrate of CYP2C19 and an inhibitor of CYP2C19, CYP2D6 and CYP1A2: A panel study, *Clin. Pharmacol. Ther.* **1995**, *57*, 670-677.
61 Koyama, E.; Chiba, K.; Tani, M.; Ishizaki, T. Reappraisal of human CYP isoforms involved in imipramine N-demethylation and 2-hydroxylation: a study using microsomes obtained from putative extensive and poor metabolizers of S-mephenytoin and eleven recombinant human CYPs, *J. Pharmacol. Exp. Ther.* **1997**, *281*, 1199–1210.
62 Setiabudy, R.; Kusaka, M.; Chiba, K.; Darmansjah, I.; Ishizaki, T. Metabolic Disposition of proguanil in extensive and poor metabolizers of S-mephenytoin 4'-hydroxylation recruited from an Indonesian population, *Br. J. Clin. Pharmacol.* **1995**, *39*, 297–303.
63 Mankowski, D. C. The role of CYP2C19 in the metabolism of (+/–) bufuralol, the prototypic substrate of CYP2D6, *Drug Metab. Disp.* **1999**, *27*, 1024–1028.
64 Ko, J. W.; Desta, Z.; Soukhova, N. V.; Tracy, T.; Flockhart, D. A. In vitro inhibition of the cytochrome P450 (CYP450) system by the antiplatelet drug ticlopedine: potent effect on CYP2C19 and CYP2D6, *Br. J. Clin. Pharmacol.* **2000**, *49*, 343–351.
65 Wedlund, P. J.; Aslanian, W. S.; McAllister, C. B.; Wilkinson, G. R.; Branch, R. A. Mephenytoin hydroxylation deficiency in Caucasians: frequency of a new oxidative drug metabolism polymorphism, *Clin. Pharmacol. Ther.* **1984**, *36*, 773–780.
66 Kaminsky, L. S.; de Morais, S. M. F.; Faletto, M. B.; Dunbar, D. A.; Goldstein, J. A. Correlation of Human Cytochrome P4502C Substrate Specificities with Primary Structure: Warfarin As a Probe, *Mol. Pharmacol.* **1993**, *43*, 234–239.
67 Kobayashi, K.; Kogo, M.; Tani, M.; Shimada, N.; Ishizaki, T.; Numazawa, S.; Yoshida, T.; Yamamoto, T.; Kuroiwa, Y.;

Chiba, K. Role of CYP2C19 in stereoselective hydroxylation of mephobarbital by human liver microsomes, *Drug Metab. Disp.* **2001**, *29*, 36–40.

68 Nadin, L.; Murray, M. Participation of CYP2C8 in retinoic acid 4-hydroxylation in human hepatic microsomes, *Biochem. Pharmacol.* **1999**, *58*, 1201–1208.

69 Yamazaki, H.; Shibata, A.; Suzuki, M.; Nakajima, M.; Shimada, N.; Guengerich, F. P.; Yokoi, T. Oxidation of troglitazone to a quinone-type metabolite catalyzed by cytochrome P-450 2C8 and P-450 3A4 in human liver microsomes, *Drug Metab. Disp.* **1999**, *27*, 1260–1266.

70 Baldwin, S. J.; Clarke, S. E.; Chenery, R. J. Characterization of the cytochrome P450 enzymes involved in the in vitro metabolism of rosiglitazone. *Br. J. Clin. Pharmacol.* **1999**, *48*, 424–432.

71 Becquemont, L.; Mouajjah, S.; Escaffre, O.; Beaune, P.; Funck-Brentano, C.; Jaillon, P. Cytochrome P-450 3A4 and 2C8 are involved in zopiclone metabolism, *Drug Metab. Disp.* **1999**, *27*, 1068–1073.

72 Yasumori, T.; Nagata, K.; Yang, S. K.; Chen, L.-S.; Murayama, N.; Yamazoe, Y.; Kato, R. Cytochrome P450 mediated metabolism of diazepam in human and rat: involvement of human CYP2C in N-demethylation in the substrate concentration-dependent manner, *Pharmacogenetics* **1993**, *3*, 291–301.

73 Ohyama, K.; Nakajima, M.; Nakamura, S.; Shimada, N.; Yamazaki, H.; Yokoi, T. A significant role of human cytochrome P450 2C8 in amiodarone N-deethylation: An approach to predict the contribution with relative activity factor, *Drug Metab. Disp.* **2000**, *28*, 1303–1310.

74 Lopez Garcia, M. P.; Dansette, P. M.; Valadon, P.; Amar, C.; Beaune, P. H.; Guengerich, F. P.; Mansuy, D. Human-liver cytochrome P-450 expressed in yeast as tools for reactive-metabolite formation studies: Oxidative activation of tienilic acid by cytochromes P-450 2C9 and 2C10, *Eur. J. Biochem.*, **1993**, *213*, 223–232.

11
Rapid ADME Filters for Lead Discovery*

Tudor I. Oprea, Paolo Benedetti, Giuliano Berellini, Marius Olah, Kim Fejgin, and Scott Boyer

11.1
Introduction

The Molecular Libraries Initiative (MLI) at NIH [1] is aimed at increasing the availability of small molecules as chemical probes for basic research, via the NIH Small Molecule Repository (NIH_SMR). Aimed at bridging the cultural divides between public and private sectors, MLI is primarily focused on the early stages of drug development, which encompass target identification, assay development, screening and hit-to-probe analysis [1]. Envisioned as "public sector science", these activities are followed by lead identification and optimization, then clinical trials (private sector science). Current proposals at NIH indicate that approximately 500 000 molecules will be physically included in the NIH_SMR. This substantiates the need for developing fast, yet accurate *in silico* technologies to evaluate large numbers of compounds.

MLI aims to bridge, at least in part, the "innovation deficit" [2] faced by the private sector that is leading to increased pressure to deliver new chemical entities (NCEs) and "best-in-class" drugs on the market. Drug discovery scientists have been forced to develop computational tools that facilitate the identification of novel molecular scaffolds (chemotypes) moving across (local) chemical spaces. In the area of *in silico* drug discovery, the public sector is about to face the same challenges as the private sector.

The entire drug discovery process, in particular at the cellular and animal level, has its own challenges [3] that contribute to the "innovation deficit". It is thus imperative that computational tools deliver rapid and accurate models. At the molecular level, drug–receptor interactions continue to be too complex to provide failsafe *in silico* predictions [4]: Entropy and the dielectric constant are but two examples of properties still under debate. The challenges of *in silico* drug discovery include the evaluation of multiple binding modes, accessible conformational

---

*) This chapter is dedicated to the memory of John ("Jack") L. Omdahl (1940–2005), vitamin D_3 pioneer and cytochrome P450 biochemist.

Molecular Interaction Fields. Edited by G. Cruciani
Copyright © 2006 WILEY-VCH Verlag GmbH & Co. KGaA, Weinheim
ISBN: 3-527-31087-8

states for both ligand and receptor, affinity and selectivity vs. efficacy, absorption, metabolic stability (site of reactivity and turn-over), distribution and excretion (ADME), as well as *in vivo* vs. *in vitro* properties of model compounds, while in the same type seeking a favorable IP position (see Fig. 11.1).

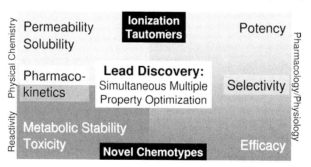

Figure 11.1. Lead discovery requires integrative approaches to address a multiple response surface optimization problem: properties related to physical chemistry, chemical reactivity, pharmacology and physiology require iterative optimization for novel chemotypes (favorable intellectual property (IP) position), if the desired outcome is a marketed drug. The increased difficulty of finding optima is suggested by darker backgrounds.

11.2
The Rule of Five (Ro5) as ADME Filter

Computational methodology shortcomings were soon discovered by the computational, combinatorial and medicinal chemistry communities. Permeability and solubility, key properties required for orally available compounds [5] are now routinely screened using computational and experimental methods [6] prior to candidate lead selection. Having a high impact on the successful progression of drug candidates towards the launch phase [7], ADME/Tox profiling has become an essential step in early drug discovery. Fast ADME filters are likely to be required (and iteratively refined) because of the high-throughput technologies being deployed throughout the industry, as well as in connection with the NIH_SMR.

The first rapid ADME filter was developed by Chris Lipinski et al. at Pfizer [6]. They analyzed a subset of 2245 drugs from the World Drug Index (WDI), in order to understand the common property features of orally available drugs. The QSAR (Quantitative Structure–Activity Relationship) paradigm for structure-permeability,[1] first suggested [8] by Han Van de Waterbeemd et al., provided the descriptor framework for the "Rule of Five" (Ro5): poor absorption or permeation are more likely to occur when molecular weight (MW) is over 500; the calculated [9] octanol/water partition coefficient (clog$P$) is over 5; there are more than 5 H-bond donors (HDO – expressed as the sum of O–H and N–H groups); and there are

---

1) The QSAR paradigm for structure-permeability expresses the passive permeability as a function of hydrophobicity, molecular size, and hydrogen-bond capacity.

more than 10 H-bond acceptors (HAC – expressed as the sum of N and O atoms) [6]. Any pairwise combination of the following conditions: MW > 500, cLogP > 5, HDO > 5, and HAC > 10, may result in compounds with poor permeability (actively transported compounds and peptides excepted). The Ro5 compliance scheme had a major impact in the pharmaceutical industry due to its simplicity, and was rapidly adopted by lead discovery research teams.

Ro5-compliance, from the perspective of putting a drug on the market [10], highlights the importance of appropriate solubility and permeability. However, one can extrapolate to other PK properties such as metabolic stability, excretion and toxicity. Based on screening results from Merck and Pfizer, Lipinski argued [10] that it is easier to optimize PK properties early on in drug discovery, and optimize target affinity at a later stage. Ro5-compliance is already considered when selecting "diversity library" plates, a significant subset of the NIH_SMR. The four parameters included in the Ro5 filters have one possibly significant shortcoming: Designed as a rapid "computational alert" [6] aimed at oral absorption, they cannot offer a comprehensive picture when it comes to understanding ADME models. Thus, we explored the possibility of using molecular interaction fields (MIFs) as the basis for advanced filters for ADME properties.

11.3
Molecular Interaction Fields (MIFs): VolSurf

GRID [11] is a molecular mechanics-based program, developed by Goodford, that estimates the interaction energy between the (macro)molecule atom types and specifically designed chemical (GRID) probes placed at regular lattice points. These interactions are parametrized on the basis of detailed information derived from crystal structures. GRID energies are the sum of the Lennard-Jones, electrostatic and H-bond interactions between the target and the probe. In the case of the DRY probe, that represents hydrophobic interactions, the energy is computed as $E_{DRY} = E_{entropy} + E_{LJ} - E_{HB}$ where $E_{entropy}$ is the ideal entropic contribution towards the hydrophobic effect in an aqueous environment, E_{LJ} is the Lennard-Jones term that accounts for the induction and dispersion interactions, and E_{HB} is the H-bond term that estimates hydrogen bond interactions between the ligand and the GRID water probe. Thus, GRID effectively estimates the MIFs between certain chemical probes and the target (small) molecule. GRID is credited [12] as the computational basis for designing Relenza™ (Zanamivir), an anti-influenza drug marketed by GSK [13].

Regardless of the encoded property, MIFs are typically tabulated energy values that are not conducive to rapid ADME property evaluation. Cruciani and his coworkers [14] developed a descriptor system aimed precisely at capturing MIF properties in a standard format. VolSurf [14] is a program that converts MIF information into fixed types of descriptors, effectively extracting 3D information (see Fig. 11.2). Using PLS [15] (partial least squares) as the statistical engine, VolSurf establishes a relationship between MIF-related information and the target property, seeking differences and similarities in the descriptor values that can be matched with differences and similarities in the target property values.

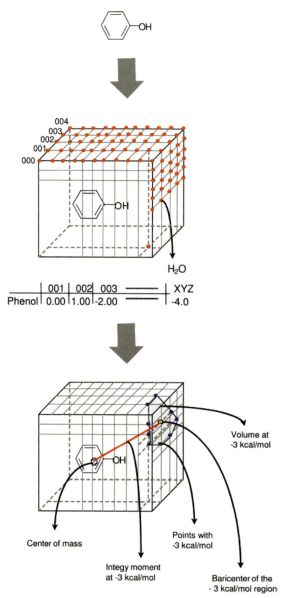

Figure 11.2. The 3-step process of VolSurf descriptor calculation:
1. The 3D structure of the input molecule (e.g., phenol) is generated;
2. MIFs are obtained using GRID; 3. surfaces, volumes and other descriptors are derived at different energy levels.

11.4
MIF-based ADME Models

VolSurf was initially validated on oral absorption [16, 17] and blood–brain-barrier permeation [18] models (see below). VolSurf has continued to be developed to improve *in silico* predictions for ADME properties, although its use has also been extended to receptor-based evaluation of binding affinity [19, 20]. While other software tools for ADME modeling are available (see, e.g., [21]), the MIF-based collection of software and models available from Molecular Discovery (MD) is both extensive and well validated by the private sector. Three programs from MD, VolSurf, MetaSite and Almond, are particularly suited for rapid evaluation of large compound sets [22] in connection with ADME/Tox related properties:

- *Cytochrome P450 (CYP) inhibition / substrate prediction.* Binding for the five most important isozymes in drug metabolism (CYP1A2, CYP2C9, CYP2C19, CYP2D6 and CYP3A4) can be evaluated with MetaSite 2.0. Using over 1000 known substrates and inhibitors for each isozyme, as well as 3D models built from crystallographic data [23] and homology modeling, MetaSite is conceptually similar to the CYP2C9 [24, 25] model described by the same authors. Substrate hydrogens are ranked in terms of the site of metabolism probability [26] using heme Fe proximity, while a docking-based procedure evaluates the ability of potential inhibitors to bind anywhere in the CYP binding site (see Chapter 12).
- *P-glycoprotein (P-gp) substrate predictions* are evaluated using Almond 3.2, a pharmacophore pattern analysis software that combines MIF distances with energies [27], using a model initially derived from P-gp ATP-ase activity [28]. The model contains about 100 drugs (60 known substrates, and 40 nonbinders), each evaluated over 100 diverse conformations (G. Cruciani, personal communication).
- *Blood–brain barrier (BBB) permeability* and all the other PK property predictions discussed here are performed using VolSurf 4.0, as reviewed elsewhere [29]. BBB permeability is predicted from a PLS discriminant analysis model [18]. This model was confirmed with GPSVS.
- *Caco-2 permeability prediction,* based on an experimental model (Caco-2 cells monolayer) that evaluates the intestinal absorption of drugs [30], is derived from known literature datasets – see [31] for a review. This model was confirmed with GPSVS.
- *Water solubility (thermodynamic) prediction,* based on various literature datasets [32, 33], is comparable to other models [29]. This model was confirmed with GPSVS.
- *DMSO solubility prediction* is based on experimental determinations from the University of Perugia [34].
- *Plasma Protein Binding (PPB) prediction* is based on collected plasma protein binding values (PPB) percentage values for therapeutic drugs from literature [35]. In this VolSurf model, values predicted to be 95% or higher are equivalent

(due to model inaccuracy for high PPB values), whereas lower values (a desired feature in drugs) are more accurate [29].
- *Volume of distribution (VD_{ss})*, a drug-disposition parameter relating the amount of drug in the body to the concentration of the drug in the blood (or tissue), tries to address the 'how often' question in the therapeutic dose regimen. The VD_{ss} VolSurf model, based on 118 drugs, is similar to that from Pfizer [36, 37].
- *hERG (Human ether-a-go-go related gene)*, a K^+ channel that is possibly implicated in the fatal arrhythmia known as *torsade de pointes*, appears to be the molecular target responsible for the cardiac toxicity of a wide range of therapeutic drugs [38]. The hERG binding VolSurf model is based on over 200 drugs collected from the literature [39, 40, 41], and is similar to work performed at Roche [42]. An Almond model using 882 measured IC_{50} hERG inhibition values has been disclosed [43].

11.5
Clinical Pharmacokinetics (PK) and Toxicological (Tox) Datasets

To evaluate the tools developed for rapid and accurate ADME property prediction, we screened the clinical pharmacokinetics literature and developed a chemical database, WOMBAT-PK (WOMBAT-Pharmacokinetics) [44]. This database contains 643 drugs with known ADME properties. Currently indexed clinical pharmacokinetics and related physicochemical properties data are summarized in Table 11.1. The top 9 properties were captured from the following sources: Goodman & Gilman's 9th edition [45] (G&G), Avery's 4th edition [46] (Av), the Physician Desk Reference [47] (PDR). FDA's Center for Drug Evaluation and Research website [48] was consulted for FDA approved drug labels. Other resources (e.g., Google™) were sometimes used to compile the WOMBAT-PK database. The maximum recommended therapeutic dose [49] (MRTD) is available from the FDA [50], whereas $MRTD_U$ (MRTD corrected for f_u, the fraction-unbound) was determined by using the %PPB data in WOMBAT-PK.

Experimental $\log D_{7.4}$ and $\log P$ values from compilation tables [51] and pK_a values from Avery [46] are also included in WOMBAT-PK. *In vitro* binding information reported for these drugs in medicinal chemistry literature was extracted from the WOMBAT [44] database. Compiling clinical pharmacokinetic data requires, typically, individual examination. Often, experimental values were "greater than" or "less than" a given value. A systematic round-off procedure was implemented, whereby "<5" was attributed a higher value (= 2.5), compared to "<1" (= 0.5). Numerical values also differ, sometimes significantly, due to various factors (e.g., multiple dose vs. single-dose, children vs. healthy volunteers, etc.), thus conflicting values were sometimes reported. The "on file" values in Table 11.1 are often averages between G&G and Av data, although ≈30% of the indexed values differ by more than 20% between these two sources (data not shown). To identify trends,

Table 11.1 Experimental ADME/Tox data for model development.

Property	On file	G&G	Av
%Oral bioavailability, %Oral	633	312	277
%Urinary excretion, %Urine	326	326	
%Plasma protein binding, %PPB	502	311	434
Clearance, Cl (mL min^{-1} kg^{-1})	491	320	422
Nonrenal clearance (fractional)	442		442
Volume of distribution, VD$_{ss}$ (L kg^{-1})	515	322	453
Half-life, $T_{1/2}$ (h)	628	338	576
Terminal half-life, $TT_{1/2}$ (h)	580		580
Effective concentration (mM L^{-1})	118	118	
MRTD (mM (kg-bw)$^{-1}$ d^{-1})	433		
MRTD$_U$ (mM f_u)	433		
Elog$D_{7.4}$	277		
Elog$P$	272		
pK_{a1}	274		274
pK_{a2}	75		75
In vitro binding data	247		

we attenuated the effect of such discrepancies by implementing an incremental increase procedure to some of the ADME properties, as illustrated in Table 11.2. Incremental rank values were selected from experience whenever possible: e.g., experimental errors related to %Oral occur mostly for values between 20 and 80%; 6/7 and 12/7 represent the $1/2$ and full value of creatinine clearance (120 mL / 70 kg min^{-1}), respectively.

Table 11.2 Parent value ranking for certain PK parameters.

%Oral	Rank3Oral	Rank5Oral	%Urine	RankUrine
0–5	0	0	0–1	0
5.1–19.99	0	1	1.01–5	1
20.0–79.99	1	2	5.01–20	2
80.0–95	2	3	20.01–50	3
>95.1	2	4	50.01–80	4
			>80	5

%PPB	RankPPB	Cl (mL min^{-1} kg^{-1})	RankCl	VD (L kg^{-1})	RankVD
0–5	0	0–(6/7)	0	0–1	0
5.01–20	1	(6.01/7)–(12/7)	1	1.01–3	1
20.01–80	2	(12.01/7)–5	2	3.01–5.5	2
80.01–95	3	5.01–10	3	5.51–12	3
95.01–99	4	10.01–15.5	4	>12	4
>99.1	5	>15.5	5		

11.6
VolSurf in Clinical PK Data Modeling

Ninety two VolSurf descriptors were computed for $N = 623$ (out of 643) drugs from the WOMBAT-PK database. An additional 12 compounds in this dataset could be computed with VolSurf. Some of these drugs, however (e.g., sodium nitroprusside, Na[Fe(CN)$_5$NO] and Auranofin (gold-based compound) are not expected to be of general interest when modeling ADME properties. For the drugs that had %Urine, %PPB and nonrenal clearance data, and low VD$_{ss}$ and low Cl (i.e., VD$_{ss}$ < 5.5 L kg^{-1} and Cl < 10 mL min^{-1} kg^{-1} or LoVLoC values), we obtained reasonably consistent models using VolSurf (data not shown). In its current implementation, VolSurf treats all molecules as neutral species. Therefore, we included estimated log$D$ values from ACDLabs [52] at 5 pH values (5.9, 6.9, 7.4, 7.9 and 8.9), as well as log$S$, log$D$ and log$P$ values from ALOGPS [53], in order to take ionization into account. Here, "log$P/D/S$" refers to estimated partition/solubility values. The combined VolSurf & LogP/D/S model for RankUrine, RankPPB and nonrenal clearance data is summarized in Fig. 11.3. The model shown is quite intuitive (see Fig. 11.3): RankUrine is in the opposite loadings region from RankPPB and nonrenal clearance. Higher RankUrine values (high renal elimination of unchanged drug) correlate well with high solubility and higher metabolic

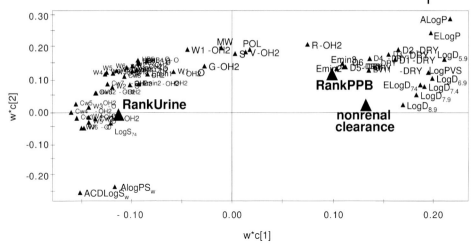

Figure 11.3. Loadings plot for the 2-component PLS model for RankUrine, RankPPB & nonrenal clearance for $N = 283$ drugs.

stability (paralleled by lower log$D$ values [21]). Thus, "log$S$-" and "-OH2" (the GRID water probe) descriptors make a significant contribution to the modeling of this property. By contrast, "logP/D" and the GRID DRY (hydrophobic) probe have a direct (and opposite to "log$S$"/"-OH2") effect in modeling RankPPB and nonrenal clearance. These are two phenomena where hydrophobic interactions are likely to be important (higher log$D$ indicates CYP metabolism [21], whereas large hydrophobic surface areas relate to direct protein binding). The model was further validated by Y scrambling, i.e., permuting the Y variables to safeguard against chance correlation (data not shown).

11.7
ChemGPS-VolSurf (GPSVS) in Clinical PK Property Modeling

To derive *global models* in a true sense, we defined *chemography* as the combination of chemical property rules and objects (chemical structures) that could provide a consistent map of chemical space [54, 55] (similar to the Mercator convention). Chemographic rules included, initially, simple molecular properties such as size, hydrophobicity and flexibility (see Fig. 11.4). By design, two categories of objects were included: "Satellites" that were intentionally placed outside the property space of interest (e.g., drug-like [56, 57]); and "core" objects that were, for the most part, orally available drugs. ChemGPS [54], the chemical global positioning system, comprises both the "core" and "satellite" molecules, and has been adapted to several 2D- and 3D-based descriptor systems. An initial PCA [58] (principal component analysis) model is developed using the predetermined set of molecules and a fixed set of descriptors. ChemGPS (chemographic) coordinates for any external set of molecules are then extracted via PCA, using the same (fixed) set of molecular descriptors that were used to define the chemographic. Thus, PCA-score prediction is used to project new molecules on this predefined map, providing a consistent method of systematically mapping the chemical property space.

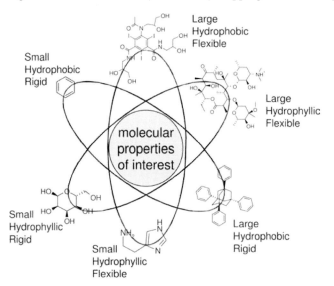

Figure 11.4. The ChemGPS concept: chemical "satellites" are specifically positioned at extreme ends in the molecular property space (as suggested by size, hydrophobicity and flexibility). When multiple objects are placed in various regions of this space, the net result is that predictions are obtained via *interpolation*, not extrapolation. This improves model predictivity, and reduces the chance of finding outliers (although chemical outliers may still produce unusual effects).

We established [54] that, in conjunction with 2D-based descriptors, ChemGPS can provide global chemical space coordinates by performing extensive comparisons with GRID-based principal properties for heteroaromatic compounds [59], principal properties ("z-scores") of a-amino acids [60], as well as by comparison to several local PCA models. The initial ChemGPS map turned out to be 9-dimensional [54].

According to ChemGPS systematics, any map coordinates derived with the same rules (dimensions) and the same objects (set of compounds) can then be used to compare large numbers of chemicals, since the coordinates no longer depend on chemistry (e.g., single vs. multiple chemical classes) and time (e.g., re-doing the model by including/excluding objects or descriptors). ChemGPS is well suited as a reference system for comparing multiple libraries, and keeping track of previously explored regions of the chemical space [61].

FDA's biopharmaceutics classification system, BCS [5], serves as a guide for *in vivo* bioavailability and bioequivalence studies for immediate-release solid oral dosage forms. BCS relies on two ADME properties, passive permeability and solubility. When combining ChemGPS objects (molecules) with VolSurf descriptors, the PCA-predicted scores from ChemGPS-VolSurf (GPSVS scores) correlate well [62] with passive transcellular permeability (1st dimension, GPSVS1) and with solubility (2nd dimension, GPSVS2).

GPSVS1 explains well the passive transcellular mechanism of absorption across the epithelial tissue in the gastrointestinal tract (as observed for a Caco-2 cell monolayer [30] system [17], and in erythrocyte ghost cells), or across the BBB (as modeled in the Crivori dataset [18]). GPSVS1 scores correlate well to Caco-2 permeability data for 22 drugs ($R^2 = 0.67$), to NMR permeability, as measured for 11 drugs in ghost erythrocyte cells ($R^2 = 0.81$) and, in combination with GPSVS2 scores, discriminate well between drugs that pass (BBB+) or not (BBB–) the blood–brain barrier (see [62] for details).

GPSVS2 explains, to a lesser degree, water solubility. Because most efforts to model solubility have been focused on its relationship with log$P$ [9], GPSVS2 was compared not only to Abraham's solubility dataset [32] (794 compounds), but also to the Pomona log$P$ dataset [63] (7954 compounds). GPSVS2 scores correlated directly to measured log$P$ values ($R^2 = 0.61$), and to water solubility ($R^2 = 0.68$).

PCA models are usually interpreted by comparing descriptor loadings (their contribution) to the latent variables, which relate to physical meaning. However, in GPSVS, the first two components are directly related to measured properties. This observation was further substantiated by comparison to PLS models derived for the same five datasets [62]. Unlike QSAR methods, GPSVS does not require biological input (i.e., dependent variables) as a training set. However, while it relies on PCA prediction, the 2 GPSVS dimensions are no longer orthogonal: as they rotate to best correlate permeability and solubility. The 2 axes form an angle [62] of approximately 43°, pointing in opposite directions, i.e., optimizing permeability has a negative effect on solubility and vice versa.

The GPSVS map in Fig. 11.5 shows WOMBAT-PK drugs color-coded according to oral bioavailability. This plot does not take into account the contribution of active processes (e.g., active efflux, intestinal metabolism, etc.) to this composite parameter. For clarity, an orthogonal view is given in Fig. 11.5. In the absence of individual evidence for each drug, low rank values for Cl and VD_{ss} ($VD_{ss} < 5.5\,L\,kg^{-1}$ and Cl $< 10\,mL\,min^{-1}\,kg^{-1}$, or LoVLoC) were used to further examine these results, as shown in Table 11.3. High rank values ($VD_{ss} > 5.5\,L\,kg^{-1}$ and Cl $> 10\,mL\,min^{-1}\,kg^{-1}$, or HiVHiC), and the alternatives (high/low) are separated in Table 11.3. If VD_{ss} and Cl are not considered, Q4 captures 75.5% of all drugs (90% of the good, 80% of the medium and 50% of the poor %Oral, respectively). When the influence of active mechanisms is reduced (LoVLoC), Q4 captures 37.5% of all drugs (63.6% of the good, ~33% of the medium and ~16% of the poor %Oral, respectively). The ratio of "good" and "medium" %Oral values drops dramatically in all the other instances, which appears to indicate that, at least with respect to passive mechanisms, GPSVS-Q4 is the target property space for orally available drugs.

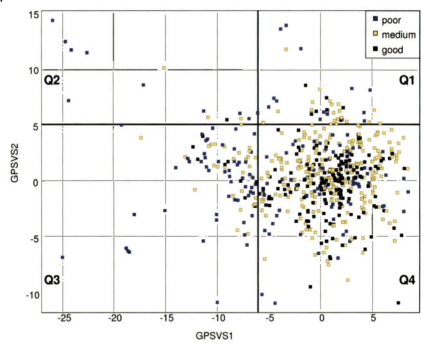

Figure 11.5. GPSVS plot for 613 drugs with known oral bioavailability. Colors codes correspond to Rank3Oral (inset and Table 11.2). Cut-off lines at GPSV1 = −6 and GPSVS2 = 5 and the four defined quadrants are shown. Q4 contains the highest number of drugs with medium (227) and good (159) oral bioavailability. For orientation, small and large MW values are located in the bottom right (Q4), and top left (Q2) quadrants, respectively; small and large clogP values are in the bottom left (Q3), and top right (Q1) quadrants.

Table 11.3 %Oraldrug counts for GPSVS1/GPSVS2 quadrants as defined in Fig. 11.5. LoV, HiV, LoC, and HiCvalues (and their combinations) are defined in text. "Unknown" indicates that one or both VD_{ss} and Cl values were not available. To the right of each quadrant (Q1–Q4) column, % refers to the percentage for that column, related to the number of drugs in WOMBAT-PK analyzed with GPSVS: 154 poor, 283 medium and 176 good %Oral values; the total row % values relate to the number (613) of analyzed drugs.

%Oral	VD_{ss} and Cl	Q1	Q1%	Q2	Q2%	Q3	Q3%	Q4	Q4%
poor	all data	14	9.1	11	7.1	52	33.8	77	50.0
medium	all data	24	8.5	4	1.4	28	9.9	227	80.2
good	all data	7	4.0	1	0.6	9	5.1	159	90.3
total	all data	45	7.3	16	2.6	89	14.5	463	75.5

Table 11.3 Continued.

%Oral	VD$_{ss}$ and Cl	Q1	Q1%	Q2	Q2%	Q3	Q3%	Q4	Q4%
poor	LoVLoC	5	3.2	7	4.5	36	23.4	25	16.2
medium	LoVLoC	6	2.1	2	0.7	15	5.3	93	32.9
good	LoVLoC	1	0.6	1	0.6	7	4.0	112	63.6
total	LoVLoC	12	2.0	10	1.6	58	9.5	230	37.5
poor	LoVHiC	2	1.3	0	0	0	0	30	19.5
medium	LoVHiC	1	0.4	0	0	0	0	31	11.0
good	LoVHiC	0	0	0	0	0	0	6	3.4
total	LoVHiC	3	0.5	0	0	0	0	67	10.9
poor	HiVLoC	0	0	1	0.6	1	0.6	1	0.6
medium	HiVLoC	3	1.1	0	0.0	2	0.7	2	0.7
good	HiVLoC	0	0	0	0	0	0	7	4.0
total	HiVLoC	3	0.5	1	0.2	3	0.5	10	1.6
poor	HiVHiC	2	1.3	1	0.6	4	2.6	4	2.6
medium	HiVHiC	3	1.1	2	0.7	2	0.7	33	11.7
good	HiVHiC	0	0	0	0	0	0	8	4.5
total	HiVHiC	5	0.8	3	0.5	6	1.0	45	7.3
poor	unknown	5	3.2	2	1.3	11	7.1	17	11.0
medium	unknown	11	3.9	0	0	9	3.2	68	24.0
good	unknown	6	3.4	0	0	2	1.1	26	14.8
total	unknown	22	3.6	2	0.3	22	3.6	111	18.1

11.8 ADME Filters: GPSVS vs. Ro5

Having established that GPSVS is compliant with the BCS system [5] regarding oral bioavailability, we further discuss its performance in comparison to Ro5 compliance [6] for the WOMBAT-PK dataset. Ro5 compliance for drugs with respect to %Oral, Cl and VD$_{ss}$ is shown in Table 11.4. Without examining VD$_{ss}$ and Cl, 0 and 1 Ro5 violations capture 80% and 15% of the total number of drugs, respectively.

When Ro5 = 0, 94% of the good, 79.5% of the medium and 65% of the poor %Oral drugs are captured. For LoVLoC, Ro5 = 0 includes 42.3% of all drugs (66.5% of the good, ~36% of the medium and ~26% of the poor %Oral drugs, respectively). The ratio of "good", "medium" and "poor" %Oral values appears to be less sensitive to VD_{ss} and Cl properties, when comparing Ro5 violations with the GPSVS quadrants. Pfizer's computational alert [6] was designed to be all-inclusive, since all drug formulations (tablet, capsule, syrup, etc.) for oral delivery were appropriate, even if %Oral was 5 or less. Even though GPSVS, like Ro5, does not include numeric values for %Oral, it is likely that the VolSurf framework, shown to model VD_{ss} and Cl, provides additional sensitivity to changes in molecular properties.

Table 11.4 %Oral drug counts for Ro5 compliance, as defined in the text. The columns are placed in a manner compatible to those in Table 11.3 to facilitate comparison. R5.n (where n is between 0 and 3) indicates the number of Ro5 criteria violated by drugs in that column. VD_{ss} and Cl values are the same as in Table 11.3. To the right of each Ro5 value (R5.0–R5.3) column, % refers to the percentage for that column, related to the number of drugs in WOMBAT-PK: 163 poor, 291 medium and 178 good %Oral values; the total row % values relate to the number (632) of analyzed drugs.

%Oral	VD_{ss} and Cl	R5.2	R5.2%	R5.3	R5.3%	R5.1	R5.1%	R5.0	R5.0%
poor	all data	15	9.7	13	8.4	35	22.7	100	64.9
medium	all data	10	3.5	6	2.1	50	17.7	225	79.5
good	all data	3	1.7	1	0.6	8	4.5	166	94.3
total	all data	28	4.6	20	3.3	93	15.2	491	80.1
poor	LoVLoC	6	3.9	9	5.8	23	14.9	40	26.0
medium	LoVLoC	3	1.1	2	0.7	10	3.5	102	36.0
good	LoVLoC	1	0.6	1	0.6	4	2.3	117	66.5
total	LoVLoC	10	1.6	12	2.0	37	6.0	259	42.3
poor	LoVHiC	1	0.6	0	0.0	1	0.6	30	19.5
medium	LoVHiC	0	0.0	0	0.0	3	1.1	32	11.3
good	LoVHiC	0	0.0	0	0.0	0	0.0	6	3.4
total	LoVHiC	1	0.2	0	0.0	4	0.7	68	11.1
poor	HiVLoC	1	0.6	0	0.0	2	1.3	0	0.0
medium	HiVLoC	1	0.4	1	0.4	2	0.7	3	1.1
good	HiVLoC	0	0.0	0	0.0	0	0.0	7	4.0
total	HiVLoC	2	0.3	1	0.2	4	0.7	10	1.6

Table 11.4 Continued.

%Oral	VD$_{ss}$ and Cl	R5.2	R5.2%	R5.3	R5.3%	R5.1	R5.1%	R5.0	R5.0%
poor	HiVHiC	2	1.3	2	1.3	3	1.9	4	2.6
medium	HiVHiC	2	0.7	1	0.4	14	4.9	23	8.1
good	HiVHiC	0	0.0	0	0.0	1	0.6	7	4.0
total	HiVHiC	4	0.7	3	0.5	18	2.9	34	5.5
poor	unknown	2	1.3	2	1.3	6	3.9	26	16.9
medium	unknown	4	1.4	2	0.7	21	7.4	65	23.0
good	unknown	5	2.8	0	0.0	3	1.7	29	16.5
total	unknown	11	1.8	4	0.7	30	4.9	120	19.6

A direct comparison between Ro5 and GPSVS quadrants is given in Fig. 11.6. When examining Ro5 scores of 0 and 1, there are three drugs classified as "bad" by GPSVS: the antineoplastics Paclitaxel and Vincristin (both i.v. formulations), and Rifabutin, an antibiotic with 20% Oral and HiVHiC (VD$_{ss}$ = 40 L kg^{-1}, Cl = 12 mL min^{-1} kg^{-1}). The eight drugs from Q2 that have Ro5 ≥ 2 are Amphotericin B (i.v.), Cyclosporine A (30.5%), Ivermectin (60%), Octreotide (1.75%), Rifampin (92.5%), Tacrolimus (26%), Teniposide (i.v.), and Vinblastine (i.v.). Of these drugs, Cyclosporine, Rifampin and Tacrolimus are high-MW peptides that do not rely on passive transcellular mechanisms. Ivermectin, an anti-parasitic macrocyclic lactone, is the only compound that appears to be a false positive.

An additional 28 drugs with Ro5 violations are placed in the Q1 and Q3 quadrants, of which 13 have poor, 12 medium and 3 good oral bioavailability, respectively. The latter are Digitoxin, Methyldigoxin (two cardiac glycosides) and Zafirlukast (a leukotriene D$_4$ antagonist). One drug that violates two Ro5 criteria is not flagged by GPSVS, and is present in the Q4 quadrant: Fosinopril, an angiotensin converting enzyme inhibitor. This is, in fact, a pro-drug (MW = 563 and log$P$ = 5.8) that loses a propanoyl fragment (MW = 52) and has logD_{74} = 2.8. Of the 105 drugs with 0 and 1 Ro5 scores (36 in GPSVS-Q1 and 69 in GPSVS-Q3), 51 have poor, and 41 have medium oral availability, respectively.

Out of the 105 flagged (but not rejected) GPSVS compounds, almost half are indeed, problematic, yet the majority can be "salvaged" in formulation; one compound flagged by Ro5 but accepted by GPSVS is, in fact, a prodrug with otherwise good ADME properties. The majority of the drugs are not flagged either by GPSVS (Q4) nor Ro5 (0 and 1 violations): Of the 462 unflagged drugs, 71 have poor %Oral values. Among these 71 drugs, 25 have LoVLoC values. To summarize, of the 613 drugs for which GPSVS and Ro5 values were compared, the two ADME filters are in agreement for 484 drugs, of which 25 are false positives (Q4,

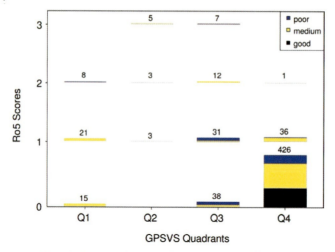

Figure 11.6. Oral bioavailability distribution comparison between GPSVS quadrants and Ro5 violations for 613 drugs. Color codes correspond to Rank3Oral (inset and Table 11.2). See text for details.

Ro5 < 2, yet poor oral availability) and 4 are false negatives (Q2, Ro5 ≥ 2, but medium/good oral availability). Thus, *the combination of GPSVS and Ro5 has a 0.7% false negative rate, and a 5% false positive rate.* We conclude that this combination provides a very effective tool for ADME filtering.

11.9
PENGUINS: Ultrafast ADME Filter

Given current hardware limitations, it takes 10–100 times longer to compute VolSurf descriptors, compared to the Ro5 properties (MW, cLogP, HDO and HAC) for the same molecules. To speed up the procedure, we have implemented a fragment-based version of GPSVS in the PENGUINS [64] software. The initial aim (2000–2001) in developing PENGUINS (Pharmacokinetics Evaluation aNd Grid Utilization IN Silico) was to perform rapid passive permeability (intestinal and brain) predictions for ca. 100 000 compounds / day / CPU, on Linux-based Pentium III machines, starting from 2D structural representation. The software was designed to perform fragment recognition on massive combinatorial libraries, then to approximate VolSurf descriptor values for whole molecules starting from stored VolSurf descriptor values for these fragments. For multiple fragment options, preference is given to larger fragments, to reduce the influence of approximations. Using approximate VolSurf descriptors, GPSVS1 and GPSVS2 scores starting from the ChemGPS/VolSurf model can be predicted, but the 3D structure of the molecule is no longer required. This proves to be a significant time-saving factor.

11.9 PENGUINS: Ultrafast ADME Filter

For comparison, the GPSVS scores from PENGUINS (fragments) are correlated with those for the entire molecules (VolSurf) for a virtual library of 1000 molecules. The reaction used to enumerate the virtual library is shown in Scheme 11.1; the 30 building fragments (R1, R2, R3) are given in Scheme 11.2. Correlation plots and R^2 values given in Fig. 11.7 show that the PENGUINS scores are 90% accurate. We note that, in the PENGUINS approach, certain corrections were introduced for, e.g., the fragmentation of the amide function which is formed between the R3-aldehyde and the R1/R2-amines. These corrections were tested on a number of small molecules before being implemented. It is encouraging that conformational flexibility (at least five nonterminal flexible bonds in the enumerated products) did not appear to influence the GPSVS scores.

The PENGUINS fragment library is not exhaustive. Fragment recognition has a built-in heuristic process, and unknown fragments can be added to the database. The current error rate is under 5% for typical, drug-like, compounds. In PENGUINS, unknown fragments are automatically converted to 3D coordinates, Vol-

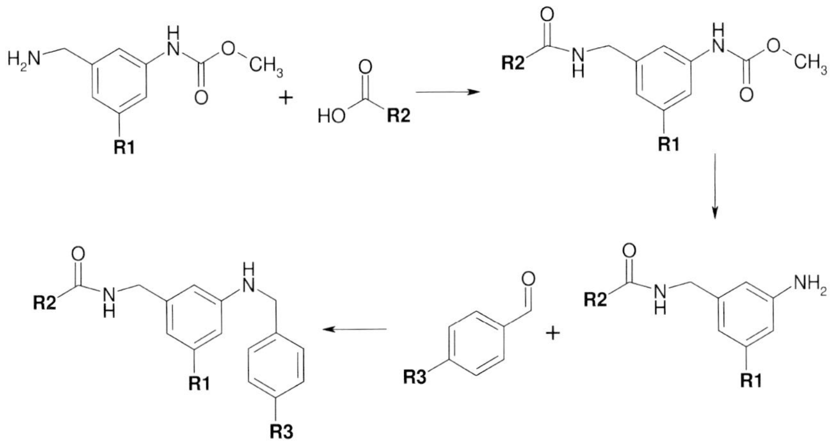

Scheme 11.1. The 3-step reaction scheme used to enumerate the 1000-compounds virtual library. The structures of the 30 building blocks are given in Scheme 11.2.

R1: A—Cl
    A—OH
    A—COOH
    A—OMe
    A—OCF$_3$
    A—NHCOCH$_3$
    A—CH$_3$
    A—NH$_2$
    A—I
    A—SH

R2: A—CH$_2$CH$_3$
    A—CH$_2$OH
    A—CH$_2$COOCH$_3$
    A—CH$_2$OCH$_3$
    A—CH$_2$OCF$_3$
    A—CH$_2$NHCOCH$_3$
    A—CH$_3$
    A—CH$_2$NH$_2$
    A—CH$_2$I
    A—CH$_2$SH

R3: A—Cl
    A—OH
    A—COOH
    A—OMe
    A—OCF$_3$
    A—NHCOCH$_3$
    A—CH$_3$
    A—NH$_2$
    A—I
    A—SH

Scheme 11.2. The 30 building blocks used for the virtual reaction in Scheme 11.1. "A" indicates the attachment points for the remainder of each fragment, as depicted for R1, R2 and R3 in Scheme 11.1.

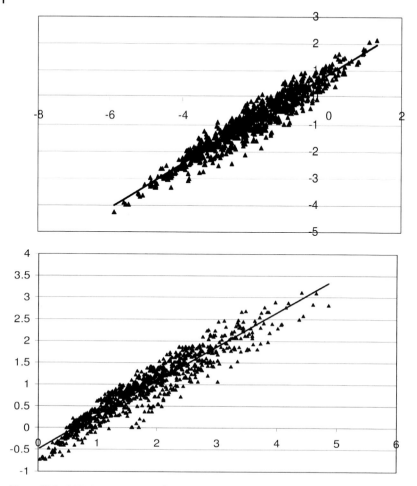

Figure 11.7. GPSVS score accuracy for PENGUINS: Scores from VolSurf (x-axis) for the 1000 structures enumerated according to Schemes 11.1 and 11.2 are compared with the scores from PENGUINS (y-axis) The R^2 values are 0.91 for GPSVS1 scores (a), and 0.9 for GPSVS2 scores (b).

Surf descriptors are computed for that fragment and stored within the fragment database.

Equally important in the PENGUINS approach is the speed factor: The fragment-based library evaluation (from 30 reactants) in PENGUINS took under 3 min, whereas the complete VolSurf evaluation for 1000 compounds took 183 min (on the same CPU, under the same operating system). Ro5 parameter computations for the same 1000 enumerated compounds required slightly more time (6.6 min) on the same machine. At current speeds (2004), running on Linux (Pentium 4, 2 GHz, 1 GB RAM, Red Hat 7.3), PENGUINS can evaluate at least 2 000 000 virtual structures / day / CPU (enumeration with ADME filtering). If the structures are already enumerated (fragment recognition only), PENGUINS can estimate over 3 000 000 structures / day / CPU. The results are fast, within 10% error (compared to GPSVS from VolSurf), and directly interpretable in terms of passive pharmacokinetic properties. We can therefore state that PENGUINS is an ultrafast, high-throughput ADME property filter.

11.10
Integrated ADME and Binding Affinity Predictions

The goal to integrate ADME property prediction with estimating binding affinity, conceptually illustrated in Fig. 11.1, is currently under implementation in PENGUINS. Our earlier studies have shown that VolSurf descriptors can effectively model ligand–receptor binding for a diverse set of X-ray complexes [19, 20]. The GRID docking procedure, implemented in PENGUINS, was applied in conjunction with GPSVS to verify if lead compounds can be optimized at the stage of *in silico* library enumeration.

The 2004.2 release of WOMBAT [44] contains 122 *para*-(R2) phenyl-(R1) methyl-ether ERα (estrogen receptor subtype a) antagonists, including Raloxifene and Tamoxifen, two drugs with %Oral > 50. We wanted to investigate whether the optimal R1/R2 combination had been found. Based on the 8 R1 and 112 R2 substituents (not shown), a total of 896 compounds were enumerated with PENGUINS and evaluated using GPSVS and GRID docking. Some of the compounds that meet the criteria for affinity, permeability and %PPB, but were not included in the initial 122 compounds are illustrated in Fig. 11.8: These compounds are estimated to have $K_i \leq 100$ nM against the receptor, as docked in the Raloxifene binding site of ERα (Protein Data Bank, PDB, entry code 1ERR), good Caco-2 (passive) permeability, and %PPB under 99%. Following a fast ADME filtering step in PENGUINS, these compounds were predicted with the built-in VolSurf model library [29], for greater accuracy. This study further substantiates the hypothesis that rapid ADME filtering prior to docking is an appropriate workflow: ADME filtering with PENGUINS remains faster (i.e., can process many compounds) compared to docking (fewer entries to process). For this particular study, the order in which these two steps were performed did not affect the final list of compounds (data not shown). PENGUINS can become effective for simultaneous ADME and binding affinity optimization, in particular when (i) the structure of the target is available and (ii) the binding mode of active molecules is known.

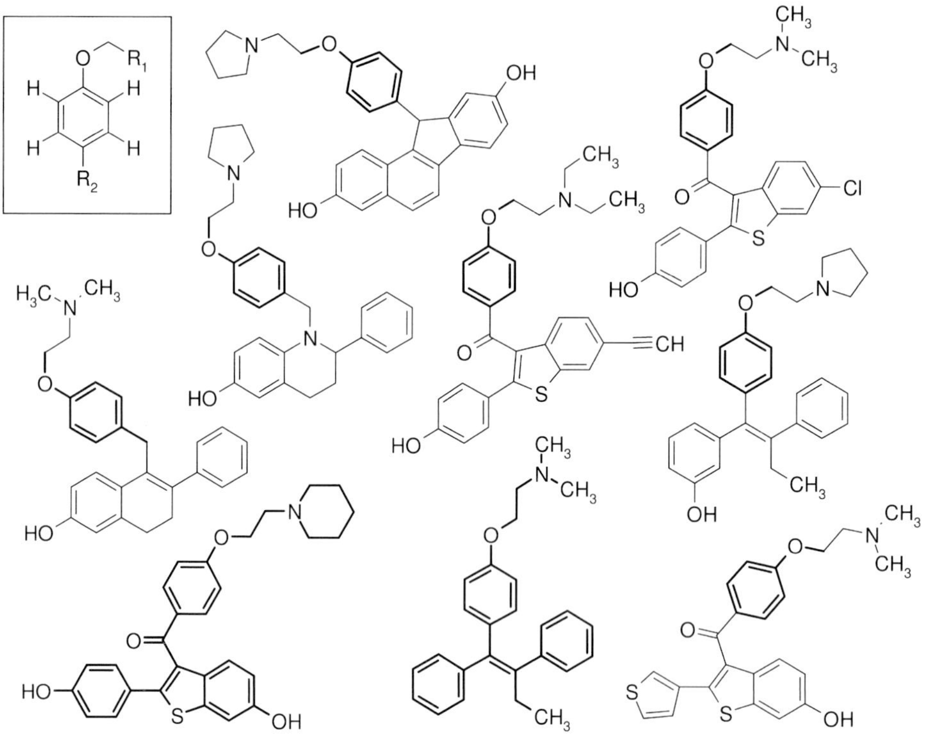

Figure 11.8. Examples of virtual compounds enumerated with PENGUINS that are predicted to be active, based on GRID docking in the ERa crystal structure (PDB code 1ERR). These compounds are also predicted to have good Caco-2 permeability and %PPB properties. The substructure in the top left (inset) is highlighted.

11.11
Conclusions

The introduction of rapid ADME filters in lead discovery is not going to replace experimental procedures; rather, it is intended to increase awareness (i.e., to provide a "computational alert" [6]) regarding potential liabilities that might occur in later stages of drug discovery [22] (e.g., lead optimization). With respect to clinical data on %Oral, we demonstrated the advantage of combining Ro5 and GPSVS, compared to using either method alone, in identifying compounds of potential interest. Although they are less accurate than the 3D-based VolSurf model, the PENGUINS-based GPSVS filters are nonetheless orders of magnitude faster. This effectively enables one to evaluate literally millions of virtual structures and, within the PENGUINS software, evaluate additional ADME properties (e.g., %PPB or nonrenal clearance) using the same descriptor paradigm. As shown in the evaluation of potential ERα ligands, one can also evaluate binding affinity when the 3D structure of the target is known.

Additional experimental profiling and appropriate adjustment of MIF-based models (to include chemotypes of interest) is part of the normal, heuristic process for model improvement. As noted, PENGUINS has the ability to recognize novel fragments, which leads to improved library (and chemotype) representation. MIF-based predictions, whether with PENGUINS or VolSurf, are intended to provide support in experimental design, planning and prioritization: For example, compounds placed in the GPSVS-Q1 region should be first tested for solubility, whereas compounds mapped in the GPSVS-Q3 region should be first tested for permeability. Compounds falling in the GPSVS-Q2 region should be avoided, whereas compounds mapped in the GPSVS-Q4 region could be considered suitable for additional optimization.

Because the current models are quite stable, we anticipate no changes in either the ChemGPS object system (currently, 525 molecules), or the VolSurf descriptors (92 descriptors to date). Thus, GPSVS – both in the full-molecule and fragment mode – is a stable system that is potentially useful in mapping large chemical probe libraries, e.g., NIH_SMR, in a high-throughput mode. For rapid estimation of combinatorial synthesis planning and for virtual screening, we consider that the significant speed gain observed for PENGUINS is well worth the trade-off in accuracy. Once the large chemical spaces are narrowed down to fewer virtual choices, the 3D-based (VolSurf) GPSVS scores, in combination with Ro5, are expected to have significantly better accuracy. One of the key challenges remains the appropriate treatment of ionic species and tautomers (as indicated in Fig. 11.1). However, in the ultrafast ADME filtering stage, this may prove to be of lesser significance.

References

1 Austin, C. P., Brady, L. S., Insel, T. R., Collins, F. S. NIH molecular libraries initiative, *Science* **2004**, *306*, 1138–1139.

2 Drews, J. Innovation deficit revisited: Reflections on the productivity of pharmaceutical R&D, *Drug Discov. Today* **1998**, *3*, 491–494.

3 Horrobin, D. F. Innovation in the pharmaceutical industry, *J. Royal Soc. Med.* **2000**, *93*, 341–345.

4 Oprea, T. I. Virtual screening in lead discovery: A viewpoint, *Molecules* **2002**, *7*, 51–62.

5 Anonymous. Waiver of in vivo bioavailability and bioequivalence studies for immediate-release solid oral dosage forms based on a biopharmaceutics classification system, **2000**. Available from http://www.fda.gov/cder/OPS/BCS_guidance.htm

6 Lipinski, C. A., Lombardo, F., Dominy, B.W., Feeney, P.J. Experimental and computational approaches to estimate solubility and permeability in drug discovery and development settings, *Adv. Drug Deliv. Rev.* **1997**, *23*, 3–25.

7 Drews, J. Drug discovery: A historical perspective, *Science* **2000**, *287*, 1960–1964.

8 Van de Waterbeemd, H., Camenisch, G., Folkers, G., Raevsky, O. A. Estimation of Caco-2 cell permeability using calculated molecular descriptors, *Quant. Struct.-Act. Relat.* **1996**, *15*, 480–490.

9 Leo, A. Estimating LogP_{oct} from structures, *Chem. Rev.* **1993**, *93*, 1281–1306. clog*P* is available from Daylight Chemical Information Systems, http://www.daylight.com

10 Lipinski, C. A. Druglike properties and the causes of poor solubility and poor permeability, *J. Pharmacol. Toxicol. Methods*, **2000**, *44*, 235–249.

11 Goodford, P. J. Computational procedure for determining energetically favourable binding sites on biologically important macromolecules, *J. Med. Chem.* **1985**, *28*, 849–857. GRID is available from Molecular Discovery Ltd, West Way House, Elms Parade, Oxford OX2 9LL, UK.

12 Taylor, N. R. Inhibition of sialidase, *Meth. Princ. Med. Chem.* **1998**, *6*, 105–119.

13 Von Itzstein, M., Wu, W.Y., Kok, G. B., Pegg, M. S., Dyason, J. C., Jin, B., Phan, T. V., Smythe, M. L., White, M. F., Oliver, S. W., Colman, P. M., Varghese, J. N., Ryan, D. M., Woods, J. M., Bethell, R. C., Hotham, V. J., Cameron, J. M., Penn, R. C. Rational design of potent sialidase-based inhibitors of influenza virus replication, *Nature* **1993**, *363*, 418–423.

14 Cruciani, G., Crivori, P., Carrupt, P. A., Testa, B. Molecular fields in quantitative structure-permeation relationships: The VolSurf approach, *J. Mol. Struct. (Theochem)* **2000**, *503*, 17–30. VolSurf is available from Molecular Discovery Ltd., http://moldiscovery.com

15 Wold, S., Johansson, E., Cocchi, M. PLS – partial least-squares projections to latent structures, in *3D QSAR in Drug Design: Theory, Methods and Applications*, H Kubinyi, (ed.), ESCOM, Leiden, **1993**, pp.523–550.

16 Guba, W., Cruciani, G. Molecular field-derived descriptors for the multivariate modeling of pharmacokinetic data, in *Molecular Modeling and Prediction of Bioactivity*, Gundertofte, K., Jørgensen, F.S. (eds.), Kluwer Academic/Plenum, New York, **2000**, pp. 89–94.

17 Zamora, I., Oprea, T. I., Ungell, A. L. Prediction of oral drug permeability, in *Rational Approaches to Drug Design*, Höltje, H.D., Sippl, W. (eds.), Prous Science Press, Barcelona, **2001**, pp. 271–280.

18 Crivori, P., Cruciani, G., Carrupt, P. A., Testa, B. Predicting blood-brain barrier permeation from three-dimensional molecular structure, *J. Med. Chem.* **2000**, *43*, 2204–2216.

19 Oprea, T. I., Zamora, I., Svensson, P. Quo vadis, scoring functions? Toward an integrated pharmacokinetic and binding affinity prediction framework, in *Combinatorial Library Design and Evaluation for Drug Design*, Ghose, A. K., Viswanadhan, V. N. (eds.), Marcel Dekker, New York, **2001**, pp. 233–266.

20 Zamora, I., Oprea, T. I., Cruciani, G., Pastor, M., Ungell, A. L. Surface descriptors for protein-ligand affinity prediction, *J. Med. Chem.* **2003**, *46*, 25–33.

21 Davis, A. M., Riley, R. J. Predictive ADMET studies. The challenges and the opportunities, *Curr. Opin. Chem. Biol.* **2004**, *8*, 378–386.

22 Oprea, T. I. and Matter, H. Integrating virtual screening in lead discovery, *Curr. Opin. Chem. Biol.* **2004**, *8*, 349–358.

23 Williams, P. A., Cosme, J., Ward, A., Angove, H. C., Matak-Vinkovic, D., Jhoti, H. Crystal structure of human cytochrome P450 2C9 with bound warfarin, *Nature* **2003**, *424*, 464–468.

24 Zamora, I., Afzelius, L., Cruciani, G. Predicting drug metabolism: A site of metabolism prediction tool applied to the cytochrome P450 2C9, *J. Med. Chem.* **2003**, *46*, 2313–2324.

25 Afzelius, L., Zamora, I., Masimirembwa, C. M., Karlen, A., Andersson, T. B., Mecucci, S., Baroni, M., Cruciani, G. Conformer- and alignment-independent model for predicting structurally diverse competitive CYP2C9 inhibitors, *J. Med. Chem.* **2004**, *47*, 907–914.

26 Boyer, S., Zamora, I. New methods in predictive metabolism, *J. Comput.-Aided Mol. Design* **2002**, *16*, 403–413.

27 Pastor, M., Cruciani, G., McLay, I., Pickett, S., Clementi, S. GRID-independent descriptors (GRIND): A novel class of alignment-independent three-dimensional molecular descriptors, *J. Med. Chem.* **2000**, *43*, 3233–3243.

28 Cruciani, G., Pastor, M., Clementi, S., Clementi, S. GRIND (GRID independent descriptors) in 3D structure-metabolism relationships, in *Rational Approaches to Drug Design*, Höltje, H.D. and Sippl, W. (eds.), Prous Science Press, Barcelona, **2001**, pp. 251–260.

29 Cruciani, G., Meniconi, M., Carosati, E., Zamora, I., Mannhold, R. VolSurf: A tool for drug ADME-properties prediction, in *Drug Bioavailability, Methods and Principles in Medicinal Chemstry*, Vol. 18, Van de Waterbeemd, H., Lennernäs, H., Artursson, P. (eds.), Wiley-VCH, Weinheim, **2003**, pp. 406–419.

30 Artursson, P. Epithelial transport of drugs in cell culture. I. A model for studying the passive diffusion of drugs over intestinal absorptive (Caco-2) cells, *J. Pharm. Sci.* **1990**, *79*, 476–482.

31 Norinder, U., Haeberlein, M. Calculated molecular properties and multivariate statistical analysis in absorption prediction, in *Drug Bioavailability, Methods and Principles in Medicinal Chemstry*, Vol. 18, Van de Waterbeemd, H., Lennernäs, H., Artursson, P. (eds.), Wiley-VCH, Weinheim, **2003**, pp. 358–405.

32 Abraham, M. H., Le, J. The correlation and prediction of the solubility of compounds in water using an amended solvation energy relationship, *J. Pharm. Sci.* **1999**, *88*, 868–880.

33 Huuskonen, J. Estimation of aqueous solubility for a diverse set of organic compounds based on molecular topology, *J. Chem. Inf. Comput. Sci.* **2000**, *40*, 773–777.

34 Meniconi, M. Solubility for potential drugs, theoretical and experimental methods. Laurea Thesis (M.Sc.), University of Perugia, Perugia, Italy, **2000**.

35 Kratochwil, N. A., Huber, W., Muller, F., Kansy, M., Gerber, P. R. Predicting plasma protein binding of drugs: A new approach, *Biochem. Pharmacol.* **2002**, *64*, 1355–1374.

36 Lombardo, F., Obach, R., Scott, R., Shalaeva, M. Y., Gao, F. Prediction of volume of distribution values in humans for neutral and basic drugs using physicochemical measurements and plasma protein binding data, *J. Med. Chem.* **2002**, *45*, 2867–2876.

37 Lombardo, F., Obach, R., Scott, R., Shalaeva, M.Y., Gao, F. Prediction of human volume of distribution values for neutral and basic drugs. 2. Extended data set and leave-class-out statistics, *J. Med. Chem.* **2004**, *47*, 1242–1250.

38 Vandenberg, J. I., Walker, B. D., Campbell, T. J. HERG K+ channels: Friend or foe, *Trends Pharm. Sci.* **2001**, *22*, 240–246.

39 Pearlstein, R., Vaz, R., Rampe, D. Understanding the structure-activity relationship of the human ether-a-go-go-related gene cardiac K+ channel. A model for bad behavior, *J. Med. Chem.* **2003**, *46*, 2017–2022.

40 Pearlstein, R., Vaz, R., Kang, J., Chen, X. L., Preobrazhenskaya, M., Shchekotikhin, A. E., Korolev, A. M., Lysenkova, L. N., Miroshnikova, O.V., Hendrix, J., Rampe, D. Characterization of HERG potassium channel inhibition using CoMSiA 3D QSAR and homology modeling approaches, *Bioorg. Med. Chem. Lett.* **2003**, *13*, 1829–1835.

41 Cavalli, A., Poluzzi, E., De Ponti, F., Recanatini, M. Toward a pharmacophore for drugs inducing the long QT syndrome: Insights from a CoMFA study of HERG K+ channel blockers, *J. Med. Chem.* **2002**, *45*, 3844–3853.

42 Roche, O., Trube, G., Zuegge, J., Pflimlin, P., Alanine, A., Schneider, G. A virtual screening method for prediction of the hERG potassium channel liability of compound libraries, *ChemBioChem* **2002**, *3*, 455–459.

43 Cianchetta, G., Li, Y., Kang, J., Rampe, D., Fravolini, A., Cruciani, G., Vaz, R. Predictive models for hERG potassium channel blockers, *Bioorg. Med. Chem. Lett.* **2005**, *15*, 3637–3642.

44 WOMBAT and WOMBAT-PK are available from Sunset Molecular Discovery LLC, http://www.sunsetmolecular.com

45 Hardman, J. G., Limbird, L. E., Molinoff, P. B., Ruddon, R.W., Gilman, A. G. *Goodman & Gilman's the Pharmacological Basis of Therapeutics*, 9th edn., McGraw Hill, New York, **1996**.

46 Speight, T. M., Holford, N. H. G. *Avery's Drug Treatment*, 4th edn., Adis International, Auckland, **1997**.

47 Anonymous. *Physician Desk Reference*, Thomson Healthcare, Montvale, **2001**.
48 The FDA labels are at the CDER website, http://www.accessdata.fda.gov/scripts/cder/drugsatfda/
49 Contrera, J. F., Matthews, E. J., Kruhlak, N. L., Benz, R. D. Estimating the safe starting dose in phase I clinical trials and no observed effect level based on QSAR modeling of the human maximum recommended daily dose, *Regulatory Toxicol. Pharmacol.* **2004**, *40*, 185–206.
50 MRTD is available from the FDA website, http://www.fda.gov/cder/Offices/OPS_IO/MRTD.htm
51 Hansch, C., Leo, A., and Hoekman, D. *Exploring QSAR*, vol. 2, American Chemical Society, Washington, DC, **1995**.
52 ACDLabs physchem suite is available from http://www.acdlabs.com
53 ALOGPS 2.1 is available from the Virtual Computational Chemistry Laboratory, http://vcclab.org
54 Oprea, T. I., Gottfries, J. Chemography: The art of chemical space navigation, *J. Comb. Chem.* **2001**, *3*, 157–166.
55 Oprea, T. I., Gottfries, J. ChemGPS: A chemical space navigation tool, in *Rational Approaches to Drug Design*, Höltje, H. D., Sippl, W. (eds.), Prous Science Press, Barcelona, **2001**, pp. 437–446.
56 Sadowski, J., Kubinyi, H. A scoring scheme for discriminating between drugs and nondrugs, *J. Med. Chem.* **1998**, *41*, 3325–3329.
57 Ajay, Walters, W. P., Murcko, M. A. Can we learn to distinguish between "drug-like" and "nondrug-like" molecules? *J. Med. Chem.* **1998**, *41*, 3314–3324.
58 Jackson, J. E. A users guide to principal components. Wiley-VCH, New York, **1991**.
59 Clementi, S., Cruciani, G., Fifi, P., Riganelli, D., Valigi, R., Musumarra, G. A new set of principal properties for heteroaromatics obtained by GRID, *Quant. Struct.-Act. Relat.* **1996**, *15*, 108–120.
60 Sandberg, M., Eriksson, L., Jonsson, J., Sjöström, M., Wold, S. New chemical descriptors relevant for the design of biologically active peptides. A multivariate characterization of 87 amino acids, *J. Med. Chem.* **1998**, *41*, 2481–2491.
61 Lipinski, C. A., Hopkins, A. Navigating chemical space for biology and medicine, *Nat. Rev. Drug Discov.* **2004**, *432*, 855–861.
62 Oprea, T. I., Zamora, I., Ungell, A. L. Pharmacokinetically based mapping device for chemical space navigation, *J. Comb. Chem.* **2002**, *4*, 258–266.
63 The Pomona Masterfile is an extensive collection of experimental log$P$ values, available from Albert Leo, Pomona College, California. http://clogp.pomona.edu
64 Oprea, T. I., Baroni, M., Zamora, I., Cruciani, G. High-throughput prediction of passive ADME properties from fragments, 224th ACS Natl. Meeting, Boston, MA, **2002**, COMP-109. See http://moldiscovery.com for additional information.

12
GRID-Derived Molecular Interaction Fields for Predicting the Site of Metabolism in Human Cytochromes

Gabriele Cruciani, Yasmin Aristei, Riccardo Vianello, and Massimo Baroni

12.1
Introduction

Metabolite identification is a crucial step in the drug discovery process. Metabolite structural information can be used to investigate phase I and II metabolic pathways, the presence of active or toxic metabolites, and to identify labile compounds. The production of this information early in the discovery phase is becoming extremely important in judging whether or not a potential candidate should be eliminated from the pipeline, and in improving the safety of new compounds. Knowledge of the location in which functional groups are metabolized is helpful in designing drugs with optimized safety profiles because stable groups can be added at metabolically susceptible positions.

Researchers have recently focused on developing faster sample preparation methods, robotic systems, and more sensitive analytical metabolite identification tools [1–4]. LC-MS-NMR is one of the latest commercially available techniques for the characterization of metabolites [5]. However, such techniques are usually highly resource-demanding tasks, consuming a considerable amount of compound. Moreover, due to the increasing abundance of potential candidates, experimental metabolite identification remains, at the time of writing, a huge challenge.

It is common opinion [5] that the use of *in silico* methods to predict a hypothetical metabolite structure, combined with the most recent experimental techniques, can speed up the process of metabolite identification by focusing experimental work on specific target structures, thus improving the method of metabolite structure confirmation and elucidation.

The aim of the present chapter is to describe a recent *in silico* method. It is fast, easy and computationally inexpensive, and able to predict human cytochrome regioselective metabolism using *ad hoc* developed 3D homology models for the enzymes and the 3D structure of the potential substrates. The method uses GRID flexible molecular interaction fields as well as the 3D structure of the potential substrates, automatically providing the site of metabolism (i.e. the place where the metabolic reactions occur) in graphical output.

Molecular Interaction Fields. Edited by G. Cruciani
Copyright © 2006 WILEY-VCH Verlag GmbH & Co. KGaA, Weinheim
ISBN: 3-527-31087-8

The fully automated computational procedure is a valuable new tool in virtual screening and in early ADME-Tox, where drug safety and metabolic profile patterns must be evaluated in order to enhance and streamline the process of developing new drug candidates.

11.2
The Human Cytochromes P450

The superfamily of P450 cytochrome enzymes is one of the most sophisticated catalysts of drug biotransformation reactions. It represents up to 25% of the total microsomal proteins, and over 50 cytochromes P450 are expressed by human beings. Cytochromes P450 catalyze a wide variety of oxidative and reductive reactions, and react with chemically diverse substrates. Despite the large amount of information on the functional role of these enzymes combined with the knowledge of their three-dimensional structure, elucidation of cytochrome inhibition, induction, isoform selectivity, rate and position of metabolism all still remain incomplete [6].

The major xenobiotic-metabolizing cytochromes P450 in humans belong to families 1, 2 and 3, and include CYP1A2, CYP2C9, CYP2C19, CYP2D6 and CYP3A4.

The crystal structures of human 2C9 and 3A4 cytochromes were recently resolved and deposited in the PDB data bank [7, 8]. These structures were also submitted to GRID computation in order to produce molecular interaction fields. The human structure for P450-2C9 was used as a template in homology modeling of the CYP2C19 enzyme. In fact, this enzyme has a high degree of similarity and identity with CYP2C9.

The initial 3D structures of the CYP2D6 and CYP1A2 enzymes were kindly provided by DeRienzo et al. [9]. Then, 3D models were built using restraint-based comparative modeling of the X-ray crystallographic structures of bacterial cytochromes P450BM3, CAM, TERP and ERYF, which were used as templates (PDB entries 2bmh, 3cpp, 1cpt and 1oxa). After this, secondary structure predictions were obtained using the method of Rost and Sander [10]. The heme molecule, with the iron in its ferric oxidation state, was extracted from the structure of CYPBM3 and fitted into the active site of each of the two cytochromes. Lastly, dynamic runs were carried out on the starting structures, without any ligands, in order to select an average bioconformation for all the isoenzymes.

It is known that CYP1A2 preferentially binds molecules with a relatively planar moiety, with heterocyclic aromatic amines, xanthines and quinolones. It is also likely that surface amino acid residues are responsible for the recognition of and the selectivity towards specific ligands [11].

CYP2C9 and CYP2C19 bind compounds with large dipoles or negative charges. Thus oxygen-rich compounds such as carboxylic acids, sulfonamides and alcohols are substrates for 2C9-2C19 cytochromes. Site-directed mutagenesis experiments demonstrated that lipophilic interactions are extremely important for binding in the enzyme cavity [11].

CYP2D6 binds compounds with a basic nitrogen and/or positive charge. Thus nitrogen-rich compounds such as arylalkyl amines are potential substrates for 2D6. It is known that the 3D pharmacophore model of 2D6 substrates needs at least one basic nitrogen atom at 5–7 Å from the oxidation site [12]. However, several substrates exist which show a greater distance between the oxidation site and the basic nitrogen, e.g. tamoxifen (>10 Å). This example demonstrates the important role played by CYP flexibility in substrate recognition.

Finally, the most abundant cytochrome in humans, CYP3A4, tends to exhibit a broad substrate specificity. It binds low molecular weight and high molecular weight compounds and shows no pharmacophoric preferences or special substrate-structural constraints. This is probably due to its large cavity. Since its substrates probably adopt more than one orientation in the active site, it is believed that CYP3A4 attacks ligand positions mainly on the basis of the latter's chemical reactivity.

All P450 cytochromes contain a protoporphyrin group with a central iron atom that is normally hexacoordinated in the ferric form. The substrates bind reversibly to the enzyme and the complex undergoes reduction to the ferrous state. This allows molecular oxygen to bind as a third partner. Molecular oxygen is transformed into oxene, an electrophilic and reactive species, which normally pulls a hydrogen radical away from the substrate and transfers a formal hydroxy group back [13]. After release of the product, the regenerated cytochrome P450 is ready for a new cycle.

12.3
CYPs Characterization using GRID Molecular Interaction Fields

This section describes the use of the GRID force field [14, 15] to characterize and compare the most important human cytochrome enzymes. Program GRID is calibrated in an aqueous environment to obtain chemically specific information about a (macro)molecule, called the *Target*. An electrostatic potential does not normally allow favorable binding sites to be differentiated for a primary, secondary or a tertiary amine cation, for pyridinium, or for a sodium cation, and the GRID method is an attempt to compute analogous potentials which do have some chemical specificity. The object used to measure the potential at each point is given the generic name *Probe*. Many different probes can be used on the same target one after the other, and each represents a specific chemical group. A great deal of chemically specific information can therefore be accumulated concerning the way in which the target might interact favorably with other molecules.

The molecular interaction fields (MIF) in the binding site of the cytochromes were obtained using a grid step size of 0.5 Å and a self-accommodating dielectric constant [16]. The grid box size for the five isoforms was placed around the active site cavities and carefully refined using the tools available in the GRID software.

The MIFs were generated by using both the rigid mode and the flexible mode in GRID (directive MOVE = 0 or 1) [17]. With the flexible option, some of the amino acid sidechains can move automatically in response to attractive or repulsive interactions with the probe. The sidechain flexibility in GRID can mimic the amino acid movements which occur in the CYPs active site to accommodate different substrates according to their size, shape and interaction pattern. In fact, when a ligand approaches a residue sidechain, their movements are always influenced by neighbors and by the ligand. For example, the methylene group in the sidechain of lysine will tend to move towards the hydrophobic moiety of an interacting ligand. However, if the ligand contains a positively charged group, the charged nitrogen of lysine will tend to move away from it. What actually happens depends on the overall balance between the effects of attraction and repulsion, and so GRID was calibrated to simulate these movements [17]. However, it should be noted that the flexible grid map cannot take into account the large movements of the protein backbone.

Figure 12.1 shows the molecular interaction fields produced by the hydrogen probe interacting with the amino acid sidechains inside the active site cavities of the cytochromes. Such MIFs are important in evaluating the available volume in the active site cavities, and the shape of these cavities.

With the enzyme sidechains in fixed positions, the active site volumes computed by the GRID hydrogen probe ranged from 1500 $Å^3$ for CYP3A4 to 640 $Å^3$ for CYP1A2. With flexible sidechains, the accessible cavity volumes increased from 5% to 10%, probably due to the release of steric hindrance and consequent reduction in steric interaction between the chemical probe and the lateral chains. The explanation put forward is that the cytochromes may use sidechain flexibility to allocate more space when necessary, making sub-pockets or small channels accessible without changing the structure of the protein backbone.

The MIFs from the water molecule probe were generated in the subsequent analysis. The water probe was used to find hydrophilic regions where structurally relevant water molecules are bridged to the enzymes. With fixed sidechains, the active site hydrophilic volumes computed by the GRID water probe ranged from 280 $Å^3$ for CYP3A4 to 50 $Å^3$ for CYP1A2. However, when corrected by the volume of the cavities, the percentage ranking of hydrophilic regions according to this scheme is: 2D6 > 3A4 > 1A2 > 2C19 > 2C9. With flexible sidechains, the total hydrophilic volume does not change very much, but the position of the hydrophilic interaction, the hydrophilic pattern, changes considerably. In fact, due to synergic interactions of the sidechains, some cytochrome cavity positions may show reinforced interaction with the water molecule. The authors are convinced that the prediction of water location and water movement due to sidechain flexibility or substrate binding is essential in order to compile the CYP–substrate interactions correctly.

Five more chemical probes were used to characterize the cytochromes studied: the DRY molecular interaction field simulating hydrophobic interaction, the N1-amide nitrogen probe simulating hydrogen-bond donor interaction, the O-carbonyl oxygen probe simulating hydrogen-bond acceptor regions, and separate posi-

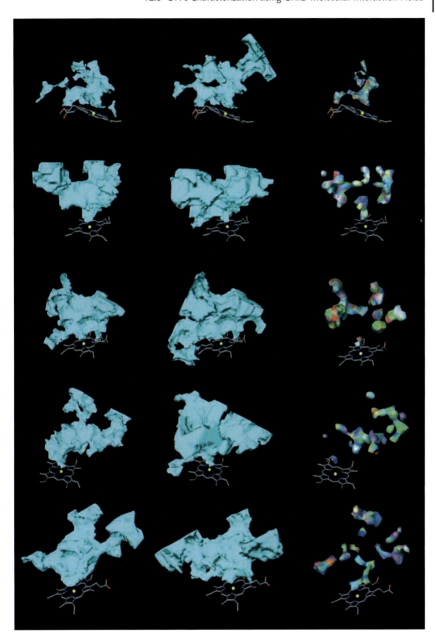

Figure 12.1. First column: molecular interaction fields, produced by the hydrogen probe, showing the volume and shape of the active site cavities, for CYPs 1A2 2C9 2C19 2D6 and 3A4 respectively. Second column: with flexible sidechains, the overall cavity volumes increase due to sub-pockets or small channels which become accessible because of sidechain flexibility. Third column: molecular interaction fields, produced by the water probe, showing strong hydrophilic regions in the active site cavities, for CYPs 1A2 2C9 2C19 2D6 and 3A4 respectively.

tively and negatively charged probes simulating charge–charge electrostatic interactions.

Figure 12.2 compares some of the MIFs obtained from the various cytochromes studied. CYP2D6 shows the highest H-bond acceptor region volume (about 50% of the cavity volume), while CYP2C9 is the cytochrome showing the largest hydrophobic regions (about 20% of its cavity is hydrophobic). CYP3A4 shows the highest H-bond donor region volume (about 25% of its cavity volume).

In agreement with the GRID findings, site-directed mutagenesis experiments demonstrated that lipophilic interactions are extremely important for binding to take place in the enzyme cavity CYP2C9. In turn, flexibility of sidechains modifies the physicochemical enviroment of the cavity, as well as the protein pharmacophoric pattern.

Figure 12.3 compares rigid and flexible molecular interaction field maps with the hydrophobic DRY probe in the active-site cavity for CYP2C9 enzyme. Some of the potential hydrophobic regions shown in the rigid structure are not present when the sidechains are free to move. With flexible sidechains, the overall hydrophobic volume does not change, but the position of the hydrophobic interactions

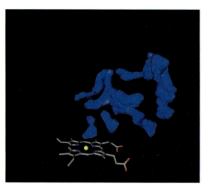

N1 cyp2D6 E= -2.8 Kcal

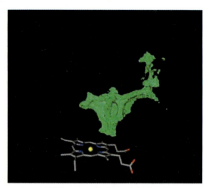

DRY cyp2C9 E= -0.2 Kcal

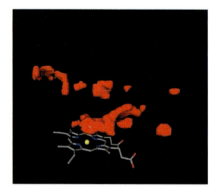

O cyp3A4 E= -2.8 Kcal

Figure 12.2. Some of the MIFs obtained from the various cytochromes studied are compared: (a) CYP2D6, (b) CYP2C9, (c) CYP3A4.

 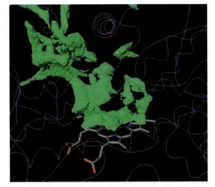

DRY 2C9 static MIF -0.2 Kcal/mol   DRY 2C9 flexible MIF -0.2 Kcal/mol

Figure 12.3. (a) Rigid and (b) flexible molecular interaction field maps with the hydrophobic DRY probe in the active-site cavity for CYP2C9 enzyme. It is noteworthy that, due to the starting 3D structure, some of the potential hydrophobic regions shown in (b) are not present in (a).

changes considerably. Due to synergic interactions of the sidechains, some positions may show reinforced interaction with the DRY probe.

CYP2D6 binds compounds with a basic nitrogen and/or positive charge, and oxidizes atoms at a distance of 5–7 Å from the nitrogen. However, several substrates exist which show a larger distance between the oxidation site and the basic nitrogen, e.g. tamoxifen (>10 Å). GRID flexible MIF in a 2D6 cavity shows that with an appropriate position of the sidechains, the basic nitrogen of tamoxifen can be well accomodated in the active site when the site of oxidation corresponds to the experimental one.

These examples demonstrate the important role played by CYP sidechain flexibility in substrate recognition.

12.4
Description of the Method

The proposed methodology involves the calculation of two sets of descriptors, one for the CYP enzymes and one for the potential substrates, representing the chemical fingerprints of the enzymes and the substrates respectively. The two sets of descriptors are then used to compare the fingerprint of the cytochrome with the fingerprint of the substrates. As discussed in the previous section, the set of descriptors used to characterize the CYP enzymes is based on GRID *flexible* molecular interaction fields (GRID-MIFs). Flexible molecular interaction fields are independent of the initial sidechain position, and are better suited to simulate the adaptation of the enzyme to the substrate structure. Similarly, the set of descriptors used to characterize the ligand substrates is based on the molecular interaction field produced around each substrate.

12.4.1
P450 Molecular Interaction Fields Transformation

The MIFs obtained from cytochrome enzymes are subsequently transformed and simplified as shown in Fig. 12.4. A three-dimensional grid map (3D map) may be viewed as a 3D matrix that contains forces of attraction and repulsion between a chemical probe and a protein. A 3D map is an image of the CYP–probe molecular interactions in which each pixel contains information about the cartesian coordinates and a physicochemical interaction. In cytochrome, where a catalytic reaction has to take place, all the 3D map information can be compressed and refers to the

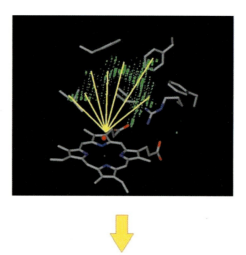

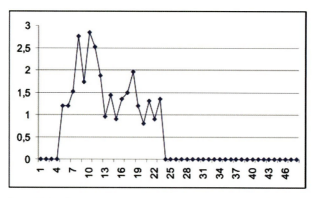

Figure 12.4. The interaction energies at a certain spatial position (the MIF descriptors) for the CYP2C9–DRY probe interaction map, are transformed into a histogram that captures the 3D pharmacophoric interactions of the flexible protein. Such a histogram is called a correlogram and represents the distance between the reactive center of the cytochrome (the oxene in the heme moiety) and the different chemical regions inside the enzyme active site. Different 2C9–probe interaction maps are computed thus generating a number of different correlograms.

catalytic center of the enzyme, that is, the oxene atom of the protoporphyrin group.

The selected 3D interaction points are used to calculate enzyme fingerprints using the GRIND technology [18]. For each CYP–probe interaction map (see Fig. 12.4), this approach transforms the interaction energies at a certain spatial position (the MIF descriptors) into a number of histograms that capture the 3D pharmacophoric interactions of the flexible protein. Such histograms are called correlograms. The correlograms represent the distance between the reactive center of the cytochrome (the oxene in the heme moiety) and the different chemical regions inside the enzyme active site.

12.4.2
3D Structure of Substrates and Fingerprint Generation

The majority of CYP substrates contain flexible moieties. Since the conformation of a substrate is relevant for CYP binding and as recognition thus has a sensible impact on the outcome of the method, the decision was taken to model each substrate by using a population of diverse low-energy minimum conformations, analysing them by running them through in-house software integrated in the computational procedure. The runs were constrained to obtain a population of conformers with 3D structures induced by the *interaction fields* and *shape* of the CYP active site.

The descriptors developed to characterize the substrate chemotypes are obtained from a mixture of molecular orbital calculations and GRID probe–pharmacophore recognition. Molecular orbital calculations to compute the substrate's electron density distribution are the first to be performed. All atom charges are determined using the AM1 Hamiltonian. Then the computed charges are used to derive a 3D pharmacophore based on the molecular electrostatic potential (MEP) around the substrate molecules.

Moreover, all the substrate atoms are classified into GRID probe categories, depending on their hydrophobic, hydrogen-bond donor or acceptor capabilities. Their distances in the space are then binned and transformed into clustered distances (see Fig. 12.5). One set of descriptors is computed for each atom type category: hydrophobic, hydrogen-bond acceptor, hydrogen-bond donor and charged, yielding a fingerprint for each atom category in the molecule. The distances between the different atomic positions classified using the previous criteria are then transformed into binned distances. In this case, the distances between the different atoms are calculated and a value of one or zero is assigned to each bin distance, respectively indicating the presence or the absence of such a distance in the substrate [19].

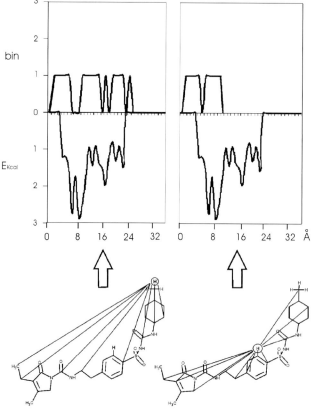

Figure 12.5. All the substrate atoms are classified into GRID probe categories depending on their hydrophobic, hydrogen-bond donor or acceptor capabilities. Starting from a randomly chosen atom their distances in the space are then binned and transformed into clustered distances. One set of descriptors is computed for each atom type category: hydrophobic, hydrogen-bond acceptor and hydrogen-bond donor, generating a fingerprint for each atom in the molecule. The distances between the different atomic positions classified using the previous criteria are then transformed into binned distances. Finally, both set of fingerprints (top = ligand; bottom = protein) are compared. Similarity is higher when the methyl hydrogen (representation on the left) is exposed to the oxene moiety.

12.4.3
Substrate–CYP Enzyme Comparison: the Recognition Component

Once the protein interaction pattern is translated from Cartesian coordinates into distances from the reactive center of the enzyme, and the structure of the ligand has been described with similar fingerprints, both sets of descriptors can be compared. The hydrophobic complementarity (see Fig. 12.5), the complementarity of charges, and H-bonds for the protein and the substrates, are all computed using Carbó similarity [20] indices. The prediction of the site of metabolism is based on the hypothesis that the distance between the reactive center on the protein (oxene atom in the protoporphyrin group) and the interaction points in the protein cavity (GRID-MIF) should correlate to the distance between the reactive center of the molecule (i.e. positions of hydrogen atoms and heteroatoms) and the position of the different atom types in the molecule [21–23].

Finally, the different atoms in each substrate are assigned a similarity score. Due to the computation mechanism, the score is proportional to the exposure of such substrate atoms to the reactive heme and represents the accessibility component.

The accessibility component, E_i, represents the recognition between the specific CYP-protein and the ligand when the ligand is positioned in the CYP-protein and exposes the atom i to the heme. It depends on the ligand 3D structure, conformation, chirality, and on the 3D structure and sidechain flexibility of the CYP-enzyme. Thus the E_i score is proportional to the exposure of the ligand atom i to the heme group of a specific CYP-enzyme.

12.4.4
The Reactivity Component

Cytochromes P450 catalyze oxidative and reductive reactions. Oxidative biotrasformations are more frequent and include aromatic and sidechain hydroxylation, N-, O-, S-dealkylation, N-oxidation, sulfoxidation, N-hydroxylation, deamination, dehalogenation and desulfuration. The majority of these reactions require the formation of radical species; this is usually the rate-determining step for the reactivity process [24].

When R_i is the reactivity of atom i in the appropriate reaction mechanism, it represents the activation energy required to produce the reactive intermediate. It depends on the ligand 3D structure and on the mechanism of reaction. Therefore, R_i is a score proportional to the reactivity of the ligand atom i in a specific reaction mechanism.

Furthermore, in this reaction mechanism, R_i does not depend on the P450 enzyme, but is only related to the molecular topology and 3D structure. There is only little experimental data available which reports the R_i component for drug-like compounds. However, the quantification of the R_i component can be approximated using *ab initio* methods. The problem is that on line calculations using *ab initio* methods take too long to be of any practical use. Therefore the authors have developed a faster procedure which has three steps. The first step involves collecting the great majority of drug-like substrates for human cytochromes and dissecting them into nonredundant chemical fragments, hundreds of fragments were selected. In the second step *ab initio* calculations[1], simulating hydrogen abstraction processes, were carried out on all the fragments and on all the fragment

---

1) Open-shell radicals were optimized at the AM1 semi-empirical level. Single point energy evaluations were performed by DFT at the B3LYP/6-311G** level of theory since correlation between experimental and calculated radical stabilities resulted in reasonable agreement for this level of theory.

atomic positions. Although this process is long and time-consuming, once done it does not need to be repeated. Fragments were classified as being stable, nonreactive, medium, moderate and strongly reactive, and were then ranked in a quantitative reactivity scale ranging from 0.5 (stable) to 1.5 (strongly reactive). In the third step, a software routine was produced that recognizes the constitutive fragments of the fragmentized substrate when a potential cytochrome substrate is given. After recognition, the reactivity component R_i can be assigned.

12.4.5
Computation of the Probability of a Site being the Metabolic Site

Once the accessibility and reactivity components are calculated, the site of metabolism can be described by a probability function P_{SM} (probability of being the site of metabolism) reported in Eq. (1), which is correlated to, and can be roughly considered to be the free energy of the overall process [25]:

$$P_{SM,i} = E_i R_i \qquad (1)$$

where: P is the probability of an atom i being the site of metabolism; E is the accessibility of atom i to the heme; R is the reactivity of atom i in the actual mechanism of reaction.

For the same ligand, and the same cytochrome, the P_{SM} function assumes different values for different ligand atoms according to the E_i and R_i components. When a ligand atom i is well exposed to the reaction center of the heme (E_i has a high score), but its reactivity is very low (R_i has a very low score), the probability of metabolism in atom i will be very low or zero. Similarly, when a ligand atom i is very reactive in the considered mechanism (R_i has a high score), but atom i is not exposed to the reaction center of the heme (E_i has a very low score), the probability of metabolism in atom i will be close to zero. Therefore, to be the site of metabolism, an atom i should possess both non-neglecting accessibility and reactivity components in relation to the heme.

Figure 12.6 illustrates the effect of Eq. (1) on a test ligand molecule, a substrate of CYP2D6 enzyme. The calculations report both the accessibility components and the reactivity components, and therefore the probability of being the site of metabolism for all the reactive atoms of the test molecule. It is important to point out that the P_{SM} value is a maximum at the experimental site of metabolism, where accessibility is high and reactivity is good, thus demonstrating that enzyme recognition and ligand reactivity play an important role, which should be considered molecule by molecule.

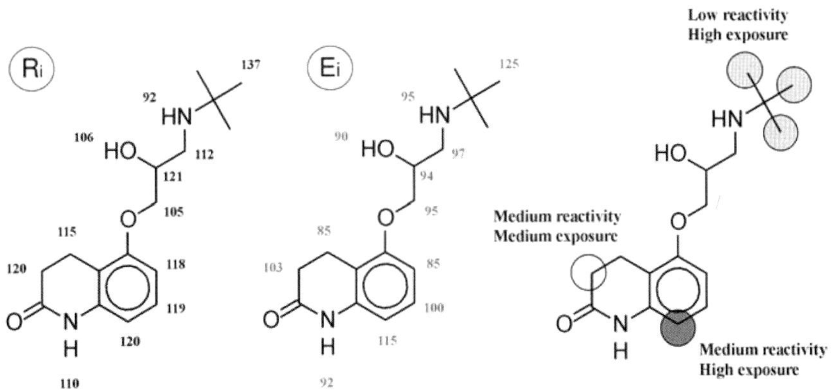

Figure 12.6. The different components affecting Eq. (1) are reported for three atoms of carteolol, a substrate of CYP2D6. (a) Reactivity component, (b) accessibility component and (c) the probability of being the site of metabolism for the first three ranked atoms of the test molecule. The probability of being the site of metabolism is a maximum at the dark gray circle. It is important to point out that the P_{SM} value is a maximum at the experimental site of metabolism where accessibility is high and reactivity is good.

12.5
An Overview of the Most Significant Results

It is important to stress that the method highlighted here requires neither training nor docking procedures and associated scoring functions, nor 2D or 3D QSAR models. The only experimental information used as input is the 3D structure of the human cytochromes. From the 3D CYP structures, GRID provides all the flexible molecular interaction fields, which in turn form the basis of the remaining calculations.

The methodology was validated on carefully selected data from the literature: 150 metabolic reactions catalyzed by CYP1A2, 160 by CYP2C9, 140 by CYP2C19, 200 by CYP2D6, and 350 by CYP3A4, together with information concerning their sites of metabolism. Table 12.1 summarizes the results obtained. It is important to note that the compounds demonstrated different metabolic pathways. Some are metabolized at only one site; others at two sites and, very rarely, some at three. Moreover, substrates show a large structural diversity, including both rigid compounds (e.g. steroids) and very flexible ones with more than 10 rotatable bonds, not to mention a wide range of molecular weights and lipophilicity.

Table 12.1 shows that in more than 70% of CYP2C9 reactions, the first option selected by the methodology matches the experimental one. Moreover, in more than 16% of cases, the second atom is that which fits the experimental one. Therefore, in considering the overall ranking list for the single and multiple sites of metabolism, the methodology predicts the site of metabolism for CYP2C9 within the first two atoms selected in approximately 86% of the reactions, independent of the conformer used.

Table 12.1. Results obtained when testing the methodology on carefully selected data from the literature. Predictions obtained from static and flexible GRID-MIFs are compared.

2C9	152 substrates	Static MIFs	Flex MIFs
Sim		70%	86%
2C19	125 substrates	Static MIFs	Flex MIFs
Sim		–	81%
Sim+React		–	
2D6	200 substrates	Static MIFs	Flex MIFs
Sim		62%	86%
3A4	340 substrates	Static MIFs	Flex MIFs
Sim		65%	78%
1A2	135 substrates	Static MIFs	Flex MIFs
Sim		–	75%

Similar results were obtained with the other human cytochromes. However, CYP 3A4 requires deeper insight. Its broad substrate specificity, probably due to its large cavity, suggests that its substrates may adopt more than one orientation in the active site. For this reason it has been reported (but not proved) that CYP3A4 attacks ligand positions mainly determined by their chemical reactivity. However, the results obtained suggest that, although the reactivity component relative to CYP3A4 is the largest, the recognition component is still the metabolic-site-determining step of the reaction.

Table 12.2 reports a similar investigation carried out by several pharmaceutical companies, using internal proprietary compounds for which information concerning their sites of metabolism in human CYPs is known [26–28]. The results show similar trends to those reported in Table 12.1 for the compounds selected from a careful search of the literature. This demonstrates that the method, unbiased by training or by local models, is generally applicable. It thus demonstrates a fundamental requirement that any method to be applied in the arena of metabolism should possess.

Table 12.2. Results obtained when the methodology was tested by various pharmaceutical companies, using internal proprietary compounds for which information concerning their sites of metabolism in human CYPs is known [26–28].

Sanofi Aventis Bridgewater USA (Drug Design I group)	2C9	90 Substrates	82%
Pfizer Sandwich (UK) (PDM group)	2D6	14 Substrates	85%
Pfizer Sandwich (UK) (PDM group)	3A4	55 Substrates	86%
Janssen Beerse Belgium	2C9 2D6 3A4	50 Substrates	85%

12.5.1
Importing Different P450 Cytochromes

This procedure was designed to work with any cytochrome structure and, for example, can be applied to humans, bacteria, fish and plant cytochromes. There are more than 120 P450 families, and more than 1000 P450 enzymes. All these structures can in theory be imported, processed and used for the prediction of the site of metabolism. The procedure is totally automatic, does not require any user assistance, and only requires the availability of the 3D structure of the enzyme.

A flow-chart of the computation is shown in Fig. 12.7. The GRID-based representations for the main human cytochrome enzymes are precomputed and stored in the software. However, as previously stated, any cytochrome structure in pdb format can be imported, with MIF also computed and stored. Once the 3D structure of the compound has been provided, the semi-empirical calculations of charges and radical abstraction energy assignments, pharmacophoric recognition, descriptor handling and similarity computation are carried out automatically. The user needs only to introduce the structure of the ligand in a smile, 2D-sdf or 3D-mol2 file.

288 *12 GRID-Derived Molecular Interaction Fields for Predicting the Site of Metabolism ...*

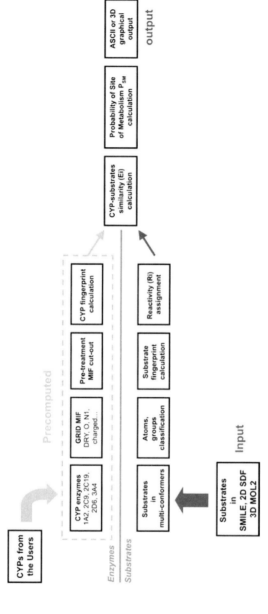

Figure 12.7. Flow-chart for the computation.

12.6
Conclusions

A methodology has been developed to predict the site of metabolism for substrates of the most important human cytochromes. On average, in about 85% of cases the method predicted the correct site of metabolism within the first two choices in the ranking list.

The methodology does not use any training set, or supervised or unsupervised technique. Conversely, the method relies on flexible molecular interaction fields generated by the GRID force field on the CYP homology modeling structures that were treated and filtered to extract the most relevant information. The methodology is easy to use and fast. The method only requires a few seconds per molecule to predict a site of metabolism for drug-like substrates. It is important to note that the method does not use any training set, statistical model or supervised technique, and it has proven to be predictive for very different validation sets examined by various pharmaceutical companies.

The 3D structure of the substrate to be analyzed (the starting conformation) has an impact on the outcome of the method. Satisfactory results were obtained using the in-house conformer generation, which is biased by the MIFs and the flexible shape of the active site of the enzymes. The latter procedure is automatically performed when a molecule or a set of molecules are provided in SMILE, 2D SDF or 3D co-ordinates.

The methodology can be applied automatically to all the cytochromes whose 3D structure is known, and can be used to suggest either new positions that should be protected in order to avoid metabolic degradation, or to check the suitability of a pro-drug. Moreover, this procedure can be used to determine potential interactions between virtual compounds for early toxicity filtering.

Although first applied in the metabolism arena, where all the enzymes contain a clear reaction center, the procedure can work equally well in all fields in which enzymes contain anchor points related to the process under investigation.

12.7
Software Package

The procedure is called MetaSite (**Site** of **Meta**bolism prediction) [25]. The MetaSite procedure is fully automated and does not require any user assistance. All the work can be handled and submitted in a batch queue. The molecular interaction fields for CYPs obtained from the GRID package are precomputed and stored inside the software. The semiempirical calculations, pharmacophoric recognition, descriptor handling, similarity computation, and reactivity computation are carried out automatically once the structures of the compounds are provided. The complete calculation is performed in a few seconds in IRIX SGI machines, and is even faster in the Linux or Windows environment. For example, processing a database of 100 compounds, starting from 3D molecular structures, takes about three minutes at full resolution with a

R14000 Silicon Graphics 500 MHz CPU, less than a minute in a Windows Pentium machine, and about 30 seconds using a Linux Pentium machine. Starting from SMILE notation, processing a database of 100 compounds (each in one of 20 conformations) takes about six minutes at full resolution with a R14000 Silicon Graphics 500 MHz CPU, three minutes using a Windows Pentium machine, and about one minute in a Linux Pentium machine.

The MetaSite software is available to non-profit organizations free of charge, and can be downloaded from www.moldiscovery.com.

References

1 S. R. Thomas, U. Gerhard, *J.Mass Spectrom.* **2004**, *39*, 942–948.
2 E. Kantharaj, A. Tuytelaars, P. Proost, Z. Ongel, H. P. van Assouw, R. A. Gilissen, *Rapid Commun. Mass Spectrom.* **2003**, *17*, 2661–2668.
3 R. Kostiainen, T. Kotiano, T. Kuurama, S. Auriola, *J. Mass. Spectrom.* **2003**, *38*, 357–372.
4 O. Corcoran, M. Spraul, *Drug Discov. Today* **2003**, *8*, 624–631.
5 A. E. F. Nassar, R. E. Talaat, *Drug Discov. Today*, **2004**, *9*, 317–327.
6 M. Riddestrom, I. Zamora, O. Fjäström, T. B. Andersson, *J. Med. Chem.* **2001**, *44*, 4072.
7 M. R. Wester, J. K. Yano, G. A. Schoch, K. J. Griffin, C. D. Stout, E. F. Johnson, http://www.pdb.org, **2004**, 1R9O entry.
8 J. K. Yano, M. R. Wester, G. A. Schoch, K. J. Griffin, C. D. Stout, E. F. Johnson, http://www.pdb.org, **2004**, 1TQN entry.
9 F. De Rienzo, F. Fanelli, M. C. Menziani, P. G. De Benedetti, *J. Comp.-Aid. Mol. Des.* **2000**, *14*, 93–116.
10 C. Sander, B. Rost, *Proteins* **1994**, *19*, 55–72.
11 D. V. F. Lewis, in *Guide to Cytochromes P450 Structure and Function*, Taylor & Francis, New York, **2001**.
12 M. J. de Groot, M. J. Ackland, V. A. Horne, A. A. Alex, B. C. Jones, *J. Med. Chem.* **1999**, *42*, 4062.
13 J. Magdalou, S. Fournel-Gigleux, B. Testa, M. Ouzzine, in *The Practice of Medicinal Chemistry*, 2nd edn., C. G. Wermuth (ed.), Elsevier, Amsterdam, **2003**, pp. 517–544.
14 P. J. Goodford, *J. Med. Chem.* **1985**, *28*, 849–857.
15 E. Carosati, S. Sciabola, G. Cruciani, *J. Med. Chem.* **2004**, *47*, 5114–5125.
16 GRID V.22, Molecular Discovery Ltd, **2004** (http://www.moldiscovery.com).
17 P. J. Goodford, in *Rational Molecular Design in Drug Research*, Alfred Benzon Symposium 42", T. Liljefors, F. S. Jorgensen, P. Krogsgaard-Larsen (eds.), Munkgaard, Copenhagen **1998**, pp. 215–230.
18 M. Pastor, G. Cruciani, I. McLay, S. Pickett, S. Clementi, *J. Med. Chem.* **2000**, *43*, 3233–3243.
19 I. Zamora, L. Afzelius, G. Cruciani, *J. Med. Chem.* **2003**, *46*, 2313–2324.
20 L. Amat, R. Carbó-Dorca, *J. Comput. Chem.* **1999**, *20*, 911–920.
21 L. Afzelius, I. Zamora, M. Ridderström, A. Kalén, T. B. Andersson, C. Masimirembwa, *Mol. Pharmacol.* **2001**, *59*, 909–919.
22 J. P. Jones, M. He, W. F. Trager, A. E. Rettie, *Drug Metab. Disp.* **1996**, *24*, 1–6.
23 S. Ekins, G. Bravi, S. Binkley, J. S. Gillespie, B. Ring, J. Wikel, S. Wrighton, *Drug Metab. Disp.* **2000**, *28*, 994–1002.
24 F. P. Guengerich, *Arch. Biochem. Biophys.* **2003**, *409*, 59–71.
25 MetaSite ver. 2.5, Molecular Discovery Ltd, **2004** (http://www.moldiscovery.com).
26 R. Vez, G. Cienchetta, Sanofi-Aventis, Drug Design I group, Bridgewater USA, personal communication (**2005**).
27 M. Howard, Pfizer, PDM group, Sandwich, UK, F. Cinato, G. Cocchiara, Nerviano Medical Science, Milano, Italy, personal communication (**2005**).
28 M. Engels, K. Ethirajulu, B. De Boeck, Validation and application of MetaSite, a computational platform for the prediction and study of metabolic hotspots, *Drug Met. Disp.*, submitted.

Index

a

Abraham's solubility dataset   259
absorption, *see also* ADME   220 ff., 250
accessibility component E_i 283 f.
– side chain flexibility   283
acetylcholinesterase inhibitor   158 ff.
ACPT   98
active transport   227
ADME (absorption, distribution, metabolism, and excretion)
– ALMOND   253
– binding affinity prediction   267
– descriptors   173
– filter   249 ff., 261 ff.
    – PENGUINS   264
    – Ro5   250 f., 261
– GPSVS   261
– GRID-derived MIF   45, 104, 219 ff.
– *in vitro* screen   173
– MetaSite   253
– MIF   197, 251
– prediction   86, 173 ff., 219 ff.
– properties   173 ff., 219 ff., 242 ff., 254 ff.
– site of metabolism   233
– VolSurf   104, 173 ff., 253
ADME filter   249 ff., 261 ff.
– PENGUINS   264
– Ro5   250 f., 261
ADME/Tox profiling   45, 250 ff., 274
– drug safety   274
affinity   60, 250
alignment-independent GRID analysis   233
ALMD   210
ALMOND   119, 197 ff., 253
– ALM directive   127
– anchor-Grind   140
– descriptor   200
– smoothing window   134
ALOGPS   256

alpha helix, amphipathic   19
amphiphilic moment   178, 221 f.
angiotensin converting enzyme inhibitor   264
anti-target   84
aqueous solubility   180 ff., 220, 253 ff.
ATOM record, *see also* Protein DataBank   14
atom type   98
atomic
– charges   17
– polarizability   17, 29, 176 ff.
– repulsion   29
ATP binding site   69
auto-correlation algorithm   198

b

BBB, *see* blood brain barrier
BCS, *see* biopharmaceutics classification system
best volume (BV)
– hydrophilic OH2 probe   176 ff.
– hydrophobic DRY probe   176 ff.
bile acid transportation system   75 f.
binding
– affinity   146 ff., 253, 267 ff.
– albumin   230
– domain   75
– fatty acid   75
– lipid   75
– plasma protein, unspecific   230, 242
– pocket   158
– phosphate   70
– purine   70
– selective   54
– site   46 ff., 78
– site characterization   50
– warfarin binding site   230
bioavailability   184, 258 ff.
– oral   73, 250 ff.
bioconformation   274

biological activity 36, 145 ff., 164 f.
biological membrane 242
biopharmaceutics classification system (BCS) 258
BioPrint database 98 f.
biotransformation, oxidative 283
block unscaled weight 56 f.
blood brain barrier (BBB) 224
– permeability 253 ff.
– PLS discriminant analysis model 253
BV, see best volume

c

Caco-2 cells 220 ff., 253 ff.
Cambridge Structural Database (CSD) 31
capacity factors 177
Carbo and Hodgkin indices 36
Carbó similarity indices 282
carrier system 227
– GRID-based model 227
Cartesian coordinates 282
CavBase 78
cdk, see cyclin-dependent kinase
charge
– complementarity 282
– negative 60, 87
– positive 60, 87
chemical global positioning system (ChemGPS) 257 ff.
chemical reactivity 233
chemography 257
chemometrical analysis 46 ff., 77, 97
– euclidean distance 105 f.
– latent variable (LV) 128, 205
– partial least squares (PLS) 120 f., 148, 174, 202 ff.
– principal component (PC) 52 ff., 128
– principal component analysis (PCA) 46 ff., 120 f., 174, 198 ff.
– SYBYL 34
chirality 85 ff., 283
– 4-point pharmacophore 90
clearance, nonrenal 256
clinical pharmacokinetics (PK) 254 ff.
combinatorial chemistry 184
combinatorial library 84 ff.
– design 129
comparative molecular field analysis (CoMFA) 34, 118, 145 ff.
– biological activity of ligand 147
– ligand-based 45, 154
– QSAR model 34
– transport 227
complementarity of charges 282
component
– accessibility 283 f.
– reactivity 283 f.
– recognition 282
CoMSiA model 207
CONCORD 199 ff.
conformational analysis 136
– Monte Carlo method 136
conformational
– change 46, 75
– flexibility 56 ff., 265
constituent of protein structure 78
contour
– isocontour 59
– map 54
– plot 59
cooperativity 215
correlogram 227, 281
– auto-correlogram 198 ff.
– cross-correlogram 198 ff.
– distance-based structure model 227
– fingerprint 235
Coulomb
– interaction 55, 146
– term 30 f.
cPCA, see principal component analysis
critical packing parameter (CP) 178, 221 f.
cross-validation method 148 ff., 201
crystal structure 17, 68, 236
CSD, see Cambridge Structural Database
cupredoxins 37
cyclin dependent kinase 69 f.
cyclooxygenase (COX) 61
cytochrome 232 ff., 253, 273 ff.
– 3D pharmacophore model 275 ff.
– CoMFA 233
– enzyme (CYP) cavity 111, 210, 275 ff.
– flexibility in substrate recognition 275
– flexibility of side chain 283
– GRID MIF 275 ff.
– GRID/GOLPE 233
– inhibition 232 ff., 253
– lipophilic interaction 275
– metabolism 257, 273 ff.
– molecular field transformation 280
– P450 family 67 f., 232 ff., 274 ff.
– recognition component 282 ff.
– recombinant 220, 234
– structure 68, 286
– substrate 274

- substrate specificity 275
- substrate-enzyme comparison 282
- xenobiotic metabolism 67, 274

d
dealkylation 283
deamination 283
dehalogenation 283
dendrogram 38 f.
descriptor 92
- 2D 258
- electrostatic forces 174
- field discretization error 121
- log$S$ 257
- GRID molecular interaction fields 279
- GRID water probe 257
- GRIND 119 ff.
- hydrogen bond 174 ff.
- hydrophilic 176
- hydrophobicity 174 ff.
- MIF-derived alignment-independent 117
- molecular 117, 136, 242
- shape 174
- variable 127
- VolSurf 119, 174 ff.
design in receptor (DiR) 84
desolvation
- lipophilic environment 225
- process 154, 225
desulfuration 283
DHFR, see dihydrofolate reductase
differential plot 75
diffusivity 178
dielectric
- constant 249
- constant value 10 ff.
- environment 25
- property 30
dihydrofolate reductase (DHFR) 61
- constitutive 61
- inducible 61
dispersion force 176 ff.
distance 282
- bin 124
distance/interaction profile 242
distribution, see also ADME 220 ff., 250
- passive 230 f.
- plasma protein binding, unspecific 228
- solubility 228
- volume of distribution 228
DMSO solubility prediction 253
DNA minor groove binding 60
DNAC 98

domain
- ligand binding (LBD) 75
- hydrophilic 178
- hydrophobic 178
- PH 37
- two tryptophan (WW) 34 ff.
DONN 98
drug
- activity 88
- orally available 250 ff.
- selectively interacting 45
- World Drug Index (WDI) 250
drug design 166, 197
- 3D-QSAR 166
- biological activity 145
- molecular interaction field (MIF) 117
- shape 103
- structure-based strategy 232
drug discovery 194, 219
- ADME/Tox profiling 250
- cytochrome P450 family 232
- in silico 249
- metabolite identification 273
- Molecular Libraries Initiative 249
- Small Molecule Repository (NIH_SMR) 249 f.
drug metabolism 220, 232
- cytochrome P450 family 232, 253
- MetaSite 253
drug property
- pharmacokinetic 179
- physicochemical 179
drug safety 197 ff., 274
drug-drug interaction 232
DRY probe 32 f., 48, 64 ff., 87 ff., 108, 121, 177, 200, 210, 221 ff., 235, 251

e
E2 ubiquitin conjugating enzyme 37
efficacy 250
electron transfer
- protein 37
- specificity 38
electrostatic
- charge 18
- interaction 30, 48, 60
- property 36 ff.
- term (E_Q) 11 f.
elongation 174 ff.
- EEFR (elongation, elongation-fixed ratio) 176
enantiomer 174 ff.
enantioselectivity 62

energy
- Coulomb 30
- electrostatic (E_{ES}) 29 ff., 91
- function 29
- hydrogen bond (E_{HB}) 29 ff., 251
- hydrogen bonding charge reinforcement (E_{RHB}) 91
- hydrophobic (E_{DRY}) 91, 251
- INTEGY moments 177
- interaction 27 ff., 46 ff., 73, 146 ff., 164 f., 176
- Lennard-Jones term (E_{LJ}) 251
- ligand-receptor interaction 146 ff.
- maxima 62
- minimization 210
- minimum 87 ff, 177
- pairwise (E_{Pair}) 11 f.
- score 124
- solvation 31
- steric repulsion (E_{SR}) 91
- van der Waals (E_{VDW}) 29
endopeptidase 70
enthalpy 16 ff., 78
entropy 32, 249 ff.
- contribution to hydrophobic effect in aequous environment ($E_{entropy}$) 251
- effect 28 ff.
- penalty 23
- term (S) 13, 29, 154
enzyme
- affinity 235
- induction 232
- inhibition 232
- selectivity 232, 242
Eph receptor tyrosine kinase 76
ephrin 76
equilibrium system, reversible
- GRID force field 16
excretion, see also ADME 250

f

FABPs, see also fatty acid binding proteins 75
factor Xa 63
fatty acid binding proteins (FABPs)
FFD, see fractional factorial design
FILMAP 48
fingerprint
- binding site 78, 90
- enzyme and substrate 279 ff.
- interaction energy 88
fingerprints for ligands and proteins (FLAP) 83 ff., 197

- 4-point pharmacophore 83, 197
- application 88
- docking 94
- GRID force field 88
- GRID map 88
- ligand based virtual screening (LBVS) 83 ff.
  - shape of ligand (FLAP LB) 95
- ligand-ligand comparison 86 ff.
- ligand-protein comparison 86
- protein similarity 95
- protein-protein comparison 86
- shape of receptor (FLAP SB) 95
- structure-based virtual screening (SBVS) 83 ff.
- term 85
- theory 86
FLAP, see fingerprints for ligands and proteins
flexibility 68 ff., 211, 257, 276 ff.
- side chain 48, 63, 78, 283
FLOG 50
- probe type 50
fractional factorial design (FFD) 148 ff., 162, 201
- multiple combinations of variables 148
free energy 16, 146, 284
- electrostatic binding 30
- enthalpy 16 ff.
- entropy 16 ff.
- interaction 29

g

generating optimal linear PLS estimation (GOLPE), see also GRID/GOLPE 54, 77, 97 ff., 145 ff.
- active plot 59
- active variable 154
- block-unweighted scales (BUW) 110
- cPCA 58
- cut-out tool 51
- irrelevant variable 154
- smart region definition (SRD) 148 ff.
GLUE
- energy scoring function (E_{GLUE}) 91
- equation (E_{GLUE}) 91
- GRID force field 89
global positioning system VolSurf (GPSVS) 257
- 1st dimension (GPSVS1) 259
- 2nd dimension (GPSVS2) 259
- PENGUINS-based filter 268
- quadrant 259 ff.

Index

GOLPE, *see* generating optimal linear PLS estimation
GPSVS, *see* global positioning system VolSurf
GREATER 88
GRID 28
– alignment-independent GRID analysis 233
– basic principles 4 ff.
– docking 267
– electrostatic attraction 25 ff.
– energetically favorable interaction 204
– energy 11 f., 29, 251
– flexibility of sidechain 276 ff.
– flexibility of target 21
– flexible option 235, 276
– hydrogen bond term 31
– interaction energy 64
– interaction fields 227 ff.
– invariant shape-descriptor 115
– ligand-based method 115
– ligand selectivity 26
– location of favorable binding sites 17 f.
– method 4 ff.
– molecular mechanics-based program 251
– noncovalent interaction 26
– output 18
– position 123 f.
– potential 147
– probe-target interaction 50 ff.
– protein cavity 89, 152
– protein-ligand complex 147
– receptor binding pocket 152
– receptor-based method 115
– standard equation (E_{GRID}) 90
– torsional rotation 19 ff., 86
– value 6
GRID force field, *see also* GRID molecular interaction fields 4 ff.
– calibrating 16 f.
– calorimetric measurement 16 f.
– crystal structure 17
– crystallographic measurement 17
– flexible molecular interaction fields 289
– interaction site 174
– ligand-target interaction energy 90
– molecular environment 16
– nonbonding interactions 104
– parametrization 98
– pharmacophore 197
– probe-molecule interaction potential 104
– structure-based drug design 197

GRID independent descriptor (GRIND) 119 ff., 198 ff., 242
– 3D 131
– ALMOND 119, 197
– ambiguity 137 ff.
– anchor 140
– application 119
– approach 214
– biological property 129
– calculation 58, 198
– conformational flexibility 137
– correlogram 125 ff.
– descriptor 138 f., 197 ff., 228
– distance range 134
– enantiomer 139
– energetically favorable interaction 214
– enzyme fingerprint 281
– highly active compound 133
– highly relevant region 120 ff.
– inactive compound 133
– insensitivity to chirality 139
– interaction with receptor 139
– internal coordinate 120
– ligand conformation 135
– limitation 135
– MIF region 138
– partial least squares (PLS) 108
– pharmacodynamic property 119
– PLS discriminant analysis (PLS-DA) 129
– PLS plot 128 ff.
– principal component analysis (PCA) 108, 131
– QSAR/QSPR model 128 ff.
– theory 120 ff.
– variable 128 ff., 198
GRID map 6 ff.
– 3D 176, 280
– flexible 276
– from DNA 13
– from protein 19 ff.
– from small molecule 24
GRID molecular interaction fields (MIF) 36, 69 ff., 103, 153 ff., 221 ff., 234 ff., 273 ff.
– ADME 242
– application 60 ff.
– DRY-DRY 206 ff., 214
– DRY-hydrogen bond acceptor 206 ff., 214
– DRY-hydrogen bond donor 206 ff.
– DRY-TIP 206 ff., 214
– flexible 279 ff.
– GRID force field 289
– hydrogen bond acceptor-TIP 206 ff., 214
– hydrogen bond donor-TIP 206 ff.

- method 46, 279
- site of metabolism prediction 273
- TIP-TIP 206 ff., 214

Grid point 5, 48 ff., 146
- fixed at 8
- free rotation 8, 28 ff.

GRID probe, see also probe 6
- anisometric 6
- C3 type 48, 64 ff.
- contour plot 59
- DRY type (hydrophobic interaction) 32 f., 48, 64 ff., 87 ff., 108, 121, 177, 200, 210, 221 ff., 235, 251
- H type 87 ff., 276
- hydrogen bond donor/acceptor capability 101
- N type 121
- N+ type 87 ff.
- N1 type (hydrogen bond donor) 48, 73, 87 ff., 108, 210, 235
- N1: type 87 ff.
- N1+ type 64 ff.
- N2: type 87 ff.
- N3+ type (positive electrostatic interaction) 235
- NM3 type 64 ff.
- O type (hydrogen bond acceptor) 48, 73, 87 ff., 108, 121, 210, 235
- O-type (negative electrostatic interaction) 87 ff., 235
- O1 type 87 ff.
- OH type 48, 73, 87 ff.
- OH2 (water) type 108, 176 ff., 225
- protein family 69
- TIP 200 ff.

GRID/cPCA 55 ff., 73 ff.
- method 46 f.
- probe 48

GRID/GOLPE 145 ff.
- application 149
- biological activity 147, 164
- combined receptor/ligand-based approach 147
- modeling of PepT1 transport system 227
- PLS model 34
- procedure 104, 118

GRID/PCA 55 ff.
- procedure 34, 46 ff.

GRIN 4 ff., 17 f.
- free moving part of target 21
- LEVL 18

GRIND, see GRID independent descriptor
GRUB 17 f.

h

HERG, see human ether-a-go-go related gene
hepatocyte 220, 234
heteroatom (HETATM) record, see also Protein DataBank 14 f.
- convention 15

heterotropic modulation 215
high-affinity molecules
- hydrophobic surface of ligand 24

histidine tautomer 28
homology model 67 ff., 233 ff., 289
homotropic modulation 215
hPCA, see principal component analysis
human ether-a-go-go related gene (HERG) 204, 254
hydrogen bond
- acceptor 87 ff., 251
- acceptor interaction (HBA), ACPT 98
- angular term 32
- donor (HDO) 87 ff., 250
- donor-acceptor center 87, 176
- donor interaction (HBD), DONN 98
- donor or acceptor interaction, DNAC 98
- function F 12
- interaction 55
- MIF 31
- term (E_{HB}) 11 f., 251
- water probe 178, 251

hydrogen bonding parameter 178
- hydrophobic (DRY) 33, 98, 153
- hydrophobic center 87 ff.

hydrophibicity 257
hydrophilic calculation 178
hydrophilic-lipophilic balance (HL1, HL2) 177, 221 ff.
hydrophobic
- cavity 215
- complementarity 282
- interaction 62, 228, 251

hydroxylation 283
- side chain 283
- N- 283

i

ILBPs, see ileal lipid-binding proteins
ileal lipid-binding proteins (ILBPs) 75
induced fit mechanism 77
inhibition
- cytochrom CYP 209, 233 ff., 253
- HERG 204, 254
- PGP 217

inhibitor 47, 62 f.
- acetylcholinesterase 158 ff.

- angiotensin converting enzyme 264
- cdk, selective 70
- MMP inhibitor 70 ff.
- NOS 74
- search 158 ff.
- serotonin re-uptake 232
interaction
- drug-drug 232
- electrostatic 30, 48, 251
- energy, negative, favorable 51 f.
- favorable location 29 ff., 160 f.
- function 29
- hydrogen bonding 33, 55, 251
- hydrophobic 54, 228
- integy moments 177
- Lennard-Jones 33, 251
- kinase-ligand 76
- motif 76
- pattern 69 ff.
- possibility 158
- probe-protein 73
- probe-target 32 ff., 45 ff., 242 ff.
- protein-ligand 47 ff.
- van der Waals 29, 54, 146 ff.
- water-bridged target-probe 28 ff.
interpretability 59
interpretation
- pharmacophoric model 202 ff.
- protein-ligand interaction 47
- single variable 133
isoenergy contour 28
isoform selectivity 74
isofunctionality 38

k
kinases
- cdk 69 f.
- eph receptor tyrosine 76
- ephrin 76
- MAP 69
- phosphate binding region 70
- PKA 69
known molecules 14

l
landscape 71
LBD, see ligand binding domain
LC-MS-NMR technique 273
lead discovery 249 ff., 268
leave-one-out (LOO) 34, 148 f., 163
Lennard-Jones
- dispersion attraction 11, 29, 251
- energy (E_{LJ}) 11 ff.

- function 29
- induction 11, 251
- interaction 33, 251
- term 11
ligand, see also FLAP 16 ff.
- 3D fingerprint descriptor 86 ff.
- 3D pharmacophore 86
- 3D structure 283
- addition of substituent to known ligand 33
- affinity 26, 89, 235
- alignment 156
- anti-receptor study 88
- binding domain (LBD) 75
- binding sites 33, 61, 84 ff.
- CoMFA 154
- critical interaction with receptor 83
- design, structure-based 27
- docking 33
- ephrin 76
- GRIND descriptor 210
- hydrophobic atom 90
- interaction with target 46 ff.
- nonbonded interaction 87
- orientation 89 f., 146
- parametrization database 87
- pharmacophore 84
- polar atom 90
- position in receptor binding site 166
- protein complex 46
- protein interaction 45 ff., 67 ff.
- receptor recognition 90
- receptor-complex 118, 146 ff.
- recognition 47
- selectivity 45, 61 ff., 78, 88
- surface-to-surface contact with target 103
lipid binding 75
lipophilic fields, 3D 177
lipophilicity 36, 45, 117, 178 ff., 221, 232
liver microsome 220, 232
loading contour map 54
loading plot 53 f.
local interaction energy minimum 177
log D 257
log P 178 ff., 194, 228, 254 ff.
log P/D 257
log $P/D/S$ 256
log S 257

m
MACC2
- encoding 124
- transform 140
MAP kinases 69

mapping of ligand binding site in protein 33
matrix 50 ff.
- generation 50
- object 50
matrix metalloproteinases (MMPs) 60 ff.
- family 71 ff.
- inhibitor 70 ff.
maximum recommended therapeutic dose (MRTD) 254
MDCK cells 220
megavariant analysis 119
membrane, biological 242
metabolic stability 97 ff., 179 ff., 192 ff., 232 ff., 250 ff.
- ADME/Tox 274
- GRID-based model 232
- in silico model 97
- *in vivo* activity 97
- rate of elimination 234
- site of reactivity 250
metabolism, *see also* ADME 198, 220, 232 f.
- human hepatocyte 234
- human liver microsome 234
- intestinal 259
- orientation of the compound inside the cytochrome cavity 233
- recombinant cytochrome 234
- site of metabolism 282 ff.
metabolite
- identification 273 ff.
- in silico method 273
- GRID flexible molecular interaction field 273
- LC-MS-NMR technique 273
- toxic 273
metalloprotein 28
MetaSite 197 f., 232 ff., 253, 289
microsome, human liver 220, 232 ff.
MIF, *see* molecular interaction fields
MINIM 48
minimum path 105 ff.
- negative-to-negative 108
MLI, *see* Molecular Libraries Initiative
MMPs, *see* matrix metalloproteinases
model
- discriminative 253
- global 257
- GPSVS 253
- GRID-based 286
- homology 67 ff., 233 ff., 289
- interaction energy 153 ff.

- ligand-based, *see also* fingerprints for ligands and proteins 45, 77, 145 ff., 227, 232
- pharmacophoric 215 f., 227
- receptor-based 77, 154 ff., 253
- structure-based 37, 166
- VolSurf & LogP/D/S 256
MOE 199 ff.
molecular diffusion 174 ff.
Molecular Discovery (MD) 253
molecular dynamics 68
molecular docking 38, 88 ff., 233
- AutoDock 151 ff.
- conformation 156
- DOCK 92
- FlexX 92
- GLUE 89 ff.
- GOLD 92
- ligand-receptor complex 146
- orientation of ligand 146
- PENGUINS 267
- steric hindrance 90 f.
molecular electrostatic potential (MEP) 28 ff.
- MIF 28
molecular globularity 176
molecular interaction fields (MIF) 117 ff.
- ADME property 251 ff.
- ALM directive 127
- application 33 ff.
- auto-correlogram 125 ff.
- calculation 27 f., 47, 77
- cross-correlogram 125 ff.
- distance 207
- drug-receptor interaction 141
- electrostatic interaction 30
- energy product 137
- flexible 67
- GRID 47
- GRID DRY hydrophobic probe 37
- GRID force field 47
- halogen 63
- ligand-receptor interaction 117
- methyl probe model 162
- rigid 67
- selectivity profile 60
- similarity 36
- structure-based ligand design 27
- transformation 119
- value 122 ff.
- VolSurf 251
- water probe model 162
Molecular Libraries Initiative (MLI) 249

molecular modelling software package 199, 210
molecular shape 87
– 3D-QSAR 109
– critical packing parameter 178, 221 f.
– descriptor 200 ff.
– globularity descriptor 179
– ligand-receptor binding 103
– metabolism 109
– molecular interaction fields (MIF) 103
– PathFinder 103
– protein-ligand shape similarity 94, 103 ff.
– shape-complementarity 103 ff.
– target-ligand selectivity 94
molecular size 257
molecular structure
– GRID force field 16
– GRID map 19
molecular surface 176
– graph 105
– grid 28
– solvent-accessible 30, 105, 153
molecular volume 176
molecular weight 178, 250
Monte Carlo method 136
motif, interaction 76
MOVE 21, 48, 62 ff., 78, 137, 235, 276
MRTD, see maximum recommended therapeutic dose
multiple binding site 215
multiple ligands binding to the same site 215
multivariant analysis 119
multivariate statistical analysis 97 ff., 198 ff.
– discriminative model 225
– partial least squares (PLS) 108, 124, 198 ff.
– principal component analysis (PCA) 108, 198 ff.

n

new chemical entities (NCE) 219, 232, 249
– ADME properties 219
NIH_SMR, see Small Molecule Repository
NIPALS 58
nitric oxide synthases (NOS) 74
– endothelial (eNOS) 74
– inducible (iNOS) 74
– inhibitor 74
– isoform 74
– neuronal (nNOS) 74
node 37, 48, 123 ff., 198
– MEP 37
NOS, see nitric oxide synthases

o

object 50 ff., 259
– core 257
– satellite 257
objective function 59
oral absorption 184, 251 ff.
oxidation, N- 283

p

3D-pattern
– interaction 76
– property 84
PA, see penicillin acylase
PAMPA, see parallel artificial membrane permeation assay
para-cellular pathway 225
parallel artificial membrane permeation assay (PAMPA) 220
partial least squares (PLS), see also chemometrical analysis 120 f., 148, 198 ff.
– coefficient 133, 228
– coefficient plot 163 f.
– discriminant analysis 191, 225
– model 180, 202, 230
– multivariate data analysis 205
– pseudo-coefficients profile 202
– regression 184
– score plot 193
partition coefficient 178 ff.
– calculated octanol/water partition coefficient (clog P) 250
– log P 178 ff., 194, 228, 254
– n-octanol/water 178 ff.
partition/solubility value 256
passive transport 221 ff.
path
– chemical entity in ligand 114
– corresponding entity in protein active site 114
path-distance frequency distribution 106 ff.
PathFinder 103
– 3D shape 109
– enzyme and receptor site 109
– functionality 109
– GRID molecular interaction fields (MIF) 104
– GRID negative MIF 107 f.
– GRID positive MIF 105 ff.
– isopotential surface 105
– mapping of active site 112
– procedure 105
pathway
– para-cellular 225

– trans-cellular 225
PC, see also principal component 52 ff., 128
PCA, see principal component analysis
PENGUINS, see pharmacokinetics evaluation and grid utilization in silico
penicillin acylase (PA) 62
– A. faecalis (PA-AF) 62
– E. coli (PA-EC) 62
– P. rettgeri (PA-PR) 62
PepT1 227
permeability 224, 242 ff., 253 ff.
– brain 264
– intestinal 264
permeation 250
peroxisome proliferator-activated receptor (PPAR) 75
PGP, see P-glycoprotein
P-glycoprotein 199, 217, 253
pharmacodynamic property 45 ff, 180, 198
pharmacological
– activity 233
– target-based study 221
pharmacokinetic (PK) property 194, 251 ff.
– MIF 27
– modeling 257
– profile 219
– study 173
– VolSurf 27, 254 ff.
pharmacokinetics evaluation and grid utilization in silico (PENGUINS) 264 ff.
– ADME property filter 267
– binding affinity 267
– fragment library 265
– fragment-based GPSVS 264
– MIF-based prediction 269
pharmacophore 84 ff.
– 2-point 85
– 3D database search query 84
– 3D model 204
– 3-point 85
– 4-point 85 ff.
– conformation 93
– descriptor 202
– feature 84
– GRID probe 93
– identification of active compounds 85
– interaction of ligand with protein target 85, 281
– ligand-based 84
– model 202, 215, 253
– molecular descriptor 85
– molecular flexibility 146
– potential 88

– protein site-derived 86
– virtual screening 84
pharmacophore fingerprint 84 ff.
– 3D property 85
– shape of ligand 88
phosphate binding region 70
physicochemical
– PAMPA 220
– property 45, 173, 194
PIPSA, see protein interaction property similarity analysis
PKA kinases 69
Plasma Protein Binding (PPB) prediction 253
plot
– contour 59
– differential 75
– grid 60
– score 52, 76, 205
– loading 53 f.
PLS, see partial least squares
pocket
– D 66
– P 64
– S1 64 ff.
– S2 73
– S3 73
Poisson-Boltzmann equation 30
polar surface area (PSA) 223
polarizability 178
polarization effect 31
– hydrogen bond 31
Pomona log$P$ dataset 259
Powell method 210
PPAR, see peroxisome proliferator-activated receptor
PPB, see Plasma Protein Binding prediction
– MIF-based 269
– site of metabolism 273 ff., 284 ff.
principal component (PC) 52 ff., 128
principal component analysis (PCA), see also chemometrical analysis 46 ff., 200 ff.
– consensus PCA (cPCA) 46 ff., 63 ff., 236 ff.
– hierarchical PCA (hPCA) 58
– NIPALS 58
– score prediction 258
– score plot 52, 205
probe, see also GRID probe 6, 27 f., 275 ff.
– 3-point 28
– acceptor 62
– acceptor/anion 50
– acid 96

- amphipathic   25
- base   96
- charge (q_1)   11, 60, 101
- donor   62
- donor/cation   50
- DRY   32 f., 48, 64 ff., 87 ff., 108, 121, 177, 200 ff., 221 ff., 251
- halogen   62
- hydrophobic   13 ff., 32, 50, 62, 101, 221
- lipophilic   96
- methyl   34
- multi-atom probe   7
- nonspherical   28
- pharmacophoric   96
- polar   50
- rotational adjustment   9
- spherical   28
- target interaction   50 ff.
- van der Waals   50
- water (OH2)   108, 176 ff., 225, 251

protein
- 3D fingerprint descriptor   86
- active site   69
- domain   21
- family   69
- fatty acid binding   75
- flexibility   73 ff.
- homology model   47, 145
- human serum albumin   230
- interaction of compounds with protein   45 ff., 242, 283
- ligand complex   46
- ligand interaction   45 ff., 67 ff.
- lipid binding   75
- kinase   69
- PH domain   37
- plasma   230, 253
- unspecific binding   230, 242
- selectivity analysis   45 ff., 232
- structure   78
- substrate specificity   221
- WW domain   34 ff.
- unspecific binding to plasmatic proteins   230, 242

protein binding site/cavity   51, 78
- complementary pharmacophore   84
- FLOG   50
- interaction points (GRID molecular interaction fields)   283

protein chain
- alpha carbon atom at N-terminus   13
Protein DataBank (PDB)   14 ff., 31, 232
- ATOM   14
- GLUE   91
- HETATM   15
- nomenclature   14 f.
- type number   14
protein GRIND descriptor   210
protein interaction property similarity analysis (PIPSA)   34 ff.
PSA, see polar surface area
purine binding   70

q

3D-QSAR   34, 84, 145 ff., 199, 227
- ADME   242
- comparative molecular field analysis (CoMFA)   34, 118, 145 ff.
- GRID force field   197
- GRID/GOLPE   34, 118, 145 ff.
- partial least squares (PLS)   34
3D-QSPR (quantitative structure property relationship)   209
QSAR, see quantitative structure-activity relationship
QSPR (quantitative structure property relationship), see 3D-QSPR
quantitative structure-activity relationship (QSAR), see also 3D-QSAR
- analysis   210, 224
- biological activity   145
- CoMSiA   118
- GRID-derived MIF   104
- GRID/GOLPE   104
- GRIND-based   130 ff.
- interpretability of statistical results   148
- ligand-based   45, 145
- MIF-based   35
- model   135 f.
- paradigm for structure-permeability   250
- pharmacophore   101, 134 f., 146
- receptor-based   145
- structure-based   166
- TOPP   97

r

rank value   259
- clearance   259
- HiVHiC   259
- LoVLoC   259
- volume of distribution   259
RCSB protein data bank   71 f.
reactivity component R_i 283 f.
receptor
- 3D pharmacophore   86
- affinity   180

- kinase 69
- type of interaction 87

recognition
- component 282 ff.
- ligand 47

region for selectivity and activity 61, 74
repulsion 29
RMSD, see root mean square deviation
Ro5, see Rule of Five
root mean square deviation (RMSD) 92, 152 ff.
rotamer 32, 93
rugosity 176
Rule of Five (Ro5) 250 f., 261 ff.

s

score plot 52, 76, 205
SDEC, see standard deviation of the error of calculation
SDEP, see standard deviation of the error of prediction
selective
- ligand 61 ff., 77 f.
- region 33, 45, 63

selectivity 52 ff., 74 ff., 250
- isoform 74
- substrate 62

selectivity analysis 51 ff., 232 ff.
- based on consensus principal component analysis (cPCA) 73, 242
- GRID molecular interaction fields 45
- GRID-based 233
- selective site of metabolism 235

serin proteases 60 ff.
serotonin re-uptake inhibitor 232
side chain flexibility 48, 61 ff., 78, 283
SIFt 78
similarity
- analysis 36
- score 283
- index (SI) 36 f.
- matrix 38
- PIPSA 34 ff.

site of metabolism
- chemical reactivity 233
- MetaSite 289
- prediction 232 f., 273 ff., 284 ff.
- probability function P_{SM} 284

site point
- favorable place for ligand atom 87
- interaction energy 87 ff.
- pharmacophoric feature 87 ff.

Slater-Kirkwood formula 29

Small Molecule Repository (NIH_SMR) 249 f., 269
smart region definition (SRD) 148 ff.
smile notation 286 ff.
solubility 45, 73, 99, 180 ff., 220 f., 242 ff., 253 ff.
- Abraham's solubility dataset 259

SRD, see smart region definition
standard deviation of the error of calculation (SDEC) 180 ff.
standard deviation of the error of prediction (SDEP) 34, 156 ff., 181 ff.
steric
- hindrance 63
- property 36

Stokes-Einstein equation 178
structure, 3D 79
substituent 33, 77
sulfoxidation 283
superimposition 46 f.
surface
- hydrophobic area 257
- planar 12
- property 34

SYBYL 199 ff.

t

target 6 ff., 27 f., 275 ff.
- charge (q_2) 11 ff.
- contour map 59
- family classification 77
- family landscape 69 ff.
- flexibility 21, 28 ff.
- immersed in water 6 ff.
- ligand complex 112
- ligand interaction 50 ff.
- ligand selectivity 94
- polar interaction with ligand 94
- probe interaction 50 ff.
- protein, kout format 88
- response to probe 8 f.
- superimposition 46
- water molecule 13

tautomer 28
test set 155 ff., 181, 207 ff.
thermodynamic
- GRID force field 16
- integration 146

thrombin 63
TIP MIF 210
tissue selectivity 232
TOPP, see triplets of pharmacophoric points
toxic effect 233, 251

toxicity   289
toxicological (Tox) dataset   254
trans-cellular
– pathway   225
– permeability (GPSVS 1st dimension)   242, 259
transport
– across cell monolayer   220
– active   227
– active transporter area   242
– bile acid system   75 f.
– Caco-2 cells   220 ff.
– CoMFA   227
– GRID/GOLPE   227
– MDCK cells   220
– passive   221 ff.
– PepT1   227
Triplets of pharmacophoric points (TOPP)   97
– drug-likeness   99
– partial least squares (PLS)   98 ff.
– principal component analysis (PCA)   98
– solubility   99
– theory   98
TRIPOS force field   199 ff.
– SYBYL   204
trypsin   63

v

van der Waals
– interaction   29, 146 ff.
– radius   17, 29
– surface   30
– term   31
variable   50 ff., 133, 200 ff., 257
– block   59
– latent (LV)   205 ff., 259
VD, *see* volume of distribution
vector   50
virtual screening
– ligand based (LBVS)   83 ff.
  – shape of ligand (FLAP LB)   95
– structure based (SBVS)   83 ff.
  – shape of receptor (FLAP SB)   95
VolSurf   36, 104, 119, 173, 197, 251
– 3D-based model   268
– application   179 ff.
– BBB discriminative PLS model   227, 253
– Caco-2 cell based model   227
– ChemGPS-VolSurf (GPSVS)   257
– clinical PK data modeling   256
– descriptors   173 ff., 221 ff., 259 ff.
– HERG binding model   254
– metabolic stability model   193
– MIF-based   173, 269
– oral absorption   253
– pharmacokinetic and physiochemical feature   119, 173
– plasma protein binding, unspecific   230
– PLS   251
– procedure   174
– VD library model   191
volume of distribution (VD)   190 ff., 220 ff., 242, 254

w

walking on the molecular surface   105
water   13, 78
– desolvation   225
– probe, *see also* GRID probe   108, 176 ff., 225, 251 ff.
water solubility, *see also* aqueous solubility   259
– prediction   253
weight root mean square (WRMS)   210
weighting procedure   56 f., 110
Wombat-Pharmacokinetics (WB-PK)   254 ff.
WOMBAT database   254 ff., 267
World Drug Index (WDI)   250
WRMS, *see* weight root mean square

y

YETI force field   152 f.
– protein-ligand complex   153

CD-ROM Information

The CD-ROM contains software and manuals described in the book "Molecular Interaction Fields" (ISBN 3-527-31087-8) edited by Gabriele Cruciani.

Installation

Softwares are separated according the Operating System they are compiled for.

Thus the CD's root directory contains the file README you are reading and the folders "irix" (for IRIX 6.5.9 or newer), "linux" (for Red Hat Enterprise Linux 3 or newer and Fedora Core) and "win32" (for Windows 2000 or newer).

Please move into your selected directory.

– Irix/Linux Users

The two folders contains respectively the following sub-folders:

irix
    grid22a_irix
    metasite_2.6.0-Wiley_irix
linux
    grid22a_linux
    metasite_2.6.0-Wiley_linux

Once moved into the specific OS and software folder, please run the installation script by typing the command:

    ./install

and follow the simple instructions will be prompted to you. A specific README file is also available within the program sub folder.

Linux distributions other than those supported would need some specific graphic libraries, please refer to your vendor for them.

– Windows Users
The "win32" folder contains the distributions of GRID and MetaSite compiled for MS Windows operating system.

– GRID
in order to install GRID, just double click on file "grid22b_win32.exe" and follow the simple instructions that will be prompted to you.

– MetaSite
the program is distributed as a ZIP compressed archive file called

"metasite_2.6.0-Wiley_win32.zip".

First of all, uncompress it in an empty folder, then open that folder and execute the installation script "install.exe" double clicking on its icon. Follow the simple instructions will be prompted to you.

How to obtain licenses

Licenses for running the programmes can be requested through the Molecular Discovery web site:

http://www.moldiscovery.com/wiley.php

Academic and nonprofit users will receive one year license and support, while commercial/profit users will be provided with three months license.

Table of Contents

Overview
    What is MetaSite?
    MetaSite features
    What MetaSite is not?
    Program citation
    Program organization
    Limitations
    Users's support
Background
    Introduction
    Protein Treatment
    3D structure of substrates, and fingerprint generation
    Substrate-CYP enzyme comparison: The Recognition component
    The Reactivity component

Computation of the probability of site of metabolism
Conclusions
References
Installation
  Install
  Installing MetaSite on a NFS server
  Editing the configuration files
Command line options
  Examples
MetaSite menu bars
  File menu
    Open/Open ascii
    Close
    Save
    Save As/Save_As_Ascii
    Export As SD File
    Exit
  MetaSite menu
  Load substrates as single conformer
  Load substrates as conformer set
  Load SDF file
  Enter SMILES
  Run site of metabolism prediction
  Evaluate inhibition
  Tabulate
  View menu
    2D View
    Substrate rendering
    Atoms color
    Sort histograms
    Reactivity correction
    Tool bar
    Status bar
    Interface Style
  CYP menu
    Import enzyme
  Help menu
  MetaSite tutorial
    Case study 1. Input of Salmeterol as single conformation file
    Case study 2. Comparing common reactivity criteria and metabolic reativity
    Case study 3. Prediction results using different cytochrome models on the same substrates structures
    Case study 4. MetaSite applications
    Case study 5. Running MetaSite from the command line
    Conclusions